The Harrowsmith Reader

Edited by James Lawrence

An Anthology from
Canada's National Award
Winning Magazine of
Country Life and
Alternatives to Bigness

Illustrated with hundreds of
photographs, including more
than 100 in full colour

Camden House

© 1978 by Camden House Publishing Ltd.

First Printing: November 1978
Second Printing: September 1979

ISBN 0-920656-00-5 (Softcover)
ISBN 0-920656-01-3 (Hardcover)
Trade Distribution by Firefly Books, Toronto
ISBN 0-920668-00-3 (Softcover)
ISBN 0-920668-03-8 (Hardcover)

Printed in Canada for
Camden House Publishing Ltd.
Camden East, Ontario
K0K 1J0

Cover Illustration by Roger Hill

The Harrowsmith Reader

Editor
James M. Lawrence

Copy Editors
Alice O'Connell, Nancy Pepper

Design & Layout
Judith Goodwin, Pamela McDonald, Lynn Dumbleton

Editorial Associates
Elinor Lawrence, David Lees, Jennifer Bennett

Typesetting
Johanna Troyer

Business & Distribution
John Blanchard, Leslie Smith, Bill Milliken

Contributors

Sharon Airhart, Kenneth Allan, Heidi Atkins, Diane Birch, Allan Bonwill, Ellen Bonwill, Jean Cameron, Barry Estabrook, Mary W. Ferguson, Jo Frohbieter-Mueller, Dick Green, Craig Gutowski, Edmund Haag, Jeffrey C. Hautala, Albert Hoffman, Stephen Homer, Peter Hutchinson, Rosann Hutchinson, Alana Kapell, Barbara Keane, G. I. Kenney, Drew Langsner, Louise Langsner, Jay Lewis, Joseph Mahronic, Robert Mariner, Nancy Martin, Donald McCallum, Janice McEwen, Shaun McLaughlin, Billie Milholand, Molly Miron, Thomas Moffat, Jurgen Mohr, Merilyn Mohr, Virginia Naeve, Magret Paudyn, George Peabody, Russell Pocock, Kathlyn Poff, Louise Price, Hank Reinink, Dr. E. Ross, R. G. Rowberry, William Rowsome, Michael Shook, David Shoots, David Simms, Kathryn Sinclair, Dr. M. V. Smith, R. Stephenson, Alan B. Stone, George Thomas, Gillian Thomas, Dorothy Wesley, Samuel Wesley, J. D. Wilson, Lynn Zimmerman

Cohorts

Mark Andrews, Frank Appleton, Gord Bagley, Mavis Bracken, Arthur Bracegirdle, Todd Bracegirdle, Elmer Boulton, Phillippa Cranston, Cheryl Empey, Ruth Geddes, Donald Goodwin, Thomas Green, Paula Gustafson, Amanda Hewitt, Margaret Hewitt, William Hutchison, Bruce Kendall, Michael T. Leo, Belle Micks, Fred Micks, Laird O'Brien, Lori Purtell, Neil Reynolds, Ian Russell, George Schültz, Glenda Smith, Mike Wells, Susan Woodend, Sylvia Wright

Contents

Introductions

Above, *Harrowsmith's* birthplace — Bellgrove Farm, near Camden East, as it appeared in 1910 shortly after completion.

Just a bit over two years ago, we typed a last-minute editorial introduction to the first issue of *Harrowsmith* that began by stating, "We find solace in Camus' observation that 'All great deeds and great thoughts have a ridiculous beginning.'"

While making no claim to greatness, we spoke with new-found knowledge of the ridiculous: We were about to publish a magazine with no paid staff, with one doubtful advertiser, 707 subscribers and a bank account that was about to slip into overdraft. Editorial headquarters were centred on one large table in a farmhouse kitchen and the sum total of our office equipment consisted of a borrowed drawing board, a T-square, several Exacto knives, a jar of rubber cement and a moribund Smith-Corona portable typewriter with a "J" key that refused to work. We had the effrontery to call this a National Magazine.

The last reserves of a *Chargex* account were drawn in cash to pay the postage, and 25,000 copies of Number One — printed at Don Mills, Ontario — were sent out to an unsuspecting public. Whether or not a Number Two was ever to appear depended entirely on the readiness of people to subscribe to an unknown magazine

with an improbable name. We had already been warned by magazine experts that a new publication must have a self-explanatory name to attract quick attention and we later learned that the absolute worst way to launch a new magazine is to mail a free copy to potential readers. (Better to send them a descriptive brochure, set their imaginations to work and have them subscribe before ever seeing the thing.)

Two weeks passed and nothing happened. One of the first letters of reaction came from a gentleman in Montreal who predicted a quick demise for such an ill-named, ill-conceived publication. The manager of a local farm and garden centre, when handed a copy and asked if we might place a few magazines at his check-out counter, simply flipped through the book and concluded, "I don't think any of our customers would be interested in *that*." We sat at home, vaguely planning a second issue but with no conviction that it would ever come to pass. A farm insurance salesman showed up one morning and, in the course of conversation, asked, "What sort of work do you do?"

"We publish a magazine."

"Oh, really. . .where?"

"Right here."

"Oh."

His silent stare was a mixture of bewilderment and disbelief — a look we soon came to recognize and encounter daily.

Finally the mail began to arrive — 25 subscriptions a day, 50, 100 — and with them a cautious euphoria. The telephone began ringing at all hours and we attempted to get a private line installed. "Impossible," said Community Telephone (Ma Bell not having reached this backwater). We argued that it was hardly fair to eight other families to have a business on the same line. They countered by saying they would not install a private line but "probably ought to start charging business rates." To get our own line we would have to find an office in the village.

In the midst of producing *Harrowsmith* Number Two, we moved hastily into the abandoned upper floor of the old Farmer's Bank building in Camden East, sweeping away many years accumulation of dust, pigeon droppings and fallen ceiling plaster. In the transition, a circulation manager was hired and government-surplus filing cabinets were bought to replace the two shoeboxes that had heretofore held our subscriber files. The second issue appeared, with contributions from British Columbia, Quebec, Prince Edward Island and the United States, and our conscience stopped twitching — the goal of publishing a truly national magazine from a village of 257 seemed suddenly plausible.

As of this writing, *Harrowsmith*'s paid circulation has reached 107,000, making it the fourth or fifth largest paid circulation magazine in this country. *The Harrowsmith Reader* has been created to preserve what we feel to be the best material from the first 12 issues and to meet an ever-increasing demand from subscribers for articles that have long been out of print.

When the first issue of this magazine appeared, media eulogies for the "back-to-the-land movement" had been heard for the better part of a decade. Many assumed that it had died with the radicalism of the late sixties, and their opinion was supported by the suicide of *The Whole Earth Catalog* in May, 1971, which ceased publication at the crest of an unprecedented success for an "alternative" publisher.

Soon after we appeared, a radio interviewer asked, "How long do you think *Harrowsmith* can last, in view of the fact that the back-to-the-earth movement has disappeared in the States and hardly appears to exist in Canada?"

Our reply was that, rather than dying, the so-called homesteading phenomenon had broadened greatly. It had passed out of the realm of turned-off university students to encompass a much more diverse section of the population. From our vantage point in Camden East, we said, it appeared that people were moving to the country in ever increasing numbers.

Along the way, expectations of what country living can offer have matured. Mercifully few are those who still believe they can hop a train to the Yukon and find instant freedom and self sufficiency with a garden plot and a dairy goat. We would like to think that those moving out of urban areas and urban occupations are better prepared for what lies ahead. One needs more than a copy of *Walden* and a Swiss Army knife to succeed out there.

We would like to think that *Harrowsmith* works to keep people skeptical, cautious, pragmatic. None of these articles, nor even the book in whole, is a complete blueprint for self sufficiency. No amount of writing can substitute for actual experience. No magazine can prevent the difficult times, the sometimes mind-bending work or the first snub from a suspicious country neighbour. Hopefully, however, this book can serve as a guide and introduction to a number of arts and skills that our generation either never had or that were somehow lost in the not-too-distant past.

In prefacing the first issue of *Harrowsmith*, we said that the magazine would be shaped by its readers, both in tone and content. To our great satisfaction, we have found an audience of readers who feel a high degree of involvement with the magazine. From them came the inspiration for, and the actual content of, much of this book. It is to the demanding, criticizing and encouraging readers who have had an active part in the evolution of *Harrowsmith* that this book is dedicated.

—JML

The Good, The Bad And The Swampy

How to find and buy land in the country

By Kathlyn Poff

The young couple had been on the road most of that Saturday, driving the 250 miles from Toronto northeast toward Pembroke. The property the agent had described to them over the telephone sounded as if it were tailored to their plans. The 70 rural acres fronted a concession road, designated as a school bus route, (assuring that it would be plowed in the winter); the heavy bush would fuel the wood stove and the rough pasture would be good enough for the goats. The land had no buildings, but that was fine; it seemed important to them to build their own house. The price, from their urban perspective, seemed right, only a few hundred dollars an acre.

The agent was busy when they arrived, too tied up with paperwork, he said, to escort them to the parcel. He described how to find the place and bid them goodbye and good luck with, they thought, surprising finality. Three hours later, the couple was back at the realtor's office.

"It's very nice," said the young man, politely, "but do you have anything above water?"

In their many previous conversations with the real estate agent, he had never used even so discouraging a word as "low." Even at the site, the couple almost talked themselves into buying the swampy land in the naïve hope that somewhere in the centre there would be high ground and in the delusion that truckloads of land fill might help in other places. If they had been green enough, the agent would have sold them this bit of bog without batting an eye, and he would have been proud of himself for being salesman enough to do it.

I was half of the land-innocent couple who had had such high hopes for that quagmire. We have since found our parcel, higher and drier, in another part of the province. To help pay for it, I've gone around the desk. I now sell real estate,

and in that capacity I've seen other young couples charge to the brink of the same pit and know of others who fell in.

News of the death of the "back-to-the-land movement," eulogized from time to time by the media in recent years, has, to borrow a phrase from M. Twain, been greatly exaggerated. Every real estate agent from Squamish to Lunenburg and from Pickle Crow to Coboconk has a list of at least 50 people, all looking for a country dream — most of them not sure if it occupies two acres or 200.

If you've got $6,000 in the bank, a good job in the city and are on the prowl for a piece of country, you are far from unique. Although your search may go quickly and effortlessly (see following articles), and although you may buy the property directly from a retiring farmer, the odds are better that you will have to work through a real estate agent. Frankly, finding the right realtor can be as difficult as tracking down the right property.

I like to think that our own land-buying experiences have helped make me sensitive to the problems common among many land seekers today. Fortunately, the country buyer has at least one advantage over the city house hunter. In small, close-knit communities, it is hard to cheat people on a long-term basis, and impossible to hide a bad reputation.

If you've decided on a general area that interests you, but have no information about local realtors, check in at the hardware store just down the street from the land agency. Make a small purchase, talk about the weather — don't hurry things. If it turns out that the agent has moved frequently from town to town or has a record of jumping from agency to agency, it is likely that he has worn out his welcome.

Once you've met an agent you feel you can

George Thomas

Spring is the best time to start searching for property, but the best prices will be had in late fall or early winter.

trust, don't be content to leave your name and address with a vague description of what you want (several secluded acres, southern exposure, fast-running stream). Chances are that he has 40 or 50 properties he's trying to move, and if you don't show an interest in one of them, your request will be filed and, temporarily at least, forgotten.

All the realtor wants to do is sell the properties he has listed. He's working for the seller, not the buyer. Simply leaving your name and telephone number will get you nowhere — dream properties sell to whomever walks through the door that day. The realtor will save the long-distance call and you for a slow day in the office when he's flogging some piece of overpriced country slum that no one else wants.

So what to do? Choose several agents and never stop telling them what you want. Visit them every chance you get. Ask to see their new listings. And someday you'll be the one who happens to walk through the door when someone has listed something special. Chances are he'll even show it to you before he shows it to a complete stranger — but don't count on it.

Always keep in mind that no rural agent will point out even the obvious drawbacks to a property if you don't see them yourself. Be sure to become familiar with local land prices. What appears to be a steal because you are still thinking in inflated urban terms may, in fact, be vastly

overpriced for the local market. And, be assured, no agent will point that out to you either.

Rural real estate seems to break down roughly into five categories: undeveloped bush (bush being anything from hardwood forest to muskeg); acreage with some pasture and some bush; abandoned farms; working farms; rural homes. The realtor is going to ask which of these you are looking for.

"Buy land. They ain't making that stuff any more."

— Will Rogers

Undeveloped bush may be the cheapest at the outset, but the job of clearing and stumping the land is of truly awesome proportions. There will be no hydro, no well, no pasture for animals (unless natural meadows exist) and no entry road. If what you want is only a place to pitch a tent and stalk the wild asparagus, untamed land is probably your best buy. Otherwise, think very carefully about the time and money you will have to invest — true homesteading involves a commitment that not one per cent of the population in North America would be able to keep.

Similarly, an abandoned farm can be a questionable investment. In parts of the Ottawa Valley, for instance, 100 acres with a falling-down shell of a house will cost very close to $20,000. The pastures have gone to milkweed, the well has probably caved-in and the driveway washed away.

However, by adding $15,000 to your budget,

10

you will also get a house that would have cost $30,000 to build and that is still warm with a life of its own. The pasture is still there, the outbuildings still standing, the driveway passable and the well useable. It won't be strictly a working farm, because there will be no machinery, no stock and a limited amount of arable land. But it can be worked at a subsistence level and improved gradually.

If you happen to be someone for whom money is not a limiting factor, and you truly intend to farm on a big commercial scale, then a working farm is your obvious choice.

But for many people, whether they know it yet or not, the last category — rural homes — is the best. Two or three acres is more than enough to plant a good-sized vegetable garden, establish an orchard and keep a few chickens and a pig or goat.

I've watched three young families come to this same conclusion. Consider, for example, the case of Linda and Dan. Fresh from Toronto and full of enthusiasm, they bought a large farmhouse with several log outbuildings on 200 acres of land. Come spring, they planted a huge vegetable patch, acquired a flock of hens, two pigs and a horse.

All of that utilized only three of their 200 acres. Two years later they realized they could no longer afford to sit on an investment so large and so idle.

They sold out for $42,000, paid all their bills and retired, mortgage-free, to a former schoolhouse on an acre of land. They still have the chickens and the garden and the horse. They ate

the pigs. (Incidentally, another couple from Toronto bought the farm.)

A good time to buy is in late summer or early fall. Properties usually come on the market in early spring, and it is extremely wise to view the land during the wettest part of the year when swampland cannot be camouflaged. By autumn, however, if a house is still up for sale, chances are good that the owner will accept a low offer since he won't be anxious to heat the place for another winter. Also, if you're looking for vacant land, the owner is likely to look favourably at your offer as the days turn crisp — he knows there will be little likelihood of selling after the snow falls.

An example: A brokerage I worked for once listed a beautiful, 200-acre farm in early spring at a price of $60,000. By midsummer, the seller dropped the price to $55,000. With the season almost over and still no buyers, he offered the house, all the outbuildings and 80 acres of land for $35,000. In this particular case, the house alone was worth almost that. It went quickly at that price, and the happy buyer is, doubtless, still congratulating himself for his patience in waiting for fall.

I cannot make too strong an argument for actually living in the area before making a purchase. This need not be lost time, either. Farmhouses often rent at prices that will gladden the hearts of those accustomed to the urban scale of things. For much less than the price of a down-

Marginal farmland: The best buy for those seeking greater self-reliance but not a career in commercial agriculture.

Donald McCallum

11

Old mills, former schools and small churches can often be bought for reasonable sums of money, but can also exact unforeseen amounts of repair and restoration work.

trodden student apartment in town, you may easily end up with an eight-room farmhouse and the use of all the land you could wish. A year or two in such quarters will almost guarantee that you buy land wisely and that you know what sort of community you are getting into.

When you come to the country or a small town, you're buying more than a house. You're buying into a whole neighbourhood.

> "I have surveyed the country on every side within a dozen miles of where I live. In imagination I have bought all the farms in succession, for all were to be bought, and I knew their price. I walked over each farmer's premises, tasted his wild apples, discoursed on husbandry with him, took his farm at his price, at any price . . . "
>
> — *Henry David Thoreau*

Tony and Anne Ruffo discovered that the hard way. Their move to the country came with his appointment as a rural area circulation manager for a national newspaper. Tony had been part of the counter-culture movement of the late sixties, but before making the transition to new home and new job, he shed the uniform; the beard went, and he acquired new and, for him, conservative clothing. When they moved into a beautiful century-old stone farmhouse, he confidently expected to charm the neighbours with his easy-going and personable manner. The community hated him.

"This is the first time I've encountered prejudice because I'm Italian," he told me. The couple says they are now looking for a home in an area less dominated by starchy Anglo-Saxons. It is often forgotten that many rural areas have strong ethnic characteristics that can work to defeat some newcomers.

If you locate a property that interests you, be sure to approach the neighbours. Ask them if that babbling, trouty-looking stream you see in May will still be there in August. They will know which fields are good, and which fields have trouble supporting burdock. They'll know if the water lines freeze in winter, if the well runs dry in summer. They may even tell you the house is known locally as a "cold house," meaning the insulation is poor (or nonexistent).

If you become seriously interested in a place, have the well tested. If the water is badly polluted, you may need to add the cost of drilling a new well to the purchase price (a job that may easily run upwards of $1,000). One thing, though, to remember about the results of the test is that well water usually comes with some bacteria. When you get the bacteria count, don't panic because you happened to have enjoyed a cool glass of water from that well. Call on the local health unit office and ask them to help interpret it.

Resist being pushed into making a hasty offer by an agent who claims he has another buyer champing at the bit. It's an old dodge. If that other buyer is really so eager, you can be sure your agent would have sold the place to him.

When you do make the offer, enter it at least 10 per cent below the asking price. List prices are routinely inflated by this amount or more. If the agent says your offer to buy isn't high enough to be worth considering, remind him that, by law, he must submit all offers, no matter how low, to his client.

Remember, you can always negotiate with the seller, and you might as well start from a low figure. He's starting high, and normally expects to come down. You can always go up, but you can never go down.

If your offer is accepted, get yourself a local lawyer who handles real estate transactions. All the local title searching has to be done in the local registry office anyway, so if your lawyer lives in the city, he's going to subcontract the work and charge you extra for doing it.

Just buying a piece of land does not, contrary to what the poets may say, make you monarch of all you can see. People fresh out of the city tend to have very wound-up feelings of territorialism and all too many of them make their first act the posting of PRIVATE PROPERTY signs.

If you didn't need — or want — neighbours in the city, you will find life extremely hard without them in the country. Make a point of introducing yourself, and be ready to do more listening than talking. New landowners can very often strike mutually advantageous deals with neighbours. One non-farming landowner of my acquaintance permits local dairymen to pasture cattle on his fields for a nominal fee and even to house some stock in his otherwise unneeded barn. His farm is well cared for while he is at work in the city and he has a constant stream of fresh eggs and vegetables left at his back door.

A half-mile down the concession road lives another new farmer. He's made it clear that he wants no trespassers on his land, including the local kids who, like generations before them, had been jumping the fence to swim and fish in the stream. He has also turned down farmers asking to rent his idle, degenerating pastures.

Last fall somebody shot his purebred Labrador retriever for running sheep. He suspects it's more because people don't like him much, and if you talk to his neighbours, you'll conclude that he may be right.

To remain on good terms with your new neighbours, change the established use of your property as little as possible and always gradually. To put it another way, remember that you are only the guardian of your farmstead. The land was there long before you ever saw it. It will be there long after you are gone.

"The land is far more important than we are. To know this is to be young and ancient all at once."

— *Hugh MacLennan*

"It's a beautiful country, all right—if you can afford it."

Drawing by Koren; © 1975 The New Yorker Magazine, Inc.

Mortgages

The Fine Print (Read It)

By David Shoots

On a home in good physical shape, or a farm with a dwelling, a new conventional mortgage can usually be arranged to cover 75 per cent of the appraised value of the property. For example, if the property is valued at $40,000, the bank (or other mortgagee) lends you $30,000 and you must come up with the rest in cash or a second mortgage.

If you are seeking a mortgage from a bank, you may find that some prefer to deal only with property on serviced lots. They will usually steer you to another bank or trust company that routinely handles rural property.

The best mortgagee is probably the person selling the property. He knows the property, he wants it sold, and he'll probably accept less than the going interest rate and may hold your note longer than a bank, trust company or neighbourhood loan shark. He might even do better than Uncle Harry, because he knows the value of the land and need not be convinced by an appraisal that his money is safe.

If you buy a successful farm you may borrow up to $100,000 (or $150,000 if you are under 35 years of age) through the Farm Credit Corporation, a crown agency. Corporation loans are set at less than the going mortgage rate but are determined not so much on the value of the land as on its productive potential. Don't go to them to talk about dreams of raising bantam chickens and a few sheep. Even serious part-time farmers are frowned upon, although if you con-vince your loan officer that you will be a full-timer within five years you might get the cash.

A covenant is any one of a number of promises made, sometimes foolishly, by the mortgagor (land purchaser) as a condition of receiving the loaned money. Most covenants are straightforward and sensible. They cover paying taxes, which take priority over mortgages, insuring the property, usually for fire, and maintaining the place.

All of this is reasonable. The lender is concerned primarily with being able to recover his investment regardless of your misfortunes.

The most important, and somewhat threatening, clause is the personal covenant by which the mortgagor promises to pay the principal and interest for the life of the mortgage. Thus, even though you later sell the property, the legal responsibility could fall back upon you if the next owner fails to make his payments. You can be sued by the original mortgagee for the missing payments as well as lose your remaining interest in the property through foreclosure or judgement.

A nasty paragraph well-hidden in the large grey areas of the fine print will discuss Power of Sale. Basically, it yields the right to the mortgagee to sell your property — in some provinces and states without need for court action — if you ignore his notice that you have fallen 15 days late in your payments. If payments are not caught up within 35 days, you may wake up in the midst of an auction sale.

Registry

The paperwork and documentation that are critical to any property purchase are all kept on file at the county registry office. For a small fee for each volume used, anyone is entitled to see copies of deeds, mortgages, surveys or any other such documents. An "Abstracts" book contains the master list and the file numbers of all documents.

All transactions that pertain to each property within the county are there, including liens on property (which can paralyze purchasing procedures) or outstanding mortgages that you might end up having to pay off.

The township clerk will tell you the annual taxes on any property and whether or not they are paid up. This is public information but I will never forget the lady in the clerk's office who didn't agree with that. She not only would not tell me, but wanted to know who was asking. Impatient county officials are an occupational hazard of property buying.

This is the information that you pay your lawyer to collect. He will want one to two per cent of the purchase price to do it and there is no doubt that the investment is usually worth it. The lawyer knows his way around the filing cabinets, knows the clerks and knows the jargon. What's more, if he makes any mistakes he is legally accountable to you. When you do it all yourself you have no one else to blame.

Right-Of-Way

If the ad says "near" find out how near. If it says "access" that means you might have no legal right to get to your property. If the ad says "right-of-way" find out where it is, how long and how wide it is and whether or not it is useable the year-round.

To have any legal standing, a right-of-way must be registered and if it is, it will be plotted on your deed with a survey of the land it passes through. If your neighbour sells out, the new owner must continue to honour the old agreement.

Water

The term "fecal coliforms" may run lightly off the tongue, but you don't want any of them in your lemonade. They are usually the result of septic tank leakage into your supply of drinking water, and their presence, in any quantity, may mean that a new well is in order.

Local health departments will analyze the results of your water samples and explain the consequences. Often, pouring liquid household bleach down the well may temporarily set things right, but a well that repeatedly becomes contaminated is largely worthless.

Testing well water at a prospective property is a must, if only because drilling now costs $15 to $20 a foot — with no guarantee that water will be struck or that it will be fit to drink. The lack of a well or the inadequacy of the existing facilities should be reflected in the purchase price. The present owner might agree to drill a well prior to

the sale or share in the costs of the project. You might wish to make the finding of drinkable water a condition of sale.

Newer wells are issued with a well driller's certificate. Ask to see it. It describes the diameter of the well casing, the depth of the reservoir and the rate of flow in the casing. Four-inch casings are common, but most houses have a six-inch casing which will handle twice the volume of water. This factor becomes critical in emergencies, as when the house is burning down. Eight-inch casings contain twice the volume again. The larger diameters, not surprisingly, are more expensive to drill.

A four-gallon per minute flow from the reservoir into a four-inch casing is considered minimally acceptable. This flow and casing size will supply a household and small numbers of livestock.

Dug wells are common, as are shore wells, which often have been

blasted into the rock near watercourses. You will find people that swear by them, but I consider neither to be reliable. Although they may have been adequate when constructed, these older wells are often too shallow to escape contamination from surface run-off. Our dug well runs year-round — it runs over and floods the basement in winter and runs dry for two months every fall.

A cistern, despite any negative connotations the word may have for you, is another kettle of fish. Cisterns are not wells, but reservoirs for reasonably clean water, usually rain collected from the roof by downspouts.

I would never destroy an old cistern. You can hook the hot water heater to it to supply hot water for baths, laundry and washing dishes. If you are contemplating solar, the cistern could serve as a heat sink.

Insurance

The simple and understandable homeowner's insurance policy that insures the city home is left behind at the town limits. In the country, insurance forms enter into new realms of complexity.

Farmers insure their homes, their barns, their livestock and their crops all separately and all for different potential catastrophes. Most important for the farmsteader are fire, wind and storm damage insurance. It is essential to shop around and compare the rates offered by different companies; it may take some looking to find a single company that will insure all your areas of potential risk.

Some firms will give no coverage to homes heated by oil space heaters, although wood stoves are usually smiled upon. Others will not insure an old farmhouse simply because it is what it is: an old farmhouse.

Frame construction, whether covered with clapboard or insulbrick, costs more to insure than brick veneer, and brick veneer costs more than solid brick. If the local volunteer fire department is stationed more than five or six miles away, depending upon the company, your rates increase: on a brick house, for example, by about ten cents per $100 value.

One senior country insurer remarked that no country home is uninsurable. Some agent with some company will always come through. At some price. He added that the only properties he considered high risk were those owned by poor housekeepers.

"There isn't any place I wouldn't insure if it were neat and tidy," he said. "It isn't the property you insure so much. It's the person."

Go East, Young Man

Some discouraging words from God's Country

By Jay Lewis

The dream is so real you can touch it. The log building overlooks a fir and spruce-lined lake. Waterfowl are dabbling in the cove and those thrust-up western mountains tower snow-capped above it all. Fields of vegetables and alfalfa are patched up and down the valley. For millions across the country and around the world, this clear, crisp image is British Columbia. Rather than buy a piece of the Rock, many would hock their souls to make a down payment on this vision of life as it should be. But before you grab your life savings and run for the door, consider the reality.

British Columbia is largely up and down. This province is mostly one mountain range pushed up against the next. The people cower in the narrow valleys and river deltas. It's great country for roaming around lakes and peaks, but for the farmer, land that will grow food is at a premium. Of the total land mass, only five per cent will support crops, less than one per cent has the highest Class 1 agricultural capabilities and only 1/100 of one per cent will produce tree fruits. Most of this acreage is now in production. There is another five per cent of marginally arable land which could produce some crops but probably not on a commercial scale. The economic implications of land acquisition are clear. The supply is limited, the demand is great and, as one would expect, the prices are high.

Good farmland occurs in segregated areas of the province. The major regions are: along the eastern coast of Vancouver Island; the Lower Mainland, south and east of Vancouver; the

17

Okanagan Valley; the Creston Valley in the south Kootenays; the Thompson Valley around Kamloops; parts of the Cariboo and Chilcotin regions; and the Peace River country east of the Rockies. The northern lands are usually suitable only for hardy grain and forage crops. The exception to this rule is the Peace River valley, which, due to a moderate micro-climate, can produce excellent field crops.

The good news for farming in British Columbia is that the NDP government passed the Land Commission Act in 1973. This landmark piece of legislation was designed to protect the province's meagre amount of farmland from urban encroachment. The major productive lands have been placed in the Agricultural Land Reserve and cannot be used in any way which would preclude a farming option. The pressure to convert every field into a residential subdivision has now largely been removed. Farmers can once again be farmers and not land speculators.

The bad news for those coming into British Columbia is that the Land Commission often will not allow the subdivision of present farms. It tends to frown on the dividing of the bigger spreads into smaller units. The new (Farm) Income Assurance Programme is also attracting highly trained people back into farming. This creates more competition for the limited land and this promotes even higher prices. It is not easy or cheap to get into farming on a commercial or even subsistence level in British Columbia.

So what does it cost? Let's look at a minimum-size orchard in the Okanagan Valley. It will take about 10 acres to support a small family. Orchard land is currently selling for about $10,000 an acre. If lucky, you can get an operation with a house and some equipment for around $130,000. By quickly calculating the interest rates involved, it will become apparent that the whole place can be paid for in — well — maybe four generations. The banks are the real farm owners in British Columbia just now. They have a lot of sharecroppers working for them.

Perhaps you would rather wrangle cattle in the Cariboo. The land is cheaper per acre, but the spreads are large. An 840-acre "dream" near Quesnel is listed at $130,000. A 450-acre ranch with sheds, but no house, near Anahim Lake goes for $105,000. Obviously, it is not absolutely necessary to go over $100,000 to get into farming in this province, but serious commercial operations need at least that. It is small wonder that the multi-national agribusiness companies are taking over.

It is still possible to find those old homesteads in the 10-to-100-acre range which go for $20,000 to $50,000 and often have some tumble-down log buildings which can be salvaged. The problem is that these farms are usually on very marginal land and will probably not produce a living income, even at a subsistence level. While they usually have been neglected and require much work before anything is produced, the advantage of these homesteads is that they are located in areas which often attract kindred spirits. The

David Stewart

Where farmland is dear: dairy country near Salmon Arm, British Columbia.

"Post-Urban Neo-Ruralists" who grabbed their savings and fled from their professional jobs in the city, found that the price and the pace was right in these marginal regions. Strong alternative communities currently exist in the Kootenays, the Kettle Valley and the Bulkley Valley, which form a support base to ease what can be a difficult transition.

Even so, transience is not uncommon as many move on to become "Post-Rural Neo-Urbanists" when such stresses of country living as limited job opportunities, conservatism, sexism and lack of excitement take their toll.

CROWN LAND

Periodically, rumours sweep the country that the vast Crown Lands (93 per cent of all land is in Her Majesty's name) of British Columbia are about to be opened up for settlement. Don't hold your breath; it is not going to happen. If you write to the Land Commissioner, Lands Branch, Ministry of the Environment, Parliament Buildings, Victoria, B.C. and request Land Series Bulletin Number 11, you will learn all about the "Disposition of Crown Lands in British Columbia."

The first point that they make emphatically clear is that people can no longer preempt land simply by moving in. The bureaucracy surrounding Crown Land leases and grants is truly formidable. First, you need to find some vacant land that no one else has a use for. That is like trying to find the proverbial needle in our vast forests. If you should locate some unwanted property, the application procedures require the patience of Job and many months. Our economic system is not particularly interested in having some people acquiring land virtually free, while others pay the going rate. The Crown Lands lease is not the best way of obtaining property if you are planning the project for your lifetime.

When most people drive up to the land on the listing sheet, they see the sparkling stream, the towering pines and the rustic cabin. But what about the road you are on? Is it publicly maintained all the way to that property or does it cross the adjacent land without a proper easement? A friend just finished pouring a great deal of money into a new approach road after a neighbour put an apple bin bearing a NO TRESPASSING sign in the middle of his long-established driveway. The easement was not properly registered. Ah, country living. It is more than prudent to check carefully to be sure that the rural land you are looking at is properly surveyed and that the boundaries are not disputed. People often buy one property and find that they actually own something quite different.

Other important considerations are water (in the dry Okanagan, farmers have been known to sit by their water intakes with a gun across their knees), power, fencing and, perhaps most important in the long run, an area with a sense of community.

It is handy to have a telephone (it saves energy), a school system, and access to living essentials (food, clothing, hardware, building supplies and implements) not too far away. British Columbia does not have a well-developed public transportation system in rural areas and the cost of going anywhere in the car or truck is skyrocketing. These factors should be considered carefully before making that offer. Unfortunately, many buyers still think of the stream through the pines and not the fencing at $2,000 a kilometre or the irrigation pipe at $6.75 a metre.

When it comes right down to it, buying farmland in British Columbia is a downright extravagant act. But what of that cabin by the lake and those lush alfalfa fields? Well, everyone needs a dream like that. It maintains our sanity in this crazy world. We must realize, however, that visions, when brought into reality can be transformed into nightmares of pressing mortgage payments, back-breaking work for little money and crushing isolation. The current economics in this province make getting into farming at this time nearly impossible for everyone but the highly financed. Come to British Columbia to walk along the lakes and into the mountains, but if you want to buy farmland, look elsewhere.

Decisions, Decisions

Do you take the no-see-ums in summer or the screaming winds of winter?

By Lynn Zimmerman

Not so long ago, Danny Mike Chiasson of Belle Côte, Nova Scotia was the artificial inseminator for area cattle and *the* neighbour to call when farm animals were ailing. I still vividly recall his large muscular hands, developed by years of farmwork, deftly delivering our exhausted ewe Blanche of her first-born lamb. Now all of his own animals have been sold, his barn torn down, the telltale mounds of rich manure bulldozed flat. Daniel Chiasson has become a rural real estate salesman.

"Now you take one old place," recalls the Cape Breton agent. "There were 100 acres with an old house in fair condition. It was nicely set on a southern slope — and that's best for light and the warmth of the sun — and a lovely pond, two acres or so — a lovely pond — there at the bottom of the slope behind the house. There was lots of lumber, both hardwood and softwood, and maybe 10 acres of cleared land.

"They were only asking $10,000, and that's really low. I must have shown it more than 25 times before one couple at last decided to buy as soon as they saw it. Now, there was no electricity, and that's a dollar a foot to run in, and the road is a bit rough, but I still can't understand it.

"That place was on the market for more than a year before it was sold."

A salesman such as Danny Mike Chiasson who comes from the land he is selling can be an invaluable asset for the newcomer. Such agents can talk expertly about soil fertility (but be sure to take along your own pH kit and *use* it), the quality and availability of water, realistic possibilities for pasturing and haymaking and the types and quality of timber on the land.

We asked him for a description of land for sale right now in western Cape Breton, and he quickly listed the following: a 100-acre piece without buildings for $9,000 and including a maintained dirt road and 1,000 feet along the river, making it valuable for gaspereaux fishing in the spring; four and five-acre lots along the Gulf of St. Lawrence for $1,000 per acre; an old farm with 200 acres total area, 50 of them cleared, a mile of riverfront, a big house in good condition, 40-by-25-foot barn and — thrown in for good measure — some horse-drawn farm equipment, all for $52,000; another farm with a good house, barn, several cabins on 565 acres of forested mountainside, including 50 acres of meadow, listed for $62,000; and, finally, 150 acres of forest near the coast with 400 feet facing on paved road, once listed for $25,000 and now down to a more reasonable $13,000. An average 100-acre farm with an old house and barn now lists between $25,000 and $45,000, he said.

TEMPORARY ROOTS

Even with the best of realtors, of course, the buyer must be ready with questions and persist until answers are forthcoming. Is the title clear? What is the legal status of the access road? Where, exactly, are the boundaries? What are the taxes?

Investigate the terrain — and not when it is under snow. Is that meadow really a marsh most of the time? Will the brook flood in spring and dry up in the summer? Try to camp on the land for several days and explore. If possible, locate the previous owner and talk about the land; at the very least, call on neighbours and inquire about the place.

The very best plan is to live in the area and try to get a sense of feeling for the community. Are cultural events available, including movies or a library? (Our intellectual sustenance comes with the monthly visit of the bookmobile.) How distant are kindred souls? How far away is a good car mechanic? We finally had to admit that owning a Volvo here is impractical because of the servicing.

By putting down temporary roots, you can determine the vagaries of the local climate. How long is the growing season? When does most of the rain fall? How bad are the flies and how strong the wind that might blow them away? Even a full year does not tell the whole story, but will give a good indication of things to come.

Relatively inexpensive land can sometimes be found by searching the records at the county tax office for property that can be bought for back taxes owed. (This route has many pitfalls of its own, and you must be prepared to investigate all the ramifications of buying such land.)

We found our own property in Nova Scotia not by that route nor through a realtor. We just happened to be in the right place at the right time.

In 1968, my husband, George Thomas, was tenting along the Margaree River on a fishing expedition. Sometime during the second continuous week of drenching August rain, he fled the tent and desperately sought the dry warmth of a small guest house right on the coast. He stayed more than a week. George had been casually looking for land and was so taken by the small farm that he eventually broached the subject of its purchase with the owners, a couple in their 60's who firmly denied any interest in selling.

But the long Cape Breton winter in a large, old, drafty farmhouse can turn one's thoughts to the cozy warmth of a smaller, better-insulated home. I know, for I now live in that same farmhouse.

In the spring the first letters were exchanged and, after a year of ongoing correspondence, we eventually bought the 30 acres, barn, shed and house for $22,500. The former owners moved into the clean, snug, pre-fab house they had built in preparation on two acres they retained.

Their land is a corner farthest from the sea and includes some of the only dense trees (spruce) on the farm. I remember saying on the June day we moved into the old house (which stands on a bluff defiantly overlooking the sea) that it was too bad they sacrificed such a spectacular view. The wise old couple knew their land better than we. While winter winds frigidly scream by our exposed and trembling house they are nestled snugly out of the way among the trees barely a quarter-mile away.

THE NO-SEE-UM FACTOR

Deeper in the forest, 30 miles up the frozen Margaree River, winter is hushed. There, even toward the end of the snow season, a young couple ski the three miles from the end of the plowed road to their lovely two-storey, octagonal log home.

Peter and Heather purchased the 30 acres of land five years ago, buying from a previous settler at $50 an acre. Most was forested, with the exception of about one-half of an acre of partially cleared land where they planted their garden.

Not including their own labour and that of their friends, they calculate that the cost for everything above the price of the land was about $5,000 (including such items as large thermopane windows, wood stoves, dimensional lumber for floors and even their antique iron bathtub). Today, with the increasing price of pulpwood, this type of land is selling for $75 an acre.

Exceptional buys can now be found in Cape Breton — the land boom of the mid-seventies having subsided — but it takes perseverance and some luck. In searching for property, it is wise to remember that every piece of land has hidden assets and, particularly, liabilities. Pursue your dream, but don't be possessed by it.

Our own exposed perch has its compensations, and they are more than aesthetic. While our inland friends are gardening amongst the black flies, no-see-ums and mosquitos, our cursed winter wind has been transformed into a balmy breeze that mercifully keeps our days and nights insect-free.

While such aspects may seem trivial, you should consider them and assess your particular ability to adjust to the peculiarities of a piece of land. *Before* you put your name on the line, try to anticipate every possible problem. This may well be the biggest investment of your life and it can mean the difference between success and failure in the country.

Quebec Farmstead

"Surely this isn't the only valley hidden away from the ravages of inflation."

By Gerry Kenney

"**L**and. The only way we can do it is to buy a piece of land."

It was maple syrup season — that most energizing of times when the earth is emerging from its winter dormancy and when human vitality quickens a notch or two as well. Elaine and I had found an old abandoned one-horse stable surrounded by sugar maples and had appropriated it for a month to boil off sap. Our evaporator was a 45-gallon drum on its side, with an aluminum pan sitting in a cut-out section atop. Twenty-five wounded maples dripped their sap into our buckets.

That month of syrup making was the turning point in our lives. We had carried in all the equipment on our backs, and, while tending our steaming contraption, we sat in the doorway of the stable, basking in the sun and the privacy of the place.

The heady spring air, redolent with the delicate maple vapours was more than our once resolute Protestant ethics could stand. We began to see that we couldn't continue working in the bureaucratic labyrinth that is the national capital, eating food we increasingly saw to be lethal, breathing air as dead as a bureaucrat's soul. We saw that we had to break our conformity and regain our self-respect. But how?

"Land. The only way we can do it is to buy a piece of land," reasoned Elaine. She was right, of course, and our search began.

We decided that our land would have to meet a few firm criteria:

1. It had to have a sugarbush. That was of prime importance. We would have to derive some income from our land and could think of no better way than by making maple syrup and sugar.

2. It had to have a good house. We would be prepared to make a few repairs, but it had to be basically sound.

3. It had to be in mountainous country (Westerners would call them hills).

4. It had to be within a reasonable distance of

21

Ottawa. During the transition period, I would have to travel to my city job.

5. We had to be able to afford it.

Apart from these requirements, we considered all options open. Where to begin?

We sought the advice of friends who already lived on a farm and their suggested approach worked beautifully.

First, we searched for a general area that would fit our plans. This took only two Saturdays of driving around. On the second foray, ending about 55 miles northwest of Ottawa, we discovered the Little Nation Valley.

The valley snaked through rugged mountains just bristling with sugar shacks. It was approaching the limits of commutability, but we reasoned that this would be a temporary factor until we were solidly on our new rural feet. Having found the area we liked, selecting the right farm proved to be just as easy.

We rattled around the dirt roads until we saw a neatly kept farmhouse with a sugar shack at the base of a mountain that rose behind the house. Our knocking summoned a grizzled old farmer to the door. After remarking on the neatness of his farm and its beautiful setting, we asked if it happened to be for sale. He said, "No."

This didn't surprise us a bit — we were, in fact, expecting it. The next question was our real reason for stopping.

"Do you know of any nice farms with a sugar bush for sale in this area?"

If farms are up for sale, our friends had told us, the local farmers will know.

"Yes, there's the old Seguin place about three miles down the road," offered the old-timer. "He's getting too old now to run the place, and I think he would sell."

And, sure enough, old Mr. Seguin wanted to sell, but his sugarbush was too small for our intended purposes. Still, he knew of another place, and so one farm led to another, until late in the afternoon we stopped at the old Sabourin farm. It was for sale, covered 100 acres, had a 2,000-tap, fully-equipped sugarbush, a solid barn, a sound 100-year-old log house (hidden under a shell of clapboard), and it was within commuting distance of my job.

Everything was perfect, even to the rugged mountains rising on all sides. But could we afford it? The asking price was $30,000 and — after a week of nerve-wracking haggling — it grudgingly dropped to $27,500. We decided that we could afford it and so we bought a new home and a new life style.

Happiness Is A Rundown Farm

There are bargains to be had — but try to remember the dry rot when that 40-mile view is stirring your soul

By Molly Miron

When we bought our first farm six years ago, we did the title search ourselves at the county record office, had the deed notarized by the Commissioner of Oaths (who was also the local fertilizer and insurance dealer) and paid in cash. We counted out our hundred-dollar bills onto the seller's kitchen table, shook hands all around and moved in.

It seemed like a bargain at the time: $4,500 for 100 acres in central New Brunswick, half of the land cleared and half covered with mixed woods, with a shabby drafty house and two rickety barns.

I should point out that I am a captive of the agrarian dream. In fact, I come from a long line of people who cultivate big gardens and raise backyard flocks of chickens in suburbia; people who, when confined to apartments, stubbornly plant vegetables along the periphery of their urban parking lots — all the while wishing for a spread of their own.

When we bought the farm, we had no intention of being hobby farmers. We sought a livelihood and were pleased to find that the land was good and responded well to cultivation. The buildings, however, proved a constant drain on our resources and morale.

The house that was airy in August was drafty in December and frosty in February. During a windy cold snap, bedroom temperatures dropped to minus 20 degrees Celsius. The plumbing froze. (It is more than invigorating after a bath to have to go down into the cellar to thaw frozen drain pipes with a blow-torch.) And the electric kettle steaming first thing in the morning could make it snow in the kitchen.

Not surprisingly, the farm failed. More accurately, we failed as farmers, being totally inexperienced and unaware of the years and dollars it takes to establish a commercial farm. We ran out of money and retreated to the city but the time had not been a complete loss; we had learned a great deal, eaten well, worked hard and made good friends in the country. Our new rented quarters in Halifax seemed cramped by comparison, but the next several years were

Employing the most useful of faculties, hindsight, we can now offer a few pointers to others searching for land. Perhaps the most important is to have a clear idea of what you want in a piece of country property. Do you want to earn a living on the land, as well as calling it home? Or do you want to live in a healthy environment while continuing to work in a nearby city or town?

Remember that even the most attractive property should be carefully scrutinized. Check with the local roads department about future development. There is nothing like a new four-lane highway to alter drastically the character of an area.

Poor fences may become an ongoing headache. You may have no need for them, but if your neighbour decides to keep animals, he can legally bill you for half the cost of repairing or erecting a common fence. That proposal can be extremely expensive.

If you have serious intentions of using wood as a primary heat source, look carefully at the woodlot. A general guide is that seven to 10 acres of mixed woods are needed to provide a perpetual supply of firewood for one home.

Perhaps the best advice, though it may be easier said than done, is to live in the area before buying. Once settled in the region, it becomes a simple matter to watch for bargains. We discovered a lake on the mountain behind our farm and went there often to swim. There was never a soul in sight, and we had no idea to whom it belonged. After about a year we sought out the owner and, as a result of our investigations, we now own 300 thickly-wooded acres surrounding a private lake — all for the sum total of $5,500.

The kind of bargain we found may be a thing of the past in Little Nation Valley. One neighbour bought his 100-acre farm-cum-sugar shack, barn and house for $10,000, eight years ago. Four years later, we paid $27,500 for our own 100. Last year an 80-acre farm two miles away sold for $50,000.

The prices here are escalating faster than can be explained by the general rise of the overall land market. Farmers in the area seem to have learned that city people can, and will, pay rates that only a decade ago would have been judged insanely exorbitant.

For the seeker of a new life style at bargain rates, this valley has drawn too close to the big city. Still, it can not have been the only valley hidden away from the ravages of inflation. There are certainly others waiting to be discovered.

First come, best served.

made easier by our sense of purpose. We were saving to buy another farm.

SWEET INNOCENCE

Eventually, we found a defunct, 100-acre property, whimsically named Ferndale on the deed. Our view of the agrarian dream having been tempered somewhat, we planned to make improvements gradually while my husband continued to commute to his city job. By the time he retires, the farm may be ready to support us.

While the size of this second farm is the same, it shares few other similarities with our first place. The house here is tighter, and sturdier, there is no barn, the 85 acres of woods are mostly second-growth spruce and the 15 supposedly clear acres are growing up in alders. We had expected a jump in price after the passing of four years, but not having to pay a price of $22,000.

Too, the sweet innocence remaining from our first land transaction was abruptly snuffed out. We discovered the demimonde world of mortgage officers, insurance companies, lawyers and appraisers, each of whom collects a fee.

Property appraisers in particular are a strange breed. The one who visited us surely must practise sneering in a mirror.

"Let's see," he observed, "that's cull land you have here. Not worth $10 an acre. Basement down there? Not much clearance. It's hardly a crawl space."

Pay cash if you possibly can and avoid the rude middlemen, high fees and ongoing interest rates. We found our second farm by consorting with real estate agents, but you are much more likely to discover bargain property if it is not listed with a realtor who, of course, must collect a commission and wants the highest price the property will bring.

ENTICEMENTS

When buying an old farm, keep your wits about you and a polite suspiciousness at the ready. It may be hard to think about dry rot and shallow wells when a 40-mile view is stirring your soul, but your naivety may return to haunt you later.

Three weeks after moving into our second farmhouse, in January 1976, the taps suddenly went dry. I was in the middle of a big washday and had baskets of soapy clothes dripping around. After examining all the interior plumbing for a leak or block, I finally peeked down the well. There I saw mud, several big rocks, a toy soldier, a rusty bucket — and no water. We were 50 miles from the nearest laundromat, but somehow coped by hauling water from the neighbours' wells and finally had a deep well drilled at great expense.

Still, I am an incredible optimist — and even after struggling with two of them — think that a rundown farm is the best land bargain. You can fix it up gradually (probably perforce) as the money comes. Last year we invested in a new roof and new electrical wiring. This year we hope to phase out the old privy. There is great satisfaction in seeing the results of your time, investment and labour. And, on rainy afternoons when the children have nothing to do, they can always be sent upstairs to peel wallpaper.

Derek Case

Country Careers

---

*Leaving the city behind —
Harrowsmith profiles
of rural independence*

Some time ago, one of the more prominent news columns in the country carried an interview with a failed back-to-the-lander. After a year in the bush with some sort of ill-defined group, the young man had come out to be refitted with a well-cut suit and, the anonymous interview implied, a shave and a haircut.

Sitting with the writer in a dimly lit, well-appointed Toronto watering hole, the young man allowed that it had been "a good experience . . . but there are just so many books you can read, just so much you can say to the chickens . . . "

That the "goin' up the country" movement of the late 1960's has changed, and for some people, failed, is unquestionable. The dream of packing a sleeping bag, a Swiss army knife and a copy of Walden in a VW van and heading off to instant freedom and self-sufficiency has not died, but it has matured.

As the following profiles by Harrowsmith contributors show, the romanticism of making a fresh start in the country is now mixed with a new pragmatism, an ability to adapt and make one's own luck.

The goals, of course, remain the same: to get out of the traffic, the queue-for-everything, frantic pace of city life — to simplify one's own life and regain a measure of control over it.

No, The Doctor Does Not Play Golf On Wednesdays

The homesteading physician of Rock Creek, British Columbia

By Billie Milholand

Marc Gable M.D., who used to make $40,000 a year and live surrounded by the regular trappings of the big-city urban physician, has turned his back on the neon ladder of success to become a mountain man in the backwoods of British Columbia.

Rock Creek, many people will tell you, is nowhere: a sneeze on the road to Vancouver, a has-been mining town whose dim memory can barely recall the glitter of the 1860's when gold precipitated population explosions in the most unlikely places. It consists of a hotel, laundromat, general store and a small collection of houses scattered along the banks of the Kettle River.

Hardly the place to look for a doctor from New York who specialized in pediatrics in Los Angeles and became the faculty expert on drug abuse at U.C.L.A. But that is where I found him, seven miles up a long winding dirt road bordered by ponderosa pines and straight-down drops with nothing to stop you for 60 feet.

Up there in those dry hills Marc, Joanne and two sons have put together a very productive small homestead, complete with log cabin (owner built), large garden, yard full of chickens and a family cow.

You originally bought land and settled in the Kootenay Mountains in the midst of a rather large alternate-culture community. That, in itself, was a dramatic change from your former life style. Why did you leave for an even more remote location?

Well, for several reasons I suppose. I made the mistake of starting a private practice in my own home for one thing, which is an impossible situation.

How so?

You can't separate yourself from your job and have any time apart from it when you do it that way. I did it for two years but things were just too hectic and so many people were moving into the Slocan that it was becoming too crowded as well.

So, my friend Tom and I looked around and found this place. It is an ideal location. Rock Creek people are essentially very self-sufficient. They haven't had a local doctor for 25 years so they have learned to deal with all of their minor ailments on their own, and they don't come running to me every time someone has a sliver or the sniffles.

They have not developed a dependency on constant prescriptions and hospital care like their counterparts in more developed areas. This is the kind of community that is excellent for any doctor who is seriously considering coming to the country.

Remote isn't such an awful place to be. This isn't total isolation; we've got vehicles and a telephone, and land that is called remote is always a lot cheaper. I didn't want to be held down to a 30-year mortgage and be hooked into the system more than necessary.

You don't have hydro here, or hot running water or television. Are there any of the amenities of city living that you miss, now that you've lowered your standard of living?

Don't think for a minute that I've lowered my living standard. That is the wrong way of looking at it. We don't have colour TV or go on ocean cruises, but I have never lived so high, physically and spiritually.

I have my own food to eat in abundance and it's of finer quality than you can buy in any city for any price. I've more freedom, peace and quiet, good neighbours. The wood that we heat with is free and it helps to keep me fit as I bring it in. And we will have hot running water as soon as I hook up the pipes to the cookstove. We have a 12-volt battery to run the radio, telephone, cassette tape deck and a few light bulbs. What else do we need?

I would like to replace the battery with my

own power in the future, but wind power is not practical up here in the mountains, and there isn't a lot of water here to be harnessed either. Of course, there is always solar power and we are looking into making a swamp in our upper corner useful. Right now I charge my batteries down at the office.

Where is your office? It's said that you doctor out of an old house in town.

Oh no, I have a more unique doctor's office than that. I have a couple of rooms in the oldest continuously operated hotel in B.C. and they are right next door to the pub and the restaurant. A great central place.

How many hours a week do you spend doctoring?

The clinic is open three days a week for a total of about 10 or 11 hours. Then, I probably spend three or four more hours a week on emergencies and things.

Do you feel that you provide adequate health care in so few hours?

Yes, I do. People don't need doctors and hospitals nearly as much as we are all taught to believe. The great gains in medicine in this century have been in public health and sanitation, not in the doctor's office or hospital.

If people are encouraged to be self-reliant, they will be. For instance, a person comes in saying that he is constipated, a common complaint; I can either give him a prescription which won't do him any good and will bind him to me and medicines forever, or I can say, "Look, what are you eating?" And then we can talk about whole grains and fresh vegetables and he can start looking to himself for getting back his health.

Hospitals too, are grossly over-used. Except for serious illness, and there's even some doubt then sometimes, there are few things that can't be looked after at home.

Take heart attack victims. They are usually rushed to the intensive care unit. Have you ever seen one? Full of machines puffing and beeping — a frightening place, enough to give someone an adrenaline rush just being there. If a person survives a heart attack after the first hour, chances are just as good, maybe even better, that he will recover at home.

Doctors aren't God; we don't hold the key to health. We have information that can help each person discover his own key.

Do you try to keep up with new developments in medicine?

Yes I do. I've become a very avid medical journal reader and I make sure that I take at least two post-graduate courses each year to get new information and to rub shoulders with other doctors. All post-grad courses are held in the winter because most doctors take summer vacations, so it works out perfectly for a homesteader. Winter is kind of an off-season on the farm anyway.

In the winter, when your road is piled high with snow, do you ever wish that you were closer to the valley?

Sometimes that thought goes through my head when I've had a rough day and I have to go such a long way down, but I don't think it for long. The advantages outweigh the disadvantages considerably.

Being remote, like this, causes people to really take stock of their situation before they arrive on my doorstep or call me to come out, so an emergency is usually really an emergency. I also don't have to listen to logging trucks all day up here or semi-trailers all night.

Does your credibility as a doctor suffer because of your unorthodox life style?

It may have, a little at first, but I took it slow and put up all of my certificates and official papers for people to check out. It didn't take long before my looks and my life style didn't matter much. I am sure that there are some people yet who are a little leery of me, but then no doctor in any situation has 100 per cent popularity.

You communicate with other members of your profession when you send one of your patients to a specialist or to the hospital. How do they relate to you?

Very well. There are few doctors who have any negative reactions to me and what I am doing. I try to be as straight and honest with them as I can and not to wave my hippie image or life style like a flag, and I find that I am treated quite well. If you don't go out of your way to antagonize people you usually get a straight deal in return.

Would you encourage other professional people to make the move to the country?

Yes, I sure would. Homesteading is the only way that I know of for people to eat properly and stay healthy anymore. You can buy very little in the supermarkets that doesn't have a list of ingredients as long as your arm — and most of them artificial and not tested for human consumption.

Some of the ones that *are* tested are only studied for their short term effects on adults. No one seems interested in finding out how much children can consume without causing damage.

Each person has to take control of his own situation, and I can't think of any profession that couldn't in some way adapt itself to country life.

Rural Arabesque

You can go back to the land but can you take your customers with you?

By Sharon Airhart

Eighteen months ago, Ron Clarke operated a successful Toronto silkscreening business and looked like any other big-city businessman: dress shoes, suit, white shirt, tie and tight collar.

Today when Ron Clarke drives into the city to do some business he's in thick-soled work boots, blue jeans, a comfortable plaid jacket and a tuque. More than one of his old customers has greeted him with a clap on the back saying, "Ron, you s.o.b. . . . I really envy you."

And with good reason. A year and a half ago, Ron, 60, and his wife Betty, 41, packed their 20-year-old silkscreening business aboard a tractor-trailer and moved it, rack, screen and press, to a 118-year-old white clapboard schoolhouse in the village of Carrying Place, 120 miles east of Toronto on the shores of Lake Ontario.

The Clarkes not only escaped the tension and traffic they had come to detest in the city, but retained all their old clients in Toronto. Today they produce the same amount of work, but do it without the expense of two $9,000-a-year assistants they had employed in the city.

Not that moving the established business did not pose its own problems: aside from the tractor-trailer load of what Betty describes as "big stuff," the Clarkes themselves moved smaller items over the course of more than 20 trips with their van.

"Reorganization was our biggest problem," says Betty, "and there is no way you can plan for it other than preparing to accept a period of utter chaos."

While an addition was being put on the schoolhouse, valuable equipment had to sit outside rusting in the rain. Other tools were lost for three or four months as the Clarkes tried to sort through their disorganized equipment.

"It was a trying time. Bills kept coming in, and production was almost at a standstill," says Betty. Costs involved with the move itself, however, posed no problem. "We had planned for over a year and were prepared to foot the bill."

After four months things had got back to better than normal and in the first year at their new location, the Clarkes' business showed the same output as it had the previous year in Toronto.

"And the lack of tension in the country means a better product," says Ron.

Freedom from distractions is the major reason why the Clarkes can produce so effectively in the country. "In Toronto we were bothered by a constant barrage of salesmen," says Betty. "Sometimes Ron would waste whole days with them." Dealings with their customers took longer than business required. Even coffee trucks proved a constant distraction for their employees.

"Now when we go to work, we can concentrate. There are no interruptions. We know that we are really producing."

Today when customers call, their calls are long-distance and therefore shorter. Salesmen are scarce and coffee trucks nonexistent.

Twice a week, Ron drives into Toronto to serve his customers. "I actually enjoy the drive. It gives me a chance to be alone, meditate and evaluate."

The arrangement seems to suit the clients, too, for even after a year and a half, all of the Clarkes' old customers still deal with them. And now when a client does manage to venture into the wilds to visit, the Clarkes are genuinely glad, and have time to socialize.

Lower overhead is another advantage the Clarkes find in their transplanted business. "It got to the point that it was costing me $100 a month just to get my garbage hauled away,"

Cliff Kenyon

Ron says. "Now I just throw it in the back of the truck and take it to the dump myself. Free."

Even though it is located in Carrying Place, Clarke-Ad is still very much a Toronto company. "We have intentionally tried not to solicit local business," says Betty. "We don't want to compete with established operations in this area."

Far from reacting in an unfriendly way to the outlandish activities that were suddenly taking place in the old schoolhouse, neighbours, according to Betty, "bent over backwards to make us feel welcome."

"But a few of our Toronto friends said we were crazy when we informed them of our intentions."

When asked if she misses anything about the city, Betty, without hesitation, answers: "Not one single thing. You couldn't drag me back to that bloody place."

Betty confesses that their first rural tax bill was so miniscule they both broke out laughing: "It was peanuts." Mortgage payments on their new workshop are more frequently butts of jokes than the cause of headaches.

The Clarkes advise anyone else contemplating transporting an urban business to more pastoral surroundings to "be ready to hang in there and persevere for the first year."

"It's a matter of faith," Ron explains. "We've learned to slow down and accept things. We're members of the local church and that has been important. We're making good friends, too. This life is so far ahead of what we had in the city that I wouldn't have believed it."

Back To Main Street

Escaping suburbia with a loaf of whole wheat bread and a bucket of natural peanut butter

By Thomas Moffatt

The first day Sylvia Walker's health food store in St. Stephen, New Brunswick (pop. 6,000) was open for business, three customers entered.

The second day no one did. Now, three years later, her small health food store in rural New Brunswick is thriving, and shows every sign of a solvent future.

For Sylvia Walker, her husband, Eric, and their family, the move to 100 acres of rolling New Brunswick farmland and a century-old house less than a mile from the sea, represented a clear instance of going "from one extreme to another."

Sylvia, who was then 40 years old, had been a housewife for 18 years in (hold your breath) a subdivision outside Richmond Hill, a Toronto suburb. "A terrible place," to use her own description. Throughout this period Sylvia held a series of dull office jobs.

But along with obvious benefits, the move back to the land posed a series of unexpected problems for the ex-suburbanites.

Not least among these were difficulties related to money. In order to meet bills amounting to $3,000 for the move down, to purchase livestock, fencing and machinery, and to pay for renovations, Eric took a job in his former field of personnel and industrial relations. Sylvia, too, cast about for ways to add to the family coffers without sacrificing her new found freedom from punch clocks and 9-to-5 days.

She found a solution in health food stores. "For 10 years I had been interested in nutrition. I even ran a health food store in Richmond Hill for a year."

But Sylvia quickly discovered it is a bumpy and often twisting road to the successful opera-

Thomas Moffat

tion of a health food store, in an area still unexposed to new trends in nutrition.

"My first step was to educate people," she told us. "They felt their food was good and nutritious: 'If it was wholesome enough for my grandparents, it's wholesome enough for me.' True, but I had to convince people that today's food is really *nothing like* what their grandparents ate."

Sylvia used a series of what the callous would call sales gimmicks to draw reluctant townspeople into her store. "There was no bakery in the area," she said. "So I made a few loaves of whole wheat bread, and gave away all of that first batch." After a few days she found herself swamped with so many orders that she could not keep up with the demand.

Homemade peanut butter and honey also served to draw in customers. "I offered good prices and people could buy in any quantity they desired. Also, it didn't take long for them to discover the difference in taste," she said.

After she had acquired regular customers for these basic items, Sylvia began coaxing them to try a more exotic health food fare.

A big break came when a group of doctors in nearby Maine began recommending unprocessed bran to patients with digestive problems. Sylvia's store was the only outlet for unprocessed bran. "This gave health food respectability here. It removed the 'hippie' connotations some people associated with it," Sylvia said.

Yes, Sylvia would do the same thing again, and, admittedly, she'd be much wiser the second time around. "Anyone considering a similar rural occupation should be certain that he is choosing something he truly enjoys," she told us. "Also, pick a field in which you have experience; you'll need it when the going gets rough."

"I've seen several health food stores fail in this area," she said. "They did so for a number of reasons."

First, they lacked capital. "I invested $2,500 in this venture," Sylvia said. "The lowest investment I could have got by with was $2,000. A comfortable amount would have been $5,000."

Promotion is a second vital ingredient of success. "Advertise," she said. "It definitely pays to advertise, and I wish now that I'd done more during the early stages.

"Remember always that you're going to have to educate people, to create your own market. People are reluctant to change their eating habits."

Although Sylvia had previous health food store experience, she immediately found that she was deficient in managerial abilities. "I'm so enthusiastic about nutrition that I can always sell health foods. Organization and accounting are equally important but, as I discovered, very difficult to master. A beginner should keep this in mind."

One clear advantage of a rural enterprise over one in the city, according to Sylvia, is that overhead, particularly rent, is cut drastically. "It wouldn't have been possible to open a store in the city on anything close to my budget."

The Walkers are now very ready to laugh at their first experiences, particularly with livestock. "We were born in the city, and so everything we have done we have had to learn ourselves and I'm afraid that some of the things we have learned have been wrong things," says Sylvia.

"Someone phoned soon after we moved in, and asked if we wanted to buy two milking cows. How could we refuse two Jerseys? So we brought home these animals that seemed like monsters, and then realized we didn't have a milking pail. The chap who sold us the cows patiently gave us advice, showed me how to milk, and said it wouldn't take long to learn all about cows, which gave us a little confidence. That night sitting around the table we decided that the only reasonable thing to do was for everyone to learn how to milk. We all took turns for the first two weeks. How those cows stood it I really don't know."

The good things in their life style? "The livestock have provided us with some of our most cherished moments, like when the first calf was born and Eric with our son Brad wheeled it into the barn in a barrow. There is something very peaceful and satisfying about sitting in a warm, fragrant barn — nothing in the city can give exactly the same experience."

The New Farmers

Self-sufficiency with a small orchard and a cider press

By George Peabody

Stephen Homer

Early every Saturday morning from September to April, Keith Helmuth sets out from North Hill Farm near Debec, New Brunswick on a 60-mile drive to Fredericton, taking 200 gallons of freshly-pressed, natural, sweet apple cider to the weekly farmers' market.

Later in the morning Ellen Helmuth makes a shorter trip to Woodstock with cider for that city's farm market.

Fifteen-year-old Eric and 13-year-old Brendan accompany their parents, helping with the marketing, just as they help with actual production. Apple cider, for the Helmuths, is a family business, and the economic mainstay of the life style they've been building since arriving in the Maritimes five years ago.

Groundwork for the Helmuth's return to the land began long before that, however. "Both Ellen and I were born into Mennonite communities," says Keith. "That gives us a deep-rooted agricultural background. We've both always been involved in gardening and producing our own food — even when we lived in cities, we always managed to find someone outside who'd let us use some land for a garden."

In 1968 while they were both on staff of Friends' World College (Keith as a librarian, Ellen as a dietitian) in Long Island, New York, they made a decision to seek a life style that didn't involve living in urban areas. After spending a year in east Africa under the auspices of the college, they settled temporarily in Vermont while looking for a permanent home.

"We felt that a farm situation would be better for the children," says Keith. This and a concern for the quality of their food were major reasons for the move.

"It was more than just wanting to grow our own food," he adds. "Ellen and I joined the Society of Friends when we married, and there's a very definite Quaker concern with the concept of 'right livelihood,' the idea of organizing your life in a way that is consistent with the cycles of the world." For them it meant earning a living from farming.

They first considered market gardening, but later opted for making cider. Keith found work in a commercial orchard by the simple expedient of inquiring at each apple farm within driving distance until one hired him.

He worked with an elderly Vermont farmhand, and Keith describes the experience as

31

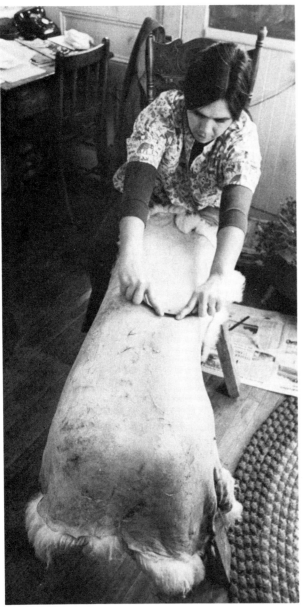

Stephen Homer

middle-aged and new ones, a dozen different varieties, though the bulk are either *Wealthies, Hobos, Cortlands* or *McIntosh*," he says.

During the first year of operation the Helmuths used a hand press and took all their cider to the Fredericton market. Response was excellent, prompting them to invest in a larger hydraulic press equipped with an electric grinder. Business doubled during their second season and is still showing steady improvement.

"Most of our sales are through the two markets," Keith says, "though three stores carry our cider, selling it to regular customers."

Of the 200 gallons that go to Fredericton each week, roughly 70 gallons are bottled in half-gallon jugs filled by the two children. The rest is brought to market in bulk and is sold to people who provide their own jugs.

The two farmers' markets have been crucial to the success of North Hill Farm, and still fit in nicely with the Helmuth's economic thinking.

Keith claims that there are a lot of what he terms "niches" — opportunities for meeting rural needs on a small scale and making a living — like this one, that can be evolved in contrast to large-scale economic agriculture.

"In anything like this," he adds, "it's marketing that's the stickler. No matter what quantity or quality you produce, it won't be much help in making a living unless you can find a way to market it.

Cider is the basis of North Hill, but it's not the farm's sole product. The Helmuths have 44 sheep and eight goats which, by the way, are quite content to derive a considerable portion of their diet from apple pomace, the pulpy leftovers after the apples are pressed.

In keeping with what Ellen calls "developing cycles that follow throughout the year," they also sell bedding plants from their greenhouse in spring, and produce from a two-acre garden.

They encountered a few problems when production in their orchard slumped after they switched to organic methods. These problems have since been overcome and they are now able to support themselves without looking for jobs away from the farm.

The key, they feel, is that North Hill Farm today is what they always intended it to be — a family enterprise.

worthwhile and recommends similar action to anyone considering a cider operation.

After wintering in Vermont, Keith and Ellen moved to Glassville where they rented a farm for a year before buying their present 250 acres, which boasts seven acres of working orchard.

According to Keith, this orchard dates from 1913. "There are a lot of old trees mixed with

Carl Pepper & His Amazing Solar Heating Machine

Alternative energy without a mind-boggling price tag

By Shaun McLaughlin

A sly smile crosses Carl Pepper's face when he recalls the day he mounted the back of his monstrous 20-by-40-foot solar collector while a frankly skeptical alternative energy entrepreneur waited at the base of the panel.

"It's too simple, too cheap. . . You're wasting your time with this thing," he'd been told, and, as a demonstration of his invention's capabilities, Pepper poured a pitcher of cool water onto the panel at its upper edge. Rushing down the steeply angled matte green surface of the heat collector, the water splashed into the waiting hands of the entrepreneur, who recoiled with surprise and slightly scalded hands.

That particular disbeliever went away convinced, as have others who have seen Carl Pepper's solar collector heat water, even in midwinter in Granton, Ontario. One February day, for example, the panel took 65 degree (F) water from the cellar storage tanks and raised it to 140 degrees. The outside temperature that day was three degrees.

A 40-year-old electrician, Carl Pepper has built the system for just $1,300 and finds that it is able to provide 55 per cent of the heating needs of his sprawling, 3,200-square-foot owner-built home. An engineering student, who has monitored Pepper's collector very carefully, estimates that, with the projected increases in fuel

33

Carl Pepper with his owner-built, 800-square foot solar panel near Granton, Ontario (total cost $1,300).

oil, it will save its inventor more than $3,000 per year by 1996.

In the short run, Pepper expects the system to pay for itself by the spring of 1978, the end of its third winter of use. Although the townspeople of Granton (pop. 365) considered Pepper and his wife a trifle crazy when the couple began work on the collector in August, 1975, some have begun to come around. One neighbour has installed a solar heater for his swimming pool and others eye the huge panel — it is the largest free-standing solar collector in Canada — with the knowledge that it is quietly proving the solar nay-sayers wrong.

"One of our greatest feelings after our unit went into operation was on December 21, 1975 (winter solstice, the day the sun is at its lowest point in the sky)," recalls Pepper.

"The day was bright and sunny and the temperature outside was 11 degrees (F). I was standing at the drain outlet of the collector, and the temperature of the water coming off the panel was 111 degrees. That day we heated 1,200 gallons of water from 65 degrees to 92 degrees in *four hours.*"

The beauty of Pepper's design is its simplicity and avoidance of expensive, hard-to-find components. The collector uses water as the heat-transfer medium in what is known as an open system. Rather than being contained in tubes exposed to the sun, the water trickles down the face of the glass-covered collector and into a trough.

The heated water flows into two metal tanks in the Pepper basement. This corner of the basement is sectioned off, and cold air ducts from the house empty into the heavily insulated room. After being warmed by passing around the hot metal sides of the tanks, the heated air is fan-forced throughout the house.

Pepper defied conventional thinking about solar heating in several respects. Rather than expensive copper plates to absorb the sun's heat, he used cheap galvanized tin and enclosed the panel with salvaged glass. Most collectors today have black surfaces to maximize heat absorption. Pepper used a matte green. This colour, according to a University of Western Ontario (London) study, is five to 10 per cent more efficient than the customary black.

As Carl Pepper says, "Just look at the fields and forests to see what colour nature uses to collect solar energy."

He saved thousands of dollars on the storage system — the tanks cost $125 — by find an alternative to the swimming-pool-sized storage chambers incorporated in other solar homes built with government funds. Ironically, the tanks had seen former use on a fuel oil truck. Getting them into the basement involved opening a section of wall and part of the house floor.

His insulation for the heat storage room and the ductwork consists of layers of corrugated board from foraged and flattened-out cardboard boxes.

Pepper admits that the panel's size is a drawback. He wanted to mount it at a 58 degree angle to make best use of the sun's position in the months of January and February, and this necessitated its free-standing, rather than roof-top, structure.

"A solar collector can be built for less than a dollar per square foot," he told a recent symposium in Toronto. "Anyone who wants to can do it, and there's no need to pay $30 to $40 per square foot for a system made by someone else." His remarks, made in the company of solar component manufacturers with displays in the lobby of the St. Lawrence Centre, may have galled them, but were warmly received by the packed house of 500.

Pepper will also tell you that solar collectors can easily be built into the roofs of new homes right now. Further, he says that the savings in shingles and plywood normally used in roof construction come close to offsetting the price of glass and metal used in a roof-mounted collector.

In his own home, Pepper hopes to improve the efficiency of the solar system to the point where wood heat will serve as the sole back-up.

He says he has no plans to patent any present or future design or innovation, or to cash in on his findings. Pepper, tough-minded and independent himself, believes people should help themselves; anyone is free to inspect his collector and ask questions. He estimates that 2,000 have already visited the house. Pepper does sell a book with details on the construction of solar energy systems and also offers his own differential thermostat for $60. This device activates the pump which feeds water to the collector any time the temperature in the collector exceeds the storage tank temperature by nine degrees. The thermostat can be reversed in the summer to feed water through the collector at night and thus cool the house. He builds these thermostats himself, including his own printed circuits.

Like others who have attempted to adapt a conventional home to solar heat, Pepper would like to start again, from scratch. He hopes to sell his one-acre lot and build a passively heated underground house with a south-facing greenhouse and several skylights.

Complete energy self-sufficiency is his goal, and his plans include a small hydroelectric generator to be powered by a fast-flowing creek. After that — you guessed it — he'll start thinking about a gasless automobile.

Dis Ole House Got Sol

Retrofitting for solar energy:
"There is no such thing as a hopeless building."

By Barry Estabrook

For decades the nondescript old farmhouse, conveniently tucked at the nether end of a seldom-travelled concession road, had served as the local emporium of after-hours libations, where those who knew of it could find a dram of home-brewed spirits on a dark night.

Recently purchased by Marilyn and Richard Hopkins, a couple in their 20's, the house no longer boasts moonshine as a topic of importance. But sunshine definitely is, as the Hopkins have recently set about the task of retrofitting their century-old, uninsulated frame dwelling to make the best use of solar heat.

They are being helped by solar designer Greg Allen, whose belief that "no house is hopeless" has already been made manifest by cash savings in the Hopkins' fuel bill.

Too often, the word "solar home" summons images of other-wordly, rhombic dodecahedrons wrapped in glistening webs of panels, pipes, ductwork and heat exchangers.

But Allen and other experts are beginning to take a look at solar energy in a new light — a perspective that suggests that the Great Solar Hope might not be a $300,000 polyhedron, but the millions of squat, one-family dwellings and drafty farmhouses that dot the continent.

Skeptics commonly say that solar heat is fine for anyone, in, say, Miami Beach, but it can never be practical in Moose Jaw. The fact is that Canada absorbs enough sunshine each year to satisfy its energy demand 7,000 times.

A study of residential energy use in British Columbia showed that the energy in the sunshine striking an average house was three times more than the total heating and lighting energy used in that house in a year.

Assuming that not everyone is going to tear down his old home and start afresh in order to capture a fair share of all this wasted energy, retrofitting must come of age.

Fortunately, the retrofitter need not look 10 years down the road to the time when the cost of modifying an existing house will repay the investment. Minor adaptations on almost any conventional home will make it a more efficient heat trap, and modest expenditures will result in fuel bill reductions of up to 50 per cent.

Greg Allen himself is no stranger to high technology solar heat. His company, Amherst Renewable Energies, already has one icosahedron to its credit and manufactures both flat plate air and water heating panels. Still, results from a passively heated solar house Amherst built last year — a house that gleaned 60 per cent of its heat from the sun with nary a panel — gave Allen pause to consider the potential of passive solar heating.

(In contrast to systems with active solar panels, a passively heated house catches its sunshine through windows and traps it in the well-insulated mass of its walls and floors.)

When Allen's friend Dick Hopkins purchased the old frame farmhouse (whose thirst for fuel oil totalled 600 gallons the previous winter), they looked for passive — and inexpensive — ways to make use of solar energy.

Anyone who has ever scratched his head over a heating bill can probably learn something from the Hopkins/Allen plan. Although the house is still being worked on, it is heartening to note that the Hopkins' driveway will not often be darkened by the shadow of an oil truck this winter.

Greg and Dick began by standing back and taking an imaginative look at the old house to see if it possessed any traits that would lend themselves to solar applications. Despite his home's uninsulated two-storey facade, Hopkins did have some factors working for him.

Illustration: Art Emery

It seemed that the original carpenters had, for once, given some consideration to modern-day energy consciousness by placing four large windows on the south wall of the house, leaving only one facing north into the cold winter winds.

A large silver maple in front of the south side would provide more than a convenient place to fasten one end of a clothesline. It would give shade in the summer and then co-operatively shed its leaves to allow the winter sun to fall upon the face of the house.

Said Allen: "The steep, south-facing tin roof seemed to be begging for a coat of black paint and a transparent covering. In short, the house's sharp gables made the perfect base for a low-priced, home-grown solar collector."

A three-step course was plotted, leading eventually to energy independence. Would-be retro-fitters will find the three phases convenient in that each is self-contained and allows progress to come at a pace amenable to individual life styles, inclinations and pocketbooks.

PHASE ONE

There is no sense even contemplating the purchase of solar panels until your house is properly equipped to retain warmth. By insulating, adding weather stripping, caulking to prevent drafts and improving windows, energy consumption can be reduced by 30 to 50 per cent. The simple, do-it-yourself steps that make up most of Phase One will begin paying for themselves immediately and will completely pay for themselves in about four years if fuel prices increase only 10 per cent each year.

Step One. On bright winter days a window serves to trap sunshine within the house, but at night it acts as a radiator for all outdoors. The addition of heavy curtains to be drawn at night and on dark days will help insulate a major area of heat loss. The Hopkins put their faith in curtains that reach the floor and which are stabilized by sewing weights into the bottom. The heavier the draping material, the better its R (insulating) value.

Step Two. Old homes are notorious for the condition of their window frames. Hopkins replaced his deteriorating frames with high-quality (if expensive) double-pane units from a builder's supply house. (A two-foot-by-three-foot window with single-pane glass gives off heat at the same rate as a well-insulated wall nine feet high and 20 feet long.)

Step Three. A caulking gun was purchased and Hopkins and Allen were not stingy with butyl caulking as they attacked all windows and doorframes. Doors received weather stripping.

Step Four. Inside storm windows were installed. The retrofitters had a choice between a clear film such as Mylar (their choice) or less attractive plastic. In either case, the storm windows are placed inside where they are protected from wind and damaging ultra-violet rays (which are filtered out by window glass).

Step Five. Insulation is the most expensive step in Phase One, but a government rebate reduced the cost by $350. Because of time limitations imposed by a busy work schedule and an approaching cold season, Hopkins had a contractor blow in cellulose insulation — four inches in the walls (R16) and six inches in the attic (R20). Hopkins could have saved $300 by leasing an insulation blower from a rental firm and doing the work himself.

Step Six. If there happens to be a stand of towering pine or spruce blocking the sun from the south side of your home, you might be well advised to add them to your stock of firewood. On the other hand, a protective windrow (perhaps fast-growing poplar or conifers) might be planted to form a windbreak on the north side.

Step Seven (optional). The Hopkins covered their home with vertical wooden siding, giving the traditional lines of the building an updated, but earthy look. At this stage, some north-facing windows and unnecessary doorways can easily be shut off permanently.

Too, south-facing windows might be enlarged, thereby creating both a more cheerful and more energy-efficient home.

Step Eight (optional). An exposed stone hearth and fireplace can mean much more than aesthetics. Rock makes an excellent storage medium for heat, and if the sun is able to strike a

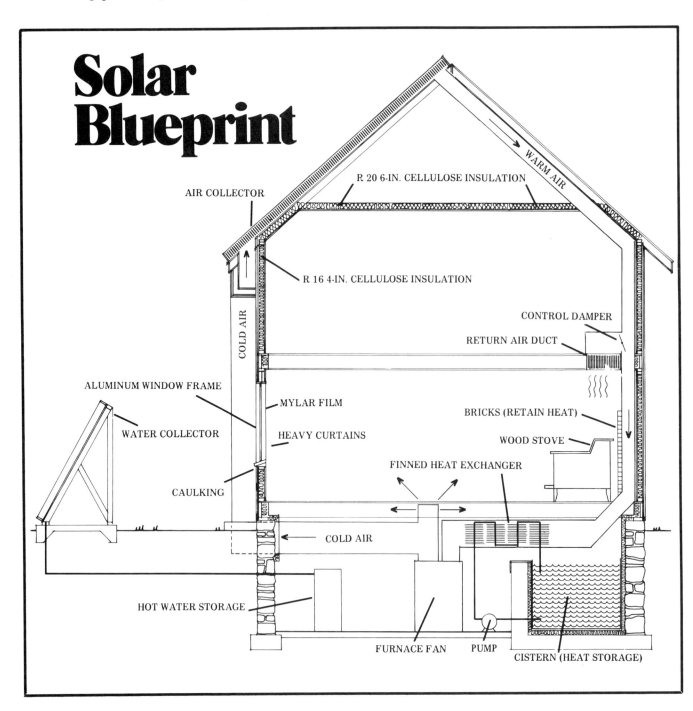

Solar Blueprint

AIR COLLECTOR

R 20 6-IN. CELLULOSE INSULATION

WARM AIR

R 16 4-IN. CELLULOSE INSULATION

COLD AIR

CONTROL DAMPER

RETURN AIR DUCT

ALUMINUM WINDOW FRAME

MYLAR FILM

BRICKS (RETAIN HEAT)

WATER COLLECTOR

HEAVY CURTAINS

WOOD STOVE

FINNED HEAT EXCHANGER

CAULKING

COLD AIR

HOT WATER STORAGE

FURNACE FAN

PUMP

CISTERN (HEAT STORAGE)

wall of rock within the house, warmth will be put in reserve to be released slowly during the evening.

Some owners of stone and masonry homes have reported good results by stripping the interior of a north wall that is struck by sun from south-facing windows (being certain, at the same time, that the *exterior* surface of the wall is well sealed and insulated.)

PHASE TWO

Step One. The standard reaction of those unfamiliar with the new airtight stove when they enter a snug, well-insulated home goes something like this: "Do you mean that little stove heats the whole place?"

Long-burning wood stoves are becoming ever more feasible, and Hopkins selected a popular airtight that he felt fitted the size of the house and the family life style. The stove was situated on the ground floor, as centrally as possible, so that heat would be dispersed to all corners of the house.

The problem of heat rising and building up in the higher reaches of the house will be solved when Hopkins carries out a plan to adapt the air ducts of the oil-heat system to circulate wood-heated air.

An air intake will be situated in the warmest part of the house, (in this case, in the ceiling above the wood stove) and warm air will be returned to the furnace and forced back through the regular household ductwork.

A brick pad was laid down under the stove and walls were constructed on two sides both as fire protection and to help retain heat from the stove. Hopkins ordered their stove with a fac-

Before the major retrofitting steps began, the Hopkins' house was no one's idea of a potential solar home.

tory-installed, metal-tubing system that, when connected to a tank above the stove, provides a steady supply of hot water whenever the fire is burning. (Working on a natural thermo-siphon principle, this wood-fired hot water heater needs no pumps or mechanical gadgetry.)

Step Two. Greenhouses, as anyone who has ever worked in one can tell you, are able to collect great amounts of heat when the sun shines. Allen and Hopkins opted for a well-designed, permanent greenhouse. (Others might choose to erect a plastic-covered model for less money, or to install double-glazed glass in a sunporch during the winter months.)

The Hopkins' greenhouse will have a floor area of 160 square feet and will, of course, be constructed on a sunny southern corner of the house. The wooden frame will be covered with two layers of Tedlar film (a strong, transparent material often used in solar panels). Ducts will allow heat to be moved from the greenhouse into the house proper, but the two will be insulated from each other, so that too-hot summer temperatures can be avoided and the greenhouse can be shut down in coldest weather.

PHASE THREE

Step One. While some argue that solar space heating has yet to become practical for northern areas, there can be no argument that solar do-

A less ambitious retrofitting plan would call for off-the-shelf components, such as these Grumman solar panels.

mestic water-heating systems are ready to go. Many companies are now offering proved designs, in the whole-package price range of $1,000 to $1,500.

With temperatures inside the operating panel cruising at 150 degrees Fahrenheit — and able to reach 350 degrees — there is no doubt that present day panels can make water hot.

Hopkins' simple system will employ two flat-plate liquid collectors situated in a never-shady spot not far from the house. Buried tubing carries the heated fluid into the cellar where it is circulated in an exchange unit in the water tank, thus transferring heat from the panel into useable hot water.

A system as basic as this can supply three-quarters of household water-heating needs and result in savings of more than $100 per year for the average family.

Step Two. The big move comes now, when the farmhouse roof is painted black and covered with a framework to support a transparent Tedlar covering. The roof has now become a built-in solar air heater.

The Hopkins' basement (see diagram), like that of many rural homes, contains a long-unused 1,000-gallon cistern. With some waterproofing and insulation, this will serve as a ready-made storage area for solar-heated water.

The forced air furnace will no longer be fired, but its fan will find service, blowing cold air from the basement up to the roof, where it will be heated as it crosses the black-painted 14-by-24-foot roof collector area. It is now returned to the basement and forced through a finned heat exchanger (in this case, a truck radiator). A separate flow of water from the cistern through the radiator is heated by this flow of air.

At night or on dark days, the process will work in reverse: hot water in the cistern will flow through the radiator, heating forced air that will be pushed throughout the house.

A temperature differential thermostat with sensors on both the roof and in the cistern assures that no air is circulated across the panel when this would result in a heat loss (at night, during periods of extended cloud). All the equipment used is available off-the-shelf from plumbing or building suppliers and/or from solar hardware sources.

The Hopkins estimate that the entire three phases of the plan will cost them close to $5,000. In return, they will have a house that is oblivious to the machinations of certain powerful Arab gentlemen and their counterparts in Alberta and Texas. There is, of course, more involvement required — pulling drapes, gathering firewood — but the deep sense of security felt at having an energy-assured future must go far toward repaying the modest investment of money and time.

What Price Energy Independence?

Step	Task	Unit Cost	Total
PHASE ONE			
One	Curtains	$5 - $20 per window	$100
Two	Windows	$40 per window	$400
Three	Caulking	$35	$ 35
Four	Inside Windows (Mylar)	10 cents per sq. foot	$ 8
Five	Insulation (contractor)	$1,280 (less $350)	$930
			$1,473
PHASE TWO			
One	Wood Stove	$250 - $500	$350
	Brick Backing	25 cents each	$ 75
Two	Greenhouse		
	Tedlar	50 cents per sq. foot	$256
	Foundation		$200
	Misc.		$300
			$1,181
PHASE THREE			
One	Hot Water System	$1,000 - $1,500	$1,300
Two	Solar Collector		
	Paint	$18 per gallon	$ 18
	Strapping	3 cents per foot	$ 10
	Cover (double thickness)	$2 per sq. foot	$ 672
	Heat Exchanger	$.00 to $200	$ 000
	Pump		$ 100
			$2,100
		Grand Total Cost	$4,754

Cord Wood House

Log home construction for areas where the tall pines don't grow

By Sharon Airhart

In the spring of 1973 John Otvos combined the unlikely ingredients of $850 in unemployment insurance benefits, some neighbourly advice from a local old-timer and 2,500 linear feet of cast-off telephone poles and old guardrails to make a 1,500 square foot woodworking shop that now supports his young family.

Log Butt, Cordwood, Stacked Wall — Otvos' building technique was known by a variety of names in its long-past rural heyday. Although very nearly forgotten, it might well be the final answer to those seeking low-cost, labour-intensive housing in the country.

The price and the construction skills needed are minimal, yet the end product can be sturdy, weather-tight and aesthetically pleasing.

"It's a real backyard way of building something," says Otvos. "After laying up the first few tiers of logs the novice can safely call himself experienced."

The technique — stacking foot-long pieces of wood butts with mortar to form walls — seems to fall naturally in a classification somewhere between stone house construction and log cabin building. John Otvos feels that building with log butts has distinct advantages over both of these.

"A traditional log cabin is quicker to build," says Otvos, "but, unfortunately, trees of appropriate dimensions have become nearly nonexistent east of the Rocky Mountains. The log butt method allows you to use much shorter lengths and you don't have to concern yourself with uniform diameters." Unlike log cabin construction, uniform diameter — even in the same wall

— is not consequential for structural integrity or for appearance.

As for stone houses Otvos is quick to point out the obvious: "Stones are heavy and wood is light. Stone is probably cheaper in some areas, because you can pick up your house rock by rock in the fields."

But Otvos contends that the log butt method is much quicker than slipform stone construction, and more suited to people new to the countryside and not yet ready for months of hauling and lifting heavy stones. Too, the log butts incorporate wood's properties.

Another important consideration is that one person can easily undertake cordwood construction, a feat impossible in log cabin building with block and tackle arrangements.

A standard chainsaw and a borrowed gasoline-powered cement mixer were the two main tools Otvos used in putting up his spacious workshop, which from a distance takes on the appearance of a stone building.

Work began with the laying of an eight-inch thick and 24-inch wide concrete foundation, poured directly on the ground, an economy that would make most builders wince.

This shortcut was made possible for Otvos by hardpan soil conditions (a combination of clay and gravel almost impervious to water and not subject to major winter frost upheavals). Such a slap-dash foundation might serve for outbuildings or workshops, but Otvos recommends that others begin their log butt homes by digging a basement or deep foundation.

"We've been through three winters, though, and our foundation isn't cracked yet."

The foundation in place, Otvos' next step was to erect large corner posts and brace them securely. Otvos used 12-by-16-inch barn beams scrounged from the remains of a neighbour's old barn in trade for a year's yield of hay from his 20-acre field. Lacking heavy beams, he says that a mock beam can be built with four one-by-12-inch boards. Three should be nailed into a U-shape and braced inside at one foot intervals with two-by-two's. When the fourth board is added to close the beam, a strong unit is formed.

Otvos stresses that corner posts really serve no structural purpose once the mortar has set, but they do act as forms while the walls are being constructed. He feels that the corner posts could be smaller but sets six-by-six as a practical minimum.

"If I were to do it again, I would consider framing the entire structure in logs," he says, citing financial and aesthetic advantages.

The ends of the upright corner posts were next set on the foundation and plumbed to form a 90-degree angle. Stakes driven into the ground served as braces for the posts during construction.

At this stage John erected doorframes (borrowed from an old house) in appropriate places along the foundation. "Remember," John said, "Basically you're building a frame and then filling in the spaces."

Top sills (boards that run along the top of the walls) may be added at this point, or you have the option of waiting until the walls approach within a foot of their final height.

Although the log butts themselves serve no greater structural purpose than fill (strength comes from the mortar) John believes that well-aged cedar is by far the best wood to use because it is resistant to decay. "If cedar is not available in your area," he advises, "find out what type of wood is being used in utility poles and guardrails. It will be the most decay-resistant wood locally available."

Aging is also vital. New wood, he explains, tends to shrink and crack badly. John's mentor went so far as to build an entire home from the mossy remains of a cedar snake-rail fence. Otvos also shies away from using hardwood because he feels the practice to be wasteful.

Your property needn't be blessed with a flourishing woodlot for you to consider log butt construction. A large number of butts for John's attractive workshop were discarded roadside guardrails. "Whenever guardrails are being replaced, the roads department usually has no objection to your making off with the old ones, and if you wait around while they are installing new ones they will usually give you the two feet cut off the tops."

Guardrails were supplemented by broken utility poles that had been left in a nearby city's public works yard. At $2.00 apiece, they were a bargain, and John estimates that there were enough discarded poles in that one yard alone to build five buildings like his.

While the quantity of butts needed will vary greatly according to the size of the structure, John estimates that 10 pickup truck loads would be ample for a modest bungalow. Although log butt construction requires almost twice as much wood as building a standard log cabin, the process makes walls twice as thick as those of a standard cabin and also enables the builder to use parts of the tree not suited to log cabins.

Before cutting the old logs into 12-inch lengths with his chainsaw, John vigorously applied a wire brush to the outer surface, removing mud and grit that would have otherwise wreaked havoc on the cutting edges of his chain. While brushing he also checked to be sure that no unnoticed nails and other metal fragments were embedded in the logs.

Otvos cut each butt separately, but if the wood is straight and untapered, the stockade system (stacking several logs in a rack and cutting through all of them in one stroke) will greatly speed the process.

Mortar was mixed in a ratio of three shovels of mortar mix to five of clean, sharp sand with one shovel of cement added per mixer load. John feels that the sand must be free of clay and pebbles, but still gritty to insure a solid bond.

Otvos recommends testing for "slump" by throwing a handful of mortar against a wall. "If it sticks," he says, "it's right. If it doesn't stick it's too runny." While runny mortar will shrink

COST BREAKDOWN:
1,500 SQ. FT. BUILDING

New lumber for rafters (2x6x24') $175
Used maple flooring from old school. 125
Cement 40
Masonry cement 115
Sand 85
Nails 30
Logs (hydro poles, telephone poles,
    old log cabin) 70
Rough lumber (floor joists, subfloor) 160
Old house for windows, etc. 50
                                    Total $850

away from the wood blocks, stiff mortar dries too quickly and becomes brittle.

An important consideration to remember when working with log butts is that the bond between the mortar and the log is of a mechanical nature. Because wood is an organic substance, mortar does not bond to it in the same chemical way that it adheres to brick or stone. Run your fingernail along a piece of wood with some mortar adhering to it and it will likely fall away. The same feat would be impossible with mortar sticking to a brick or a stone.

This mechanical bond limits the height to which log butt walls can be built. John's walls are 12 inches thick and eight feet tall. Before going any higher Otvos says he would recommend increasing the width to 24 inches. John also recommends driving 12 one and one-half-inch galvanized roofing nails partly into the corner posts so that the mortar will have a surface to grab.

Working in successive tiers, John laid two thin (two-inch) strips of mortar where the ends of the logs would rest. This method insures an insulating pocket of air inside the wall and also saves time and materials. "Filling the entire wall with mortar simply isn't necessary for strength," he says.

"I quickly learned to tap — not pound — logs into place and to scrape excess mortar from log faces as I went along." Construction should progress at a rate no greater than three or four courses per day, and mortar should be pointed about an hour after it is laid, a technique which prevents later cracking.

He suggests that a beginner practice by starting on a back wall until he has the experience and confidence gained through building two or three courses. After this brief apprenticeship, most will be expert enough to move continually around the building one tier at a time.

Because his workshop has no basement, John worked from the inside wall, thereby making it simpler to keep a perfectly flush interior surface fit for application of strapping and wallboard. With a basement, scaffolding would be needed for inside work, although the problem could be circumvented by working carefully from the outside. In either situation, logs of uniform length will prevent problems.

To install window frames John filled the course with mortar until it was level. Rough frames are built, plumbed, braced and spiked into the logs. Installation of the top sill is accomplished in the same way.

Inside, double reflector aluminum foil tacked directly to the log butts, in combination with strapping and drywall, makes a practical and climate-proof interior finish for a house. A person with abundant time and plenty of logs could build himself a very snug and uniquely attractive domicile by building a double wall separated by sheets of Styrofoam insulation, leaving natural butt ends as interior finish. But when finishing the dwelling, Otvos says, it is important to remember that, with time, cracks will develop in any wood and cold winter wind can enter through these cracks.

Floors, rafters, and the roof are built according to traditional methods, and in the case of John's workshop they accounted for half the costs.

"This is a learn-as-you-go building method," John insists, "one easily adapted to personal preferences, availability of materials, and skill of the builder."

Below, John Otvos and the wood-butt building he erected for $850 to house "Wood Synthesis", his furniture making and woodworking shop.

*A Harrowsmith Report
on the new composting and
alternative toilet systems*

Privy Redux

By Barry Estabrook

Sister Monique Gemme is a hyperactive
housekeeper. She punctuates conversation by stooping to run her disinfectant-
soaked rag along a kitchen baseboard or
pausing to polish an already gleaming toaster.

She is no less stringent in the standards of
cleanliness she demands of the five orphans who
live with her in the village of St.-Francois-du-Lac,
60 miles east of Montreal. Beds are drum-tight;
personal belongings nowhere in evidence.

It is a curious anomaly, then, that Sister Monique has allowed tons of mouldering excrement
and kitchen refuse to accumulate in her basement for the past year and a half.

No, the Sister is not guilty of an extreme exercise in "sweeping it under the rug." She and
her wards are using an experimental Clivus Multrum composting toilet installed in their home
by the McGill University School of Architecture's Low-Cost Housing Group. Far from being
an example of poor housekeeping, use of the
Clivus Multrum is a venture into an environmentally harmless manner of treating human excrement.

Architect Witold Rybczynski (rib-chin-ski),
author of *Stop The Five Gallon Flush*, is keeping a close eye on the St.-Francois Clivus Multrum. The final product of its mouldering process will be a rich, dark humus-like fertilizer
(approximate nutrient rating: 20-12-14) far removed from the tons of human excrement,
kitchen scraps, tampons, garden wastes and dis-
posable diapers that went into the composting
system.

The increasing availability of such systems
makes it clear that the day is fast passing when
people of ecological bent can sit with clear conscience on their old-fashioned flush toilets.

Once extolled as the answer to cholera-infest-
ed, fly-breeding open sewers, the water closet
and its role in polluting and unbalancing the environment is becoming more publicly known.
Every flush whisks away five to seven gallons of
formerly pure water, to transport perhaps a pint
of human excrement, or 10,000 gallons per year

for the average toilet flushing person. What happens from there is not one of society's nicer
tales. In all too many cases it goes more or less
directly into the nearest handy body of water.

Says Abby Rockefeller, Clivus Multrum's
North American distributor and a leading advocate of composting toilets, "The flush toilet,
which has had an active life of about a century,
must give way to reason. The damage it causes
far outweighs its benefits."

The St. Lawrence River, receiving all Montreal's raw sewage, now ranks as the world's third
most polluted river, following the Mississippi
and the Rhine. Eutrophication plagues even the
lakes and rivers of municipalities that treat their
sewage. Despite the great expense of such treatment plants, the end product still contains quantities of heavy metals, nitrogen and phosphorous.
Although fine for gardens, nitrogen and phosphorous are disastrous for waterways. Often
compared to amphetamines and their effect on
the human body, these elements produce wild
algal growth which causes watercourses to become oxygen depleted and filled with floating
masses of green algae and bloated, belly-up fish.

The answer to all of this, and to the intense
world-wide need for fertilizers, is very likely to
be a composting system of some sort. For attuned urban planners and macro-ecologists, the
solution may be municipal sewage salvaging centres, where tons of human "waste" is composted
into agriculturally valuable fertilizer. For Rybczynski, who heads McGill's Low Cost Housing
Group, the answer will probably be something
very much like a Clivus Multrum.

Developed 30 years ago in Sweden, the un-
disputed leader in ecological toiletry, the Clivus
(pronounced *cleave*-us) appears to be the most
successful alternative privy designed to date.

Approximately 2,000 of the units have been
sold in Sweden, and the company notes that
none has ever been removed from a home for
malfunctioning. There are currently 800 Clivus
units installed in the United States and about 20
in Canada.

A DARK GLACIER

Having no moving parts and requiring no chemicals or energy input, the Clivus Multrum, despite its futuristic image, works solely on simple natural processes. Multrum means mouldering room in Swedish and clivus is Latin for incline, the key to this composter's operation.

Basically, the Clivus Multrum is a large fiberglass box, nine feet long, seven feet high and four feet wide. It is usually installed in the basement of the house, and two pipes drop into its uppermost chamber — one from the toilet above and one from the kitchen.

Excrement from the washroom and scraps from the kitchen drop into the uppermost chamber. The floor of the Clivus is on a 30 degree angle, allowing the mass of refuse slowly to slide toward the bottom. This process may take up to two years, the biodegradable mass moving like a heavy, mouldering glacier.

By the time it has reached the lower chamber, the material has been transformed into a sanitized compost not unlike clean smelling potting soil. The biological process behind this transformation is aerobic decomposition.

Accomplished in the presence of oxygen, it is unlike the methane producing and foul smelling anaerobic composting, the process that predominates in a simple outhouse system. Within the Clivus a self-sustaining biosphere is created with a dense population of micro-organisms constantly breaking down the fecal matter and vegetable wastes.

To thrive, these micro-organisms require carbon, which must be infused into the system in the form of kitchen scraps, peat moss, grass clippings and other organic refuse. These carbonic materials absorb urine and aid in stabilizing the valuable nitrogen so it can be used by plants. Cellulose present in organic refuse is also necessary for the survival of other micro-organisms which destroy disease-producing bacteria and viruses.

Barry Estabrook

47

Ecological implications aside, perhaps the most common question to be asked about such a system is a simple, "Does it smell?"

The answer is that it shouldn't. Inventor Rikard Lindstrom delights in taking incredulous visitors into his bathroom, bending over the toilet receptacle for his own Clivus and blowing a waft of pipe smoke across the opening. His unit is working so efficiently that the smoke is sucked quickly into the chamber below.

The Clivus Multrum has a tall ventilation stack running to the roof of the building in which it is installed. Acting exactly like a chimney, it allows the venting of carbon dioxide and water vapour while using the heat produced by decomposition to create a natural draft. This draft sends odours up the chimney rather than into the house and it also keeps up the oxygen circulation needed for aerobic decomposition.

Extensive testing in North America and Sweden leaves no doubt that the final compost is free from pathogenic organisms and perfectly safe for application to the garden. According to Scandinavian studies, the lengthy process in which harmful pathogens undergo attack from other soil organisms creates the danger-free final product. Viruses which require a watery habitat, such as those causing hepatitis, are killed more rapidly in a Clivus Multrum than in sewage treatment plants.

As with solar heating, however, the composting toilet is an idea whose time has come but which is held on the verge of mass popularity by cost, technical problems and public wariness.

THE INCREDIBLE BULK

The task of marketing the Clivus to people reared on the flush toilet in North America goes to Ms. Rockefeller, 33, an English professor at the New England Conservatory of Music and daughter of David Rockefeller (chairman of the Chase Manhattan Bank). Her involvement with toilets can be traced to the February 1972 issue of *Organic Gardening & Farming* magazine, in which Robert Rodale discussed the potential of the Clivus and noted ". . . as yet, no American firm has expressed interest in the toilet unit."

He then quotes Carl Lindstrom, son of the Clivus inventor, as saying: "It is too expensive to ship the units from Sweden. Maybe someone reading this article will see here a good business and environment opportunity, and will make the Clivus toilet available to Americans."

So far the opportunity for Abby Rockefeller has been mainly environmental. Both cost and the size of the thing have hampered sales here. The composting chamber's incredible bulk (it fills a small room), coupled with the fact that it is best located directly under both the bathroom and the kitchen, makes it desirable to design a home around a Clivus, rather than just moving one in (although that is possible). The company now manufactures a conveyor tube with an internal screw design to transport refuse and excrement to a chamber not located directly below

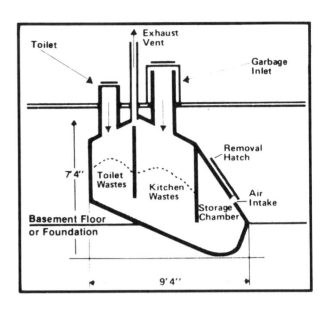

Above, *schematic diagram of the Clivus Multrum composting toilet system.* Below, *Witold Rybczynski, Canadian author of* Stop The Five Gallon Flush, *with test model of the new Toa-Throne, a Swedish design that uses less space than the Clivus.*

Barry Estabrook

Sources

CLIVUS MULTRUM
Crowdis Conservers
R.R.3, Baddeck, Cape Breton,
Nova Scotia B0E 1B0
(902) 295-2275
$1925 complete basic package

CLIVUS MULTRUM U.S.A.
14-A Eliot Street
Cambridge, Mass. 02138
(617) 491-5820
$1685 complete basic package

HUMUS TOILETS
Future Eco-Systems
680 Denison Street
Markham, Ontario
(416) 495-6450
$699 and up

ECOLET
Canadian Inventor Ltd.
Box 541
Don Mills, Ontario M3C 2T6
$675

ECOLET
9800 West Bluemont Road
Milwaukee, Wisconsin 53226
(414) 257-1830
$736

COMPOSTING PIT TOILET PLANS
Farallones Institute
Integral Urban House
1516 Fifth Street
Berkeley, California 94710
$2.50

BOOKS
Stop The Five Gallon Flush!
By Witold Rybczynski
McGill University
School of Architecture
3480 University Street
Montreal, Quebec H3A 2A7
$4.00

Goodbye to the Flush Toilet
296 pages, paperback $7.95
Rodale Books

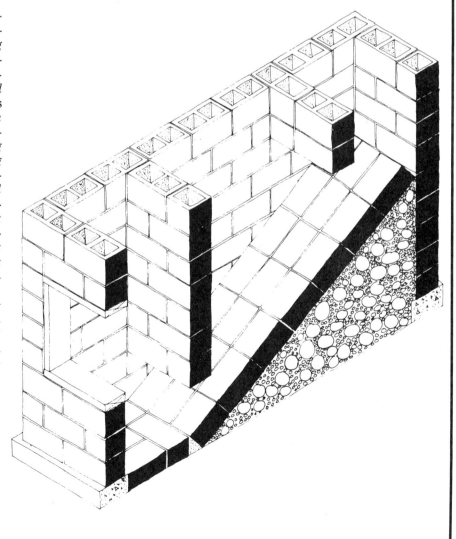

Right, *design for a low-cost, homemade composting toilet built using the same principles employed in the Swedish-designed commercial model. Dubbed* Clivus Minimus, *it is the work of the McGill University Low-Cost Housing Group, which is seeking alternate, low technology designs for third world countries. Fabricated with cement blocks, and given a substantial coat of sealant (bitumen), this unit would receive both toilet and kitchen wastes. In addition, the unit would need a chimney-like exhaust pipe and non-corrosive ventilation ducts to assure that oxygen is supplied to the mouldering heap.*

the receptacles or even outside the house. This, of course, increases the initial cost and requires electricity, thus disrupting the plan for purely natural, unmechanical operation.

Present prices range from $1685, but rise for an installed unit. To this must be added the expense of a septic tank and drainage bed still required by most rural health authorities for disposal of "greywater", the effluent from sinks, bathtubs and washing machines. Compared to "blackwater" (that contaminated by feces and urine), greywater is not as harmful to the environment, especially if the household does not use detergents high in phosphates. Greywater contains only 10 per cent of the nitrogen and pathogenic organisms found in normal household sewage water, and its lack of large particle material permits rapid breakdown. A further advantage is that total waste water is reduced by 50 per cent when blackwater is eliminated.

Various systems for handling such greywater in composting-toilet-equipped homes are now being developed, but government regulations will prove a hurdle when septic systems are threatened with elimination.

THE FLY FACTOR

Although the theories behind the workings of a Clivus Multrum are sound, in practice difficulties do arise. The McGill Clivus has experienced a build-up of liquid in the lower chamber, Rybczynski says, despite the addition of absorbent material. A further problem with the unit is one of odour, most evident on sultry days. The day we visited, the St.-Francois unit was cool and breezy and no smells were evident.

However, Sister Monique noted that on the still humid summer days common to the St. Lawrence valley, odour has become so intense that neighbours were loathe to visit. Rybczynski hopes to overcome this problem by installing a small, noiseless fan in the vent pipe.

Fruit-fly-like insects also plagued the McGill unit for a time. Sister Monique said the problem was so severe that whenever either the toilet seat or the kitchen receptacle cover was opened swarms of flies would emerge into the house.

Prior to the infestation the Sister had been preserving some 300 pounds of tomatoes, and the insects presumably entered the system on the peelings and scraps. Rybczynski feels that insects themselves are no problem as long as they remain in the chamber. Better draft through the fan should prevent them from entering the house until time creates a biological balance that would prevent freak population explosions.

Abby Rockefeller says that there is a two-year break-in period with any Clivus Multrum before such a balance is achieved in the unit, assuring trouble-free operation (the St.-Francois unit has been operative one year). Liquid build-up in the lower chamber commonly occurs during this period, according to Ms. Rockefeller, who stresses that the liquid is not urine and possesses none of its unpleasant traits.

Our examination of the liquid in the McGill Clivus verified Ms. Rockefeller's claims. It was coffee-coloured and gave off no tell-tale odour (it has been compared to the dark run-off water from a forest). Rybczynski says that the excess of liquid has diminished as the system matures. He is working on the problem by feeding the Clivus large amounts of leaves, sawdust, peat moss and other highly absorbent materials. The better draft is also expected to speed evaporation.

As for the fly problem, Ms. Rockefeller suggests the installation of screening and exercising care to keep all covers closed when not in use. She lists this as "a rough edge that hasn't been refined out of the system."

NO CLIVUS PLUMBERS

Despite the drawbacks, Rybczynski (who discusses 66 different toilet systems in his book) is convinced that the scientific principles behind the Clivus design are basically correct.

Below, *The Battery — St. John's, Newfoundland neighbourhood with medieval sewage problems.*

Dick Green

50

However, he cautions people contemplating installation of one to be prepared for active participation in the process for at least the first two years.

"Problems will arise," he predicts. "A thorough understanding of the principles of decomposition is necessary. Unlike today's flush toilets, you will find no 'clivus plumbers' at your call when something goes amiss."

Shirley Garrett, sales manager for Clivus Multrum USA (also the Canadian distributor) says the company is working to lower the price. She points to the expense of fiberglass construction and research as primary reasons for the hefty price tag.

"The company is still losing money and probably will continue to lose money for some time," says Ms. Rockefeller.

Aside from the Clivus, the Low Cost Housing Group is examining a smaller, but similar, Swedish composting toilet called Toa-Throne. At first Rybczynski had hopes that reduced dimensions of the Toa-Throne's composting chamber would overcome some problems of the Clivus. According to Rockefeller, the size of the Clivus Multrum chamber is absolutely necessary to accom-

modate natural fluctuations within the unit.

The Montreal group has also constructed a composting toilet much like the Clivus Multrum from standard cement blocks, in an effort to develop low-cost composting toilets suitable for third-world areas. A prototype "clivus minimus" is now functioning satisfactorily in Manila, according to Rybczynski.

Although being self-powered makes Clivus Multrum the most aesthetically pleasing compost toilet, a large variety of smaller composting toilets are sold in Canada with the chamber directly below the receptacle forming a single, compact unit. These toilets must rely on artificial heat to decompose and dehydrate excrement.

Cutbacks in size require rapid decomposition of excreta — electric heat assures a high, tropical humidity conducive to rapid decomposition. Electricity also powers a fan which serves the two-fold purpose of adding air to the unit and venting foul odours out of the house.

A package of prepared soil-supporting micro-organisms necessary for decomposition comes with each unit. Once a year under normal conditions a drawer is removed from the base of the toilet and compost much like that produced by the Clivus Multrum is removed.

NIGHT SOIL DILEMMA

Humus and Ecolet are Canadian composting toilets. Humus toilets have proved exceedingly popular, with over 7,000 toilets in Canada (about 90 per cent of all Canadian units). The price, $700, makes the Humus more accessible than Clivus Multrum, and the Ecolet is similarly priced.

However, reliable sources who have studied operation of this type of composting toilet have found that problems are commonplace.

Witold Rybczynski has visited sites in northern Ontario, northern Quebec and northern Manitoba where the Humus unit was used extensively. He says there were almost always problems with smell, flies and liquid accumulation and says that they are not really suitable for use by a lot of people on a regular basis, although they might be appropriate for vacation homes.

Frank Smith of Humus says that although "we do have some difficulty from time to time" the enviable sales figures point to a reliable product. But, counters Rybczynski, the reasoning that "if thousands of people are buying them, they must be good" doesn't really hold up. Just consider some automobiles.

Dr. John Evans, a professor of marine biology at Memorial University in St. John's who has embarked on a crusade to introduce widespread use of composting toilets in that province, is also dissatisfied with the Humus.

He has centered his activities around a small borough of St. John's known as The Battery. Houses in this area are built directly on the bedrock of a steep cliff, terrain preventing installation of municipal water and sewer facilities. People in The Battery must carry their night soil in pails each evening to areas where trucks wait to cart it away, a practice identical to that of medieval times.

"And The Battery is only one of hundreds of Newfoundland villages in exactly the same position," Evans says.

The nature of the terrain, financial limitations and older home designs make it impractical to use Clivus Multrum in The Battery. Evans sought a solution in Humus toilets, but found that odour was a problem which made them unsatisfactory. Evans blames this on a tendency toward anaerobic decomposition, a contention positively denied by the manufacturer, who says that odour problems arise only in units where the ventilation system has been improperly installed. The Director of Public Health Inspection Services for the Province of Newfoundland said in a recent address, "It has been our experience in Newfoundland that self-installation has resulted in some odour problems. Ninety per cent of the units are self-installed. Two complaints re odour problems have been documented." As an alternative Evans has constructed his own composting toilet that — in testimony to its lack of offensive characteristics — sits proudly in the professor's bedroom.

ROMANTIC HOPE

Building your own *is* a viable alternative for people discouraged by the high cost of commercial compost toilets. By adhering to three basic principles of composting human excrement, a satisfactory toilet can be built with little expense and effort.

First, assure aerobic conditions. Do this by providing adequate ventilation to the entire decomposition chamber. A turning mechanism (or 30-degree sloping floor in a large unit) helps. Second, add plenty of carbonic materials such as grass, peat moss and kitchen scraps. Third, ensure watertight conditions to prevent leaching. Old-style pit privies were too often direct routes for contaminants to gain access to the water table.

"The idea is to retain all nutrients," Ms. Rockefeller explains. "If they get away they can cause nothing but trouble."

However, before application of compost from a homemade unit to your garden have it checked for disease-causing organisms at a laboratory. To be perfectly safe, apply the final product only to ornamental gardens, forage fields and young orchards.

By writing to Farallones Institute (see Sources) a set of instructions on how to construct a simple, privy-type composting toilet can be obtained.

Clearly, we now know the environmental dangers of sewage, and we are beginning to have the technological ability to create alternative sanitation systems. What remains is altering our inherited laissez-faire attitudes.

The solution does not entail reinstating last century's privies (though this would be preferable to today's flush toilets in some areas).

Or, in the words of Abby Rockefeller:

"The fact is that people's thinking has something to do with the kind of technology that is used. Take nonreturnable bottles for example. People have a sense, internal and unconscious as it may be, of all the trouble the stuff is causing. They feel bad and they behave badly.

"I have a severely romantic hope that the Multrum, or something like the Multrum will give hope back. This system, even with the screw design, is practically pure and does all good things and no harm. The effect of having such a creature in your house, something that is absolutely good in your system, causes other things to come together."

Life With A Clivus

Imagine what the neighbours are thinking

By Jay Lewis

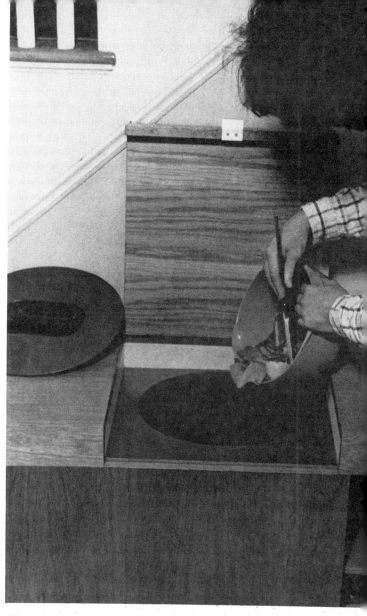

Biodegradable wastes all go into the kitchen counter-top receptacle for a quick fall into the composting chamber of a Clivus Multrum in the cellar below.

Living with a Clivus Multrum toilet for fifteen months has been an interesting experience. It has become the centre of attention in our new home. With a flourish, we flip up the toilet seat and launch into our dissertation on the virtues (no water, no electricity, a proper cycle for human and kitchen waste, no torn land for a septic drainage field) of the Clivus. This performance usually rocks people back on their heels as they peek into the black void. Dimly, down below, once your eyes adjust, is a pile of excrement just like ye olde outhouse. But there is a difference. This system does not smell (usually) and produces finished compost (after a two-year cycle) for our fruit trees. In general, we're happy with the performance of the Clivus and it's nice to know we're not fouling the waters of the Okanagan Valley any further.

There are, of course, some problems of which people should be aware. They start when you approach the health authorities for a permit to use a Clivus Multrum. We were given an experimental permit and had the full support of the local health unit, but other people have met resistance. ("What! You're going to keep it all in your basement for two years?") The next concern is for the design of the house. There needs to be six to eight feet of vertical clearance below the bathroom and kitchen to accomodate the container, depending on the model. We ran into bedrock quickly, so our first floor is four feet above grade, and not as unobtrusive as we had planned.

Because the human and organic waste enters the composting chamber through two chutes and a vent stack must run between them, the bathroom and kitchen must be back to back. This requires some forward thinking to place the floor joists and the partition correctly. If the water plumbing is included in the wall as well, the whole system can be centralized.

Once primed with peat moss, top soil and compost, the Clivus should run without attention — theoretically. In the first year of operation, some liquid percolates through the mass before the composting process gets fully cranked up. We bailed it out (the shrubs love it), and the situation has not occurred recently. Insects can be a problem. They get into the chamber on fruit and then multiply. A Clivus is a good place to watch exponential growth in action. The populations explode and crash depending on the season, temperature and our use of a mild insecticide. The situation has not been too bothersome although I sometimes wonder where the fly that is walking on my lettuce was walking last.

When the venting system is working properly, it should create a downdraft at the toilet seat which eliminates odour — an improvement over the traditional W.C. We have a 20-foot vent stack, but the drafting is sometimes marginal — yes, there can be similarities to the old outhouse if you lift the toilet cover too quickly. We are working on this problem by altering the seat cover and obtaining a more efficient vent cover.

The disposal of "greywater" (from showers, sinks and the washer) is a problem that the environmental engineers from Victoria (we've had a lot of interested people coming through) were most concerned about. We collect ours in a 500-gallon tank and periodically drain it out onto some pines and aspens. The odour, which can be quite noticeable, disappears quickly as it hits the soil. Some other solutions would be to use a dry well, an abbreviated drainage field or a municipal sewage system.

Next spring we will be putting our first compost out, and I expect the trees will enjoy it, while the creek continues to run clean. The system needs some attention, but sewage treatment plants need even more and they are costly while not doing the job very well. By not flushing, the two of us use less than 1,000 gallons of water per month in the house. The cost of the Clivus system was fairly high at about $2,000 (that includes the unit with all accessories, shipping from Maine, taxes, duty and the steel "greywater" tank), but septic systems with the in-house fixtures, tanks and drainage fields can be costly as well. The Clivus Multrum expense could be reduced by going to mass production or by building your own version using locally available building materials such as concrete. The Clivus makes a lot of sense because it works with biological systems rather than disrupting them. While we've had some problems, we're glad that we kicked the "flush" habit.

The W.C. A Brief History

Thundermugs, Three-holers & The Father of the Flush Toilet

If ecologists have their way, chances are excellent that your trips to the white porcelain receptacle mounted with such nobility in your bathroom are numbered. Before the flush toilet joins gramophones, sadirons and pedal zephyrions in the ranks of memorabilia, *Harrowsmith* would like to pay tribute to the device's long and frequently colourful existence.

History shows that the flush toilet has been with us considerably longer than the 100 years normally credited it. In ancient Rome, citizens enjoyed the benefits of both public and private flush toilets. Of the former there were 144 used in the city as early as 315 A.D.

Like so much else, the flush toilet became forgotten during the Dark Ages. Feudal folk were not inclined toward things temporal — a realm that included all aspects of personal sanitation. Not only did they scorn bathing (once a year being considered mildly excessive), but they permitted technology connected with disposal of human excreta to drop to pre-civilized levels.

Flushing during those dark days was limited to the very primitive direct-drop-in-a-handy-body-of-water method. This accounts for cathedrals and castles placed close to watercourses.

A monastery at Canterbury boasted one "*reredorter*" that measured 145 feet long. A polyholer of such size was necessary because rigid routine dictated that the Fathers visit *en masse*.

With an eye always to defense, feudal man learned quickly that his own excrement lent greater powers of repulsion to castle moats. Unfortunately, the defensive qualities of excrement often backfired. Britannia's own ships were constantly bombarded as they passed beneath London Bridge (the houses built thereon used a crude, gravity-powered waste disposal system), a fact which made more than one hearty seaman state that the passage beneath the bridge was the most dangerous part of any voyage.

Castle turrets were often sites for privies which were artfully situated next to each other around the circumference of the turret so that one could enjoy the amenities of polite conversation while maintaining a modicum of discreet privacy.

Legend tells us that at least one nobleman met his demise while mounted on a turret "throne." A zealous assassin, it is alleged, braved the moat to gain entry to the cesspool below the noble's privy, and with crossbow in hand waited for gloom to descend over the circle of light above.

Lack of a handy stream or moat caused difficulties beyond measure. Two alternatives were the pit and the removable-barrel systems. Edward III is remembered for the Great Pit he ordered dug at York Castle. Emptying these sanitary devices was no mean feat. It took 13 men five nights to empty the pit at Newgate Jail. The men, it should be noted, commanded wages three times higher than normal.

For milady, who understandably preferred not to venture into drafty privies, there were numerous ornate, lushly upholstered portable stools. One made for the royal boudoir was covered with velvet and accentuated with gilt nails, silk ribbons and fringes. Seat and arms, of course, were down-filled; and needless to say the top came equipped with a stout lock to prevent illicit use. To servants was relegated the duty of emptying these stools.

Another popular expedient was merely heaving the contents of the chamber pot from urban windows onto the street below, a method causing consternation among members of the fair sex. One Englishman lamented that "Daughters and Wives of Englishmen should encounter, at every corner, sights disgusting in every sense."

Jonathan Swift, one Augustan with an eye always on mankind's base necessities, reported that only the foolhardy would proceed along Edinburgh's streets (where fecal plummets of up to 10 storeys were recorded) during waste disposal hours. A lilting *"Gardy loo"* (look out below), was a phrase commanding utmost respect.

Desperation drove enterprising minds to seek a better device. As early as 1596 Sir John Harington constructed what he dubbed "A Privy of Perfection" near Bath, England. Harington is truly the father of the flush toilet, for his invention incorporated all the technological intricacies of 20th-century water closets, suggesting that in the past 400 years we haven't come such a long way.

That man is reluctant to alter his toilet habits is one steadfast law known to all students of human excreta. This accounts for Harington's invention failing to gain popular following, and explains why 189 years were to pass before Alexander Cummings, a British watchmaker, applied his nimble fingers to the development of a water closet that became the first patented in Britain.

Entering the scene just as the industrial revolution boomed, Cummings' invention gained a measure of acceptance, inspired the usual flock of imitators, became the baby doll of upper classes, and within 100 years had swept the land. The flush toilet became so popular that one British writer, a Mr. Eassie, said in 1874, "Nothing can be more satisfying than a good water closet, properly connected with a well-ventilated sewer."

Despite snowballing popularity of the flush toilet during the middle years of the 19th century, we must be wary not to forget the efforts of honourable scientists who, in the face of the inevitable flush, were diligently working on what could have been truly a privy in perfection. Cardinal among these unfortunates was one Rev. Henry Moule. His 1860 invention, humbly called the Earth Closet, is a forerunner of today's composting toilets. Instead of flushing five gallons of water after each use, Moule's invention quietly sprinkled a pint of fresh, dry earth over the excrement. Civil planners worked on methods to assure that fertilizer produced by widespread implementation of earth closets would be recycled to agricultural holdings in a manner which would assure some profit for the municipality.

It remained only for people of the 20th century — armed with Augustan technology and Victorian squeamishness — to coin the term "waste" in its present connection with human excrement.

Go Directly To Windmill. Do Not Collect

Energy independence with a recycled windplant

By David Simms

Our first winter in the country was the earliest and most merciless I can remember. It was, to be honest, a kind of a nightmare and an experience that has led us to believe that much of what you read about going back to "live on the land" is glorified while the inevitable problems are ignored.

Chill November winds eddied through the partly-chinked walls of our unfinished log house, straining the heating capacities of our Trolla 800 airtight stove and the kitchen wood range. The snow lay a foot thick in the woods and our firewood wasn't even in yet.

Our well left much to be desired — the water was ultra-hard and smelled of swamp. We found we weren't able to wash clothes in it and were forced to walk to a spring a half mile away just to obtain drinking water.

With extraordinary effort we finally cut and split four cords of wood and stored it in the cellar, despite the snow. Of course, snow-covered, wet wood does not burn easily and we came to think that those who write about country living are either afraid of discouraging others by mentioning the hard times, or that they've had enough money to pull out and go to Florida when the going got rough in the middle of winter.

Looking back, one of the few things that worked smoothly during this difficult transition was our refurbished Jacobs 1800 wind plant that even then supplied all of our electricity.

Dreams of the totally energy self-sufficient homestead, adorned with solar panels, windmill towers, Clivus Multrum exhaust stacks and stovepipes exuding tart wood smoke, prompt many back-to-the-landers to approach the subject of wind power with a zealous but grossly over-simplified attitude. And when this over-simplified attitude leads to yet another failure, more fuel is added to the skeptical arguments of engineers and government experts who claim that wind power is simply too costly to be practical.

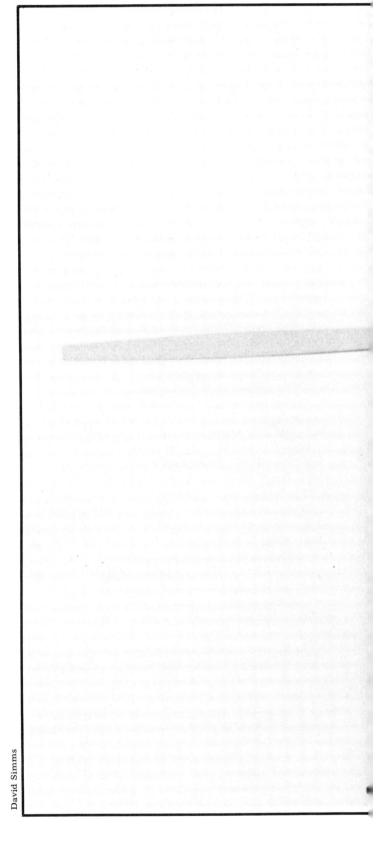

David Simms

56

Do Not Enter Public Utilities. Monthly Windbill.

But, as we've proved to ourselves, a reasonably priced, domestic wind power system is within anyone's grasp today, provided he lives in an area with adequate wind speeds and that he approaches wind power in a rational manner.

We find that the most frequent question about our system is, "How much electricity are you getting from the windmill?" Since it is our sole source of electrical power, we have no trouble answering that we run everything with wind power.

To avoid the appearance of being cagey, I should point out that our electrical demands have been pared to the point where they are far below those of an average North American home.

Right now, we are running all our lighting, a large, double door, commercially-sized fridge, a water pump, radio, iron, bench grinder, a half-inch electric drill, air compressor and a couple of small inverters which run small appliances such as an electric shaver, blender, floor polisher, etc. whenever we need them. Needless to say, we don't operate all these things at the *same time* but we do use them as needed. The only time we really begin to watch our consumption of electricity is after several calm days when the voltmeter on the kitchen wall reads less than 32 volts.

So far, with an average wind speed of less than nine m.p.h., our electrical consumption appears to be well balanced to the weather systems which prevail locally during most seasons. The $2,700 wind system carries nearly 100 per cent of the load. However, when the spells of calm, humid weather cause drops of sweat to trickle down our noses in mid-July, we sometimes fire up our standby, a gasoline-powered light plant. This, a 1944 Onan one kw unit was put together from two identical ones and some original spare parts that were still in stock in the factory.

The occasional use of a light plant is quite interesting insofar as expanding one's appreciation of wind power. Although these old-timey clunkers were of robust construction, intelligent design and quite dependable in operation, they do make considerable noise, consume surprising quantities of gas and leave clouds of exhaust around to remind you that they're running (in case you could forget). The point is that each kilowatt-hour produced by a light plant such as ours requires one hour of noisy, stinky operation, one-sixth of a gallon of gas and bits of attention in the form of oil checking and gas tank filling. By way of comparison, any kilowatt-hour coming down the wire from a wind plant demands almost no attention, presents no intrusion and is entirely free from any input of fuel. This comparison can also be extrapolated to consider how that clean, wonderful Reddy Kilowatt which buzzes around the grid is normally produced. In many cases, the plant that turns out those Reddy Kilowatts is not much more environmentally or aesthetically acceptable than a grown-up light plant, and in other (like nuclear power plants) it can create a spectre which is downright frightening.

Acrophobia is a clear deterrent, but author David Simms has found wind power cheaper than Hydro Quebec and a practical alternative for the homesteader.

As with other people who have gone in for wind power, we have been willing to cut down somewhat on our power consumption. Wind power happens to fit in well with a different life style, especially on a rural homestead, where other cheap sources of power are available.

Eliminating those major appliances which use electricity to generate heat makes a vast difference — in fact, it closes the gap between needing a system that is prohibitively expensive and one that is within the reach of most of us. For example, by getting rid of three common appliances — the clothes dryer, the electric stove and the electric water heater — your monthly power consumption can be cut by 500 kilowatt-hours. These tasks can easily be performed by other energy sources.

Wood or propane can replace the electric range, solar energy and circulation systems working in conjunction with wood stoves can heat water, and every time you hang your laundry on the line rather than toss it into the dryer, you save 6,000 watts of energy.

DEVICE FROM THE PAST

As far back as 1930, some very efficient, electric wind systems were in wide use in the western parts of Canada and the United States. These pre-dated the establishment of rural utilities and some units were easily capable of producing 800 kilowatt-hours per month. Many were trouble-free in that they regulated themselves and governed their rotor speeds automatically.

Inquisitive people are usually quick to ask why the western farmers and others using wind power did not continue in this blessed state of independence from the power utilities. It is a good question, and in my ramblings and reflections, I, too, have attempted to discover why.

It was not, I have come to believe, due to any inherent fault in the idea of using wind power, but a result of several factors — not the least of which was the boom in new inventions and increased power consumption. Many of these new appliances operated only on 110-volt AC and they became very difficult to adapt to the old DC voltages.

At this point, it is important to understand that wind power, then and now, is in the form of DC, which is the only kind that can be stored in batteries. Common voltages in the past were 6, 12, 32 and 110 DC. Unfortunately, direct current cannot be transformed as easily or as efficiently as alternating current, and more and more AC motors were produced. They began to replace the old universal motors, and so direct current gradually became outmoded — and the hydro poles eventually supplanted self-sufficient windmills.

It is interesting to note that in secluded regions of Switzerland and Australia, wind power never lost its pre-eminence and is still widely used today.

The mainstay of our own wind power system is actually a survivor of the earlier days of wind-milling. It is a rebuilt Jacobs Wind Plant which cost $1,800. The Jacobs is considered by many to be the "Cadillac" of used wind plants. Michael Hackleman, author of *The Homebuilt, Wind-Generated Electricity Handbook*, says of the Jacobs: "Only the best materials were used to make a machine which would last a lifetime. . . . The commutator is so large it could be turned down every 10 years for a hundred years . . . . " Having restored a number of Jacobs plants myself, I am convinced that the guts of the machine should last indefinitely.

Our system is 32 volts DC, which means that modern appliances must be run through an inverter or be modified (if possible) before they'll function for us.

Dealing with bizarre voltages and currents appears a frightening proposition, especially to a person unfamiliar with the complicated jargon of electricity. But a bit of grass-roots gumption combined with excursions into the darkest corners of appliance stores' back rooms and an occasional visit from the neighbourhood electrician should enable the wind power neophyte to overcome most problems.

We had to coax the manager of a local Canadian Tire store to order us special 32-volt light bulbs (they are listed in the catalogue), but he did, and the light bulbs work fine. In years gone by, non-standard voltage systems were more common than they are today, and you can still find old appliances that were designed for non-standard systems.

Automobile generators can be used unaltered as electric motors on my 32-volt system, and with generators, the handyman can convert refrigerators and freezers to operate on 32-volt DC

POWER, CURRENT, & MONTHLY KW-HR CONSUMPTION OF VARIOUS HOME APPLIANCES

Appliances	Power in Watts	Current Required in Amps		Time Used Per mo. in hrs.	Total KW-hrs. per mo.
		at 12V	at 115V		
Air Conditioner (window)	1,566	130.00	13.70	74.0	116.00
Blanket, electric	177	14.50	1.50	73.0	13.00
Blender	350	29.20	3.00	1.5	.50
Broiler	1,436	120.00	12.50	6.0	8.50
Clothes Dryer	4,856	405.00	42.00	18.0	86.00
Coffee Pot	894	75.00	7.80	10.0	9.00
Dishwasher	1,200	100.00	10.40	25.0	30.00
Drill (¼ in. elec.)	250	20.80	2.20	2.0	.50
Fan (attic)	370	30.80	3.20	65.0	24.00
Freezer (15 cu. ft.)	341	28.40	3.00	29.0	100.00
Freezer (15 cu. ft.) frostless	440	36.60	3.80	33.0	147.00
Frying Pan	1,196	99.60	10.40	12.0	15.00
Garbage Disposal	445	36.00	3.90	6.0	3.00
Heat, electric baseboard, av.- size home	10,000	832.00	87.00	160.0	1,600.00
Iron	1,088	90.50	9.50	11.0	12.00
Light Bulb, 75-Watt	75	6.25	.65	320.0	2.40
Light Bulb, 40-Watt	40	3.30	.35	320.0	1.30
Light Bulb, 25-Watt	25	2.10	.22	320.0	.80
Oil Burner, 1/8 HP	250	20.80	2.20	64.0	16.00
Range	12,207	1,020.00	106.00	8.0	98.00
Record Player (tube)	150	12.50	1.30	50.0	7.50
Record Player (solid st.)	60	5.00	.52	50.0	3.00
Refrigerator-Freezer (14 cu. ft.)	326	27.20	2.80	29.0	95.00
Refrigerator-Freezer (14 cu. ft.) Frostless	615	51.30	5.35	25.0	152.00
Skill Saw	1,000	83.50	8.70	6.0	6.00
Sun Lamp	279	23.20	2.40	5.4	1.50
Television (B&W)	237	19.80	2.10	110.0	25.00
Television (colour)	332	27.60	2.90	125.0	42.00
Toaster	1,146	95.50	10.00	2.6	3.00
Typewriter	30	2.50	.26	15.0	.45
Vacuum Cleaner	630	52.50	5.50	6.4	4.00
Washing Machine (auto)	512	42.50	4.50	17.6	9.00
Washing Machine (wringer)	275	23.00	2.40	15.0	4.00
Water Heater	4,474	372.00	39.00	89.0	400.00
Water Pump	460	38.30	4.00	44.0	20.00

systems by using belt-driven compressors (used ones available through refrigerator repairmen for $15 to $20). It pays to spend the $100 a qualified repairman will charge to disconnect the sealed motor compressors found on most modern refrigerators and freezers, and to connect the piston compressor.

The key to making the maximum number of appliances operate on a limited amount of electricity is to avoid waste. We save major jobs for periods of high winds, thereby using electricity as it is produced. It is also wise to have a substantial battery storage capacity to accommodate extended periods of calm.

The advantages in choosing a used generator become clear when you compare my total installation cost of just under $2,700 to the $7,000 you would have to pay for a similar, new system.

Used, reconditioned wind-power systems can be obtained through several outlets — most of which are in the United States. If Canadians can prove that the wind system is for farm use, they may apply for a customs ruling by asking that the equipment be entered duty-free and exempt from federal sales tax (under tariff item 40944-1). Merchandise will cross the border with greater ease if you have the sender fill out four copies of what is known as the "MA Invoice."

An excessive need for maintenance could easily render small wind electric systems impractical. However, because wind generators turn so slowly and have few moving parts the attention which they *must* be given is absolutely minimal. Insofar as the Jacobs direct drive units are concerned, an annual trip up the tower to grease the governor, turntable bearing, and to check the tension on the brushes is all that's needed. Every three or four years the blades should be repainted (without removing them) and every 10 or 15 years the brushes in the generator will have to be replaced. There cannot be many machines which give so much in return for so little.

We further economized by purchasing a scrapyard battery that had formerly powered a fork-

lift. At eight cents a pound, the one-ton battery was bargain-priced, but because of its junkyard origins, I was taking a chance on getting a ton of unfunctional lead and zinc.

I checked the outside of the battery for leaks and ran through a short check list. How were the caps on top of the cells? Were most of the screw corks in place? I checked the electrolyte level in all cells.

I immediately determined that one of the battery's 18 two-volt cells was inoperative, and at home I discovered another malfunctioning cell. But this posed no problems. In perfect condition, the battery would have had a capacity for 36 volts. Since my system was 32-volt, the two dead cells were not needed and I bridged them.

With time and tender care, our battery has actually improved with age and has given us only minor problems during periods of no sun and severe cold. It is so big and heavy that we had to locate it outdoors, where the electrolyte might freeze. To prevent this, we built a little shed around it, insulated it well and sealed it carefully. We also painted the south side of the battery case black and put a double-glazed window in that side of the shed to allow the sun to heat the battery. The whole battery thus acts as a passive solar collector and this helps to prevent it from freezing.

Of course, there is some satisfaction in knowing that if the battery had proved useless, we had bought it as scrap and could have resold it

Average Monthly Output in Kilowatt-Hours

Nominal Output Rating of Generator in Watts	Average Monthly Wind Speed in mph					
	6	8	10	12	14	16
50	1.5	3	5	7	9	10
100	3	5	8	11	13	15
250	6	12	18	24	29	32
500	12	24	35	46	55	62
1,000	22	45	65	86	104	120
2,000	40	80	120	160	200	235
4,000	75	150	230	310	390	460
6,000	115	230	350	470	590	710
8,000	150	300	450	600	750	900
10,000	185	370	550	730	910	1,090
12,000	215	430	650	870	1,090	1,310

for very close to what we had paid the junkman. In time, our battery should even appreciate.

To determine what size battery you should buy, consider what your electrical demands will be (see chart) and select a battery or set of batteries which will accommodate your demands.

Batteries are rated in amp-hours, a system which tells for what length of time a battery can discharge a given number of amps at its voltage rating. For instance, a 500-amp-hour battery can discharge five amps for 100 hours, 50 amps for 10 hours, etc.

Remember, too, that batteries are designed to handle a limited rate of charge. Exceeding this rate can overheat the battery and cause damage. Your battery system should be able to absorb the maximum charge rate that your system can produce.

The lesson to be learned here is that, if you are contemplating the purchase of a large wind plant, you must purchase batteries which are capable of storing the power it will produce. Batteries are very expensive.

Which begs an obvious question: How large a windplant do you need? First jot down a list of electrical appliances that you consider to be the least with which you can comfortably live. Next, determine roughly what your total electrical demand will be per month (see table). Your wind generator should be able to handle these projected needs at the average wind speed prevailing in your area. Manufacturers and distribu-

Sources

USED WIND GENERATORS
(RECONDITIONED)

North Wind Power Co.
Box 315
Warren, Vermont
(used Jacobs)

Roland Coulson
RFD No. 1, Box 225
Polk City, Iowa 50226
(assorted models)

Steve J. Paul
Best Energy Systems for Tomorrow
Box 819, Route 1
Necedah, Wisconsin 54646
(used Jacobs)

Paul Biorn
1123 South Lawson
Aberdeen, South Dakota 57401
(used Jacobs)

BOOKS AND PERIODICALS

Alternative Sources of Energy Magazine
Route 2
Milaca, Minnesota
($6 per year)

Wind Power Digest
54468, Cr. 31
Bristol, Indiana 46507
($6 per year)

Environmentally Appropriate Technology
by Bruce McCallum
Advanced Concepts Centre
Office of the Science Advisor
Environment Canada
Ottawa, Ontario

Wind Energy . . . Achievements and Potential
Symposium Proceedings
Universite de Sherbrooke
Sherbrooke, Quebec

The Homebuilt, Wind-Generated
Electricity Handbook
by Michael Hackleman
Earthmind
Boyer Road
Mariposa, California 95338

Wind and Windspinners
by Michael Hackleman
Earthmind
Boyer Road
Mariposa, California 95338

tors usually provide an estimate of what their machines will produce in areas of given wind speeds. Novices might be tempted to purchase the largest wind plant available to assure more electricity but a close examination of different units will show that bigger is not necessarily better.

In low wind speed areas (less than 10 m.p.h.) it's critically important to be sure that the wind machine you choose has as low as possible "cut-in" speed (wind speed at which the unit starts to generate useable power). In these areas, a cut-in speed of more than seven m.p.h. is impractical because there are many more hours of low wind speeds and with higher cut-in speeds many hours of potential output are lost. In areas of high winds, these considerations may not apply with the same vigour because there exist many more hours of high wind speeds.

The Jacobs 1800, for instance, produces 1,800 watts in a wind of 18 to 20 m.p.h. The larger Dunlite 2,000-watt machine will put out more power at its rated wind speed of 25 m.p.h., but at lower wind speeds (more likely to occur in most areas) my smaller machine puts out more power than the Dunlite.

Our backyard generation plant has been free from all but minor difficulties. As a matter of fact, we are now oblivious to blackouts experienced by neighbours tied into Hydro Quebec. There *have* been some uncomfortable moments: Our lights dimmed when the battery became frozen during a cold snap, and we now regard those normally appreciated Indian Summer days with apprehension — they are clear and warm — but windless. Today when we turn on an electrical appliance we are more energy-conscious. We have nevertheless learned an important lesson: Simple alternative energy systems are not only within the financial grasp of most North Americans, but they work. Now we wish that we had designed our home around a solar heating system as well as our wind-generated electrical system.

If you think wind power is for you, begin by subscribing to informative periodicals like *Alternative Sources of Energy Magazine*, talking to wind power veterans, and overcoming your fear of heights.

I find that the more independent, economical and ecological an alternative looks, the more it should be matched with ingenuity and care by the person using it.

When you generate your own electricity, you are free of monthly power bills. You are also your own lineman, engineer, electrician and repairman. To be independent, you should be your own trouble-shooter, and that's not easy for everyone. The one thing you will never be able to do for yourself, however, is make the wind blow, so check out how much you have — *first*.

COST BREAKDOWN

WINDMILL	
Used Jacobs 1800-watt 32-volt plant	$1,800
Used 50-foot, four-post steel tower	400
Scrap-yard fork-lift battery	148
Wiring and fixtures	229
Miscellaneous (transportation, etc.)	100
Total Costs	$2,677

HYDRO QUEBEC	
Installation of line	$1,095
Wiring by electrician (required by Hydro Quebec)	1,500
Total Costs	$2,595

WoodHeat

"Everything," said Goethe, "has been thought of before. The difficulty is to think of it again."

By Craig Gutowski

Consider the typical reader of a flashy newspaper article that proclaims, "Wood stoves are In and Back!" Suddenly struck by the inherent good sense of heating with wood — and well aware of what has happened to his winter heating bills — he hops into the car and rushes out to buy a stove.

The first stop is a hardware store or Simpsons-Sears where he is given a highly supportive sales talk by a high school kid whose concept of the word Alternative goes no further than lunch at Burger King rather than McDonald's:

"Yessir, these stoves really *throw* the heat. Yessir, these stoves really *hold* the heat."

One hundred and sixty-six dollars and sixty-six cents changes hands and, as if it is a very expensive pet, he takes it home. The family buys it food, cleans up after it and soon the novelty wears off. The Fire-Eating Franklin is simply too much work; it requires constant tending and the woodpile just doesn't seem to last. Soon it sits cold and ignored, except for the New Year's Eve party when a 99-cent BLAZE-O supermarket log is thrown on for good cheer.

The fact is that the namesake of today's most popular stove must be gyrating in his unrestful grave. The Made-In-Taiwan versions that pass for Franklin stoves today bear little or no relationship to the efficient, controlled-draft version that the 18th-century inventor envisioned. If he were alive today, we are sure that Ben would be warming his shins in front of one of the new airtight stoves that make woodburning a rational proposition once again.

Although many people are happy with their old-fashioned stoves or will swear by the open fireplace, theirs is a bliss built upon ignorance. Today's efficient stove is capable of giving the same level of heat from the same amount of wood, but for twice as long. It produces far fewer ashes and of a notably finer texture — an indication of more complete combustion of the wood.

The full impact of this more complete combustion comes when you realize that it is no longer necessary to arise in the night to rekindle and stoke the fire. Many stoves that are readily available today — although not yet on any mass scale — allow the woodburning family to get a full night's sleep and awake the next morning to walk barefoot across warm floors to rake the ashes slightly and plop in a few new logs.

No kindling, fanning, matches or rib-knocking shivers — you fill the firebox in the morning, again in the afternoon and once before going to bed. The average user of an airtight stove can always count on six straight hours of heating to a filling, with up to 12 or 14 hours easily achieved with careful filling and good, dry hardwood.

I have met more than a few seasoned users of woodburners who get 24-hour fires from a single filling. It is, of course, best not to count on this extreme — better to be happily surprised than disappointed.

Some would mount the soapbox to extol the efficiency of wood stoves and their ability to help conserve our firewood resources. I would merely point out that one of these efficient stoves is simply a whole lot easier to operate and costs less (in both time and money).

There is no mystery about how these stoves work, although some do become quite complicated. My own definition of an efficient wood stove is one which is designed to burn not only the wood itself, but also some portion of the smoke and gases that just go up the chimney of other stoves.

The goal is to get a slow, even, controlled burn that releases and burns the maximum amount of wood gases and creates a bed of long-burning charcoal. This is only accomplished in an airtight firebox.

64

Some will be charmed by the chrome-trimmed parlour stove or Quebec heater, but you are enjoying the luxury of a primitive device that sends most of its heat up the chimney.

AIRTIGHT CASE

In a truly airtight stove, the doors, seams and controls should be tight and positive enough to kill the fire by oxygen starvation when the controls are closed.

The other day a prospective airtight-stove buyer recounted for me the heating abilities of his sculpted, cast-iron farm-auction special once the ashes start to mount. Unfortunately, this system is not patentable — when the ashes get halfway past the air intake grate, with the air intakes closed, enough of the cracks and seams are plugged to give a positive control over the rate of burn.

For anyone considering the purchase of an airtight stove for the first time, or even to replace an old Ashley, the number of variables among stoves today is both exciting and a bit confusing.

Approached differently, from manufacturer to manufacturer, is the system for bringing volatile gases and smoke back down to the hottest part of the fire to be mixed with oxygen and burned. Most stoves have one or more of three basic solutions to this problem (a few, however, defy pigeon-holing); the three systems are *baffled*, *secondary air* and *downdraft*.

A *baffle*, of course, is an obstruction around which the smoke and gas must flow to get out of the stove and up the chimney. This is typified by the Jøtul stove, in which gases rise to the underside of the baffle, are forced to travel laterally near the fire until reaching the end, after which they travel back to reach the exit.

The *secondary air system* works by introducing a stream of air above the firebed to create a swirling, circular and/or downward motion of gases inside the firebox. Carried back down to the hot coals, wood gases are burned off before passing into the flue.

Downdraft is the last of the three basic gas combustion methods. The simplest way to explain its workings is to point out that the gases cannot rise directly out of the stove, but must first be drawn down through the firebed and grates and thence allowed to exit up a flue at the back of the stove in a normal updraft manner.

The theory of downdraft stoves makes good sense by directly addressing the problem of getting hot (and therefore rising) gases as close as is physically possible to the hottest part of the fire.

Thermostatic controls are available on many stoves. They are mechanical rather than electrical and consist merely of a bimetallic coil (two different metals, with different expansion and contraction rates, bonded together). One end of the coil is attached to a small hinged door over the primary air inlet to the wood stove, and the other end to a control dial (HIGH-MEDIUM-LOW).

The thermostat is constantly comparing the setting on the dial to the temperature of the room air entering the stove, allowing more air into the stove when the room starts to cool, shutting down when the temperature gets to the right level.

Whether or not a stove has a thermostat should not be the sole deciding purchase factor — the device will help slightly in maintaining a steady heat plateau, but other thermostat-less models will surprise you with the evenness of their operation.

One last internal feature to check is if the combustion air entering the stove is preheated. This is normally done by allowing the air to enter through a piece of channeled iron or steel which is exposed on one side of the fire. There are various reasons offered for doing this, but in my opinion it is simply to preheat the air and maintain the high temperature of the firebox.

Before buying a stove, consider how it is loaded. Toploaders tend to be cleaner, but can be difficult when it comes time to remove ash. Check the size of the door-opening on the stove — a diminutive mouth means you will have to feed it small-diameter wood. Remember that every extra inch of loading-door size will, over the years, be worth many hours saved from the task of splitting.

COOKING WITH (WOOD) GAS

Occasionally people will ask if you can cook on an airtight stove and the answer is a qualified "yes" in most cases. The question is what kind of cooking and how often — if your only concern is for blackouts, any stove will serve to brew tea.

If you're serious about making double use of a wood stove, consider three things: how much useable space there is on the surface, the working height (most are quite low but can be raised on a hearth) and the thickness of the top.

If a cookstove is what you need there are a few airtights available. However, a regular old cookstove has so much mass, surface area and baffling that it is not all that vulnerable to criticism. The firebox is usually so small that you won't be able to waste much wood, but forget any notion of holding a good fire overnight. (Beware: Some of the modern-looking wood "ranges" have prohibitively small fireboxes, unless you truly enjoy making little pieces of wood out of big ones.)

EFFICIENT FRANKLINS

There are several airtights which try to offer the best of both worlds: efficiency and an open fire. When the double doors are open you have,

in effect, a fireplace, but it is no more efficient than a bonfire in an open field.

The advantage of fireplace/airtights is that, once you have satisfied the need to watch open flames, you can shut the doors and once again have an efficient heater.

(As with any open fireplace, the amount of room oxygen it constantly draws up the chimney is tremendous. This creates a vacuum in the house which is filled only by outside air entering through cracks and around windows and doors. There are cases of too-tight houses where people have lost consciousness when the fireplace sucked away the oxygen.)

There are great debates over the question of stove materials: cast iron or sheet metal. Cast iron is more brittle, but is more stable under high temperatures. Although steel can be continuously welded for a very tight stove, cast iron can be moulded into any number of beautiful designs.

Heaviness is another question — admiration for a stove that "holds the heat" is a vestige of the days when a stove was expected to go out in the night. If the stove was heavy and thick enough and it retained heat from its long-dead fire, that was good.

Wanting an efficient wood stove to hold the heat, however, is ludicrous. Wood heat should be produced and used — the best stove transmits its energy into the room quickly and does not store it or send it up the chimney.

There is, of course, the other side of the coin: durability. Some stoves are too light and underbuilt, others, in my opinion, overbuilt and a waste of metal.

I am familiar with a number of stoves and am not prepared to say that one is "best." Because there is no standard household, with standard wood, standard chimney and standard operators, there is no foundation for recommending one stove to everyone.

A good stove dealer will ask as many, if not more, questions than the buyer, and will attempt to sort out the many variables before making a recommendation. To be considered are the characteristics and size of house to be heated, the type of chimney, the size and hardness of fuel woods to be burned.

To me, this is the beauty of wood heat and the reason it has become the only widespread solution, as of yet, to the high price of standardized (and therefore easily monopolized) fuels for the average unstandardized person.

Take the opportunity to sharpen your wits and use your common sense in selecting a stove. Then sharpen your axe. It will be the first of a series of steps to cut your dependence on the fuel companies. You conduct the fuel exploration surveys, you direct the extraction of fuel and your vehicle replaces the super-tanker. You become the deliveryman, the meter reader and the thermostat. An efficient stove will be your refinery.

Woodlot Strategies & Tactics

Twelve acres and freedom, if you manage wisely

By Billie Milholand

The tree is the most beautiful solar collector yet perfected, needs no man-hours, money or fossil fuel to build, and is a renewable resource. It quietly gathers in the sun's energy through photosynthesis and easily releases it through combustion, cheering both body and soul of the user. It is probably the most aesthetic of fuels and with careful management the most ecological. There lies the only hitch — careful management. Wood is only a renewable resource if it is harvested and used with forethought and planning, so, before we all rush off to install a fireplace or an airtight stove in every room, we need to consider several factors.

First of all, what is the availability and source of supply for you? If you live on a small acreage or in a city and would have to purchase cordwood, you would be well advised to think carefully before you decide to depend solely on wood for fuel.

If, however, you are among the fortunate who have 12 acres or more of useable woodlot per household, then you can be confident that with careful, sensitive management you can rely on wood as your major source of fuel. I know that seven acres per household is the traditional figure, but I feel that that can only be sufficient under the most ideal conditions; if, for example, it is a good, mixed hardwood stand in an area that receives plenty of moisture and moderate winters.

The figures on the B.T.U. charts look depressing for evergreens, poplar and willow, but they do have their place. You can use your conifers first thing in the fall and last in the spring, saving the hardwood for the coldest part of the winter. Poplar is fairly low on the totem pole for heating, but it is excellent for the smokehouse, especially for fish.

The only time you should cut down a healthy tree is when it is crowding other healthy trees. If you thin discriminately you make more light available to the remaining stand which will respond with faster growth. Proper thinning is very important, but it needs to be done in stages because too much sudden sun puts a physiological strain on trees that have adapted to shade, causing them temporarily to suspend growth and become more susceptible to disease.

The fastest way to cut your trees into stove lengths is to pile them horizontally between two rows of posts and go through the whole pile at once with a chainsaw. We find our chainsaw a noisy nuisance for many jobs and usually use a Swedish bow saw, but for bucking firewood we feel the advantages outweigh the disadvantages.

Burning wood is easy, but burning wood efficiently is a problem that needs careful study. The first consideration here is a tight house so that you are not sharing your hard-earned warmth with the great outdoors that has little

An actively harvested and managed woodlot is ecologically more sound than one which is allowed to become a tangle of stunted, diseased trees. **Above right:** *A well-tended woodlot. Proper thinning makes more light available and results in rapid growth.* **Below:** *A perfect example of a lot that would benefit by cutting — and yield well-cured deadwood at the same time.*

need of it. When I was a kid, I sometimes visited an old man who would holler whenever I left the door open too long: "God only expects us to warm ourselves, not all of creation." He thoroughly understood the dynamics of efficient heating. He even had what he called his "winter sashes," simply heavy curtains made from old overcoats that he drew across the windows at dusk to keep the night heat from escaping through the glass.

There are some of us who may not be too pleased with old overcoats hanging around the walls all winter, but the basic idea is a good one and is an illustration of the kind of ingenuity we need to foster in order to make the best use of our resources.

Twentieth century houses are deplorably designed in general, but it is hard to believe that such a technical age as ours could produce housing so inefficient. In the '60's, Professor Harold Clark of Columbia University, did a study in which he discovered that in the 40 countries covered in his survey. so few *private* dwellings met even basic heating or cooling efficiency standards that they were not worth mentioning.

Areas to be checked are your insulation, weatherstripping, windows and the draft area around the perimeter of the house. If you cannot bank up your house, consider insulating your ground floor, and if winter curtains do not appeal to you, put one-half inch insulation board panels up against your windows at sundown on cold evenings, and up to 20 per cent more heat will be kept inside.

The wide mouth of a fireplace sucks warm air from the room up the chimney causing a great waste of heat. If you put a ventilator duct from the back of the fireplace to the outside or to the cellar, some of this heat loss can be eliminated. Or, you can install a register in the floor in front of the fireplace that can draw air from the basement instead of from your living room. If your fireplace does not have a damper, always cover the opening when it is not in use to prevent warm air escaping up the chimney.

One of the problems of heating with wood that has challenged designers for centuries is the build-up of creosote in the chimney, resulting in possible chimney fires. The airtight stoves are the worst offenders and this is their only drawback. What happens is this: At night these stoves are customarily packed full of logs and the primary draft is turned down. The temperature in the firebox drops and since there is only enough oxygen coming through to permit minimum combustion, the chimney cools also. Therefore, all night long, water vapour and volatile gases distill in the cold chimney and pyroligneous acid is formed, which eventually turns to solid creosote in the pipes. As the creosote layer builds up in the chimney, the danger of fire increases.

Friends of ours who are very content with their Valley Comfort airtight stove use this trick to solve the problem. In the morning when they stoke up the fire and add more wood, they open all the dampers and let a hot fire roar up the chimney for 15 to 20 minutes in order to burn up the night's accumulation of creosote. They insist that as long as they make a habit of this, there is no creosote build-up in their chimney.

Cut large quantities of firewood by stacking logs in a rack then cutting down through the pile at the desired intervals.

Robbie McCallum

Heating Values Of Different Woods

Species	Wt. Per Cord (Density)	BTU Per Air Dried Cord	Equivalent In Heating Oil	Cost Equivalent At 61 Cents Per Gal.
Hickory	3595	30,600,000	219 gal.	$133.59
Red Oak	3240	27,300,000	195	118.95
Beech	3240	27,800,000	199	121.39
Hard Maple	3075	29,000,000	207	126.27
Yellow Birch	3000	26,200,000	187	114.07
Ash	2950	22,600,000	161	98.21
Elm	2750	24,500,000	175	106.75
Soft Maple	2500	24,000,000	171	104.31
Tamarack	2500	24,000,000	171	104.31
Cherry	2550	23,500,000	168	102.48
Spruce	2100	18,100,000	129	78.69
Hemlock	2100	17,900,000	128	78.08
Aspen	1900	17,700,000	126	76.86
Basswood	1900	17,000,000	121	73.81
White Pine	1800	17,100,000	122	74.42

Strange Bedfellows

The Voodoo and Science of Companion Planting

By Lynn Zimmerman

I still remember the summer of our first venture into companion planting. I'd read somewhere that beets, carrots and onions all grow well near each other. I recall reading at about the same time that beans and onions definitely do not cohabit successfully. Our local agricultural extensionist would surely have regarded this as just so much voodoo, but I thought, why not?

When early May brought some warm, dry weather, I chose a section of our garden and sowed first a row of onion sets, next a row of beets, then a row of carrots; I repeated the sequence three times and ended with an extra planting of onion sets.

My dad, who comes from Pennsylvania "Deutsch" country and has gardened all his life, visited us that summer, and I still picture his puzzled expression when, during the inevitable daughter-turned-homesteader garden tour, he spied my mixture of onions and beets and carrots. He has always been a one-of-a-kind together gardener, and he could not possibly imagine any sense in companion planting. My explanation of how plants interact biochemically as well as physically did not at all impress him.

I wish he could have been there for the harvest. Those mixed-up vegetables proved to be outstanding — beets, carrots and onions all seemed larger, tasted wonderfully sweet and appeared insect-free. My experiment was hardly scientific — there should have been control group plantings — but the practical results were impressive. I was beginning to comprehend the enormous similarity between companion planting in the garden and the interplanting observable virtually everywhere in nature.

Consider, for example, the case of "soft chaparral," a unique association of evergreen shrubs and trees that is found in the semi-arid land of western North America, in central Chile and in countries surrounding the Mediterranean Sea. Reaching no more than eight feet in height, these thickets of broadleaf evergreens and stunted shrubs and trees have the startling ability to invade grasslands and encircle themselves with dry moats of absolutely bare soil three to six feet wide.

Scientific study of this phenomenon shows that, while these species thrive together, other plants that have close encounters with soft chaparral have little chance of survival. These xerophytic — arid climate loving — plants have been found to release chemical compounds called terpenes from their leaves into the surrounding air, creating a characteristic fragrance. (Commonly known terpenes include camphor, rosin, natural rubber and turpentine.)

The soil around the shrubs absorbs these naturally occurring terpenes, which accumulate in sufficient amounts during the dry season to inhibit the germination and growth of other plants in the surrounding area.

Companion planting is still very much a backyard science, but there is increasingly solid evidence proving that at least some species of green plants can actually affect or change the environment in which they grow. Instead of the immobile, silent, passive bumps in the garden row you may have always assumed them to be, plants can be aggressive, competitive, active participants in assuring their own survival.

DEADLY CUCUMBERS

The case of soft chaparral is an example of allelopathy, the natural ability of some plants to inhibit the growth of competing "weeds" by releasing toxic substances into the soil.

Patricia Doucet of Petit Rocher North, New Brunswick with her organic, diversified garden by the sea.

Seeking to test this theory in the laboratory, Michigan State horticulturist Alan Putnam and Cornell agronomist William Duke pitted two weed-like plants — millet and mustard — against 41 varieties of wild cucumber. The researchers found that about three per cent of the cucumbers inhibited weed growth by more than 75 per cent.

This suggested that the plants were releasing some sort of natural herbicide. Root drippings collected from the "toxic" varieties of cucumber proved to have the power to stunt the growth of mustard seedlings in the laboratory, thus proving the theory.

Dr. F.W. Went of the University of Nevada noticed an example of positive interaction between plant species in the case of certain desert flowers, such as chicory and fiddleneck, which grew at least two and as much as 10 times better in locations near shrubs, even if the shrubs were dead. Dr. Went offers a mundane explanation for this natural association of companionable plants: organic matter blown by the desert wind is caught by the shrubs and then dropped or washed into the soil, providing nutrients for the flowers.

While those who practise it are often met with raised eyebrows and good-humoured skepticism, companion planting often takes advantage of such simple relationships. One common-sense example of putting two or more species together in the same bed for their mutual benefit involves the planting of corn and pole beans. The corn stalks provide support for the growing bean vines and the beans, which are planted later and have the ability to take nitrogen from the air and introduce it into the soil around their roots, help feed the nitrogen-greedy corn.

UNKNOWN FACTORS

Other companion planting relationships are less simply understood and occasionally appear to border on the magical, but some of the causes and effects are finally coming under scholarly scrutiny.

Dr. Richard Root of Cornell University has been experimenting for years with collards and the flea beetles that often infest them. He found that nasturtiums, the old stand-by of some companion planting advocates, produced no benefit when interplanted with the collards. However, his experiments showed that planting potatoes with the collards definitely discouraged the insects, as did interplantings with tobacco and tomatoes.

74

"There *is* something there, but all I can tell you at this point is that it is complicated," he says. It is dishearteningly difficult, in an outdoor setting, to record all relevant variables — or even to determine what the variables are. Too, scientific results are only considered valid when an experiment (with the same variables) can be duplicated.

To add to the problem, Root explains, insects have been shown to respond not only to plant odour and colour, but also to plant spacing and to the background against which plants are set. Furthermore, some pests, such as the cabbage worm, are attracted to isolated plants, whereas others, such as the flea beetle, tend to infest whole plantings.

The key to many observable companion planting effects comes in the form of secondary chemical compounds, including the essential oils, that some plants release into the air, soil or even into rainwater. Some of the aromatic chemicals have the ability to selectively attract or repel insects; they can stimulate or inhibit growth of nearby plants, as with the shrubs of the chaparral; and they are capable of affecting the kinds and numbers of microorganisms living on plants and in the surrounding soil.

Garlic oil, for instance, is a powerful larvicide and is shunned by virtually every pest in the garden. Two Bombay scientists isolated the active ingredients in the oil, diallyl disulphide and diallyl trisulphide and found them, even in minute quantities, fatal to mosquito larvae, potato tuber moths, red cotton bugs, red palm weevils and houseflies.

The roots of the black walnut tree give off a toxin that wreaks havoc on tomato plants, while an exudate that has been isolated from the roots of carrots is now known to have a beneficial effect on the growth of peas. Mustard oil released from the roots of the mustard family plants will help sweeten acidic soil and inhibit nematode cysts.

A research team at the University of Maryland has isolated a chemical from asparagus that, when sprayed on tomato leaves, will — apparently — travel through the plant and kill nematodes attacking the tomato roots. Plain asparagus juice reportedly will do the same, and the home gardener should pour water from cooked asparagus around his tomato plants.

NEMATODE ATTACK

Most gardens support at least some of the many different species of nematodes, microscopic worms which infest and parasitize the roots of many kinds of plants from lawn grasses and eggplants to zinnias and tomatoes. While they seldom actually kill the plants (after all, it is in the best interests of the parasites that the host plants survive), nematodes do often noticeably stunt growth. Even the most vigilant gardener may not readily identify this foe because the pest cannot be seen by the naked eye and concentrates its attack in the roots.

A remarkable, "organic" solution to this prevalent problem was investigated at the Connecticut Agricultural Experiment Station by P.M. Miller and J.F. Ahrens. These researchers took soil samples from beds of different species of plants, including beans, marigolds, petunias and tomatoes.

The number of nematodes in each soil sample was counted, and the population of these miniscule pests was measurably lower in the soil from the marigold bed. Next, zinnias were planted in the various soil samples and grown for six weeks. Significantly, the soil that had previously grown marigolds still showed less than half the nematode population seen in the other soils. Follow-up experiments confirmed these findings.

The active ingredients in marigolds turn out to be sulphur-containing substances known as thiophenes, isolated by Dutch scientists from the roots of the African marigold. The best method of nematode control is to plant wide rows of marigolds in several different locations of the garden and rotate them each year. At the Connecticut station it was also discovered that the beneficial effects of sowing marigolds last about three years. They won't show much effect during the year they are grown and will compete with other young plants for growing space. Thus they should either be transplanted into the garden after the other plants are established, or, better yet, sown directly in their own rows. Avoid the new, fancy ornamental marigolds — some of which have had much of their "smelliness" bred out. William Dam Seeds sells a cheap, "old-fashioned smelly" variety specifically for companion planting and nematode control.

PHYSICAL WAYS & MEANS

Each summer, I plant sunflowers all around the garden, and last year scattered marigolds, cosmos and nasturtiums among the vegetables, primarily for their colour. According to Dr. Kring of the Connecticut Agricultural Experiment Station, some insects are repelled by the colour of the nasturtiums. I therefore planted them in the rows of cabbages to see if there would be any effect on insect pests. Unfortunately, the experiment ended abruptly when our two cows broke their tethers early one July morning and selectively devoured all the cabbages, cauliflowers, broccoli and beet tops. But the nasturtiums flourished, and the flowers finally accented a tossed salad created by an adventurous friend. The vivid yellow, orange and red flowers have a semi-hot and tangy taste and make for a very companionable salad.

Even horticulturists who feel these biochemical relationships smack of mysticism will see that vegetable species growing closely together in the confines of a garden cannot help but interact in physical ways. Consider, for instance, that the roots of a single corn plant will reach down four feet and around the plant in a radius of five feet. It is easy to see why corn plants spaced too closely or planted in poor soil will suffer. Some plants, such as the familiar and edible pigweed, develop deep and extensive root systems, crumbling compacted soil and making available the nutrients and minerals which were trapped in the subsoil. The roots of comfrey reach down 10 feet to bring nutrients to the surface for shallow feeders.

Some vegetables need an abundance of rich fertilizers — and I mean rotted manure, seaweed,

Companionable Plants

Close associations of the species in the two columns is generally believed to enhance growth and/or result in better insect protection.

Anise	Coriander
Asparagus	Parsley; Basil; Tomatoes
Beans	Celery; Marigolds; Potatoes (repel Mexican Bean Beetle); Summer Savory (known as the bean herb in Germany)
Beets	Kohlrabi; Carrots; Onions; Brassicas
Brassicas (Cauliflower, Kale, Broccoli, Brussels Sprouts, Rutabaga, Turnip, Cabbage, Collards)	Onions; Beets; Radishes; Lettuce; aromatic herbs like Dill which help repel White Cabbage Moth
Carrots	Parsley and Onions which repel Carrot Fly; Sage; Peas; Beets
Corn	Squash; Cucumber; Beans; Lettuce; Spinach; Soybeans to repel Cinch Bug
Cucumbers	Corn; Radish, repels Cucumber Beetles
Lettuce	Radish; Cabbage; Onions, to repel Rabbits
Marigold	Cut Nematode populations, plant throughout garden
Nasturtiums	Squash; Brassicas; Potatoes
Onions	Lettuce; Carrots; Radish; Brassicas; Beets
Peas	Carrots; Corn; Beans; Turnips; Potatoes
Pepper	Basil; Swiss Chard
Potatoes	Corn; Peas; Eggplants; Beans repel Colorado Potato Beetle
Radish	Lettuce; Carrots; Parsnips; Tomatoes; Squash; Brassicas
Squash	Corn; Radish; Beans
Strawberries	Bush Beans; Lettuce; Spinach; Borage
Tomatoes	Marigolds repel Nematodes; Radish repels Two-Spotted Spider Mite; Carrots; Spinach; Nasturtiums; Lettuce; Basil
Turnips	Peas

hay and other organic additions — while others have far lower requirements for good growth and production. Heavy feeders, which should be grown in freshly fertilized soil, are the leafy *Brassicas* such as broccoli and cauliflower, all vegetables with heavy foliage such as lettuce, Swiss chard, squash, tomatoes and corn.

These should be supplanted the following season by legumes — peas and beans — for these valuable vegetables harbour nitrogen-fixing bacteria within their root nodules, and they greatly enrich the soil by making atmospheric nitrogen available to themselves and other plants. Light feeders, including such root vegetables as turnips, radishes, carrots and beets, should be rotated with the heavy feeders and legumes in the home garden.

In another example of physical interaction, tall plants may shade smaller species either to their detriment or benefit. Sunflowers and pole beans, for example, make poor companions, as they compete for sunlight. On the other hand, lettuce can be kept from bolting for a time during the hot summer by transplanting it into the luxurious shade of a sprawling tomato plant, and pumpkins and squashes grow well within the cool shelter of a corn patch, while peppers would be at a disadvantage growing in even partial shade.

One sort of companion planting which works consistently has solely to do with timing. I always sparsely sow radish seeds in the same drills with all the carrots and parsnips. Since radishes are the fastest of maturers while carrots and parsnips germinate ever so slowly, they are ideal companions. Quick to sprout, the radishes are up above the ground within days to mark the rows, which, if they depended on the lagging carrots and parsnips, would remain invisible for days and days more. And, within the month, the plump pink radishes are mature and ready for harvest just as the delicate wisps of carrot and parsnip tops need their first thinnings.

Last summer I planted radishes along with lettuce, which seemed to enhance the taste of the radishes, and amongst many of my *Brassicas* which include cabbages. This coming summer I plan to sow radishes between the tomato transplants, where they supposedly repel the two-spotted spider mite, and in the squash hills, where I'll let one or two in each hill go to seed to ward off cucumber beetles.

One neighbouring Cape Breton gardener with a reputation for cultivating a most bountiful and beautiful vegetable-flower-herb garden is Norma Taussig. When we spoke about companion planting she emphasized again and again the importance of growing many kinds of plants together to create a well-balanced, healthy environment. For that reason, Norma advocates companion planting, though she has "experimented with many of the conventional do's and don't's without always getting convincing results."

She did discover that dill and carrots do not grow well when planted near each other, and she reported that when she grew wormwood amongst *Brassicas* to repel cabbage moths it did not work, and in fact, appeared to stunt the growth of cabbages around it. Also, quite by accident, Norma, one year, sowed two identical rows of pole peas near her onions. The row of peas immediately next to the onions was about 18 inches shorter than the other row and bore only half the crop.

The biological basis for companion planting becomes especially clear when one is asked to name a single natural habitat which isn't a mixture of various plant species. Certainly, there are predominant plants such as spruce trees in a coniferous forest or beach grass on the dunes, but close scrutiny always reveals an incredible assortment of less conspicuous species well-integrated with the major plants. Ecologists explain that this vast variety of species, all of which are to some extent interacting, is the very essence upon which our delicate life system depends.

In marked contrast stands monoculture, which is responsible for the great bulk of commercial foods consumed in North America. The practice of planting one crop on acres and acres of land, and often year after year, can put a tremendous stress on the ecological balance of an area. The same nutrients are removed; the same pests and diseases are attracted season after season (and are better equipped each year to resist the pesticides). An artificial balance must be maintained by massive applications of herbicides, pesticides, fungicides and fertilizers. Aside from the environmental and health considerations, this type of agriculture is extremely costly. In fact, these chemicals and their application comprise much of the expense of modern vegetable production. For instance, it costs more than five or six hundred dollars per acre each season to spray mono-culturally-grown Florida tomatoes.

One of Canada's best-known organic growers is John Harrison who feels that the importance of companion planting is sometimes "blown all out of proportion."

For 30 years Harrison used entirely organic practices in farming 80 acres on Lulu Island near Vancouver, and his experiences are recorded in *Good Food Naturally* (J. J. Douglas, 1972). He says that natural pest control is mostly a factor of healthy soil and adequate moisture, and stresses that the latter is more important than most gardeners realize. (Even a slightly wilted plant can quickly fall prey to insects.)

"Avoiding the horrors of monoculture is really the answer," says Harrison, who is a firm believer in crop rotation. His own vegetable crops were rotated with plantings of green manures to be tilled under for composting in the field, and he reports virtually no pest problems in all his years of farming.

Harrison does feel that other forms of companion planting are practical, desirable, and to some extent unavoidable, for home gardeners. He points out, however, that the interplanting of marigolds or other labour-intensive practices are not feasible for most commercial growers and would like to see more research done on the subject.

Incompatible Species

The following combinations have been repeatedly observed to result in retarded growth when planted in close proximity.

Asparagus	Onions; Garlic
Beans	Onions; Garlic; Shallots; Gladiolus
Beets	Pole Beans
Brassicas	Strawberries; Tomatoes
Carrots	Dill
Corn	Tomatoes (The corn earworm will attack both, if handy.)
Cucumbers	Potatoes; all aromatic herbs
Onions	Peas; Beans; Pole Beans
Peas	Onions; Garlic; Shallots
Potatoes	Tomatoes; Squash; Sunflowers
Tomatoes	Dill; Fennel; Corn; all Brassicas
Most species	Wormwood; Fennel; Sunflowers; Walnut Trees; all known to inhibit growth of nearby plants

While the literature on chemical ecology is expanding, companion planting is far from being a high priority at most research institutions. It is not seen as practical in agribusiness and the experimentation being done is still comparatively slight, although both Richard Root of Cornell and Jean Forget of Agriculture Canada believe that more research money will be directed this way in the future.

"We're doing research," says Dr. Root, "but there needs to be a lot more done. Organic gardeners just need to keep on truckin'."

Indeed. Companion planting alone will not make for great gardens, and what works well for one person may not necessarily succeed as conclusively for another under different conditions. Too, until well-designed, controlled experiments are done, we must recognize that a certain amount of the advice on companion planting is based on hearsay and observations that cannot pass scientific muster.

Fortunately, the exaggerated monoculture of the Florida tomato fields is unlikely to seduce the home gardener, who will have little inclination to plant his whole spread in *Big Boys* every year. Neither is it likely that anyone will duplicate truly natural conditions in a space where the Mediterranean cabbage and the South American tomato are close neighbours. Nonetheless, it pays to plan your "domesticated polyculture" as intelligently as possible.

The accompanying lists of companion and antagonistic plants can serve as a guide -- and they are an appropriate beginning — but each garden is unique. The challenge for each of us is to determine what works best under your own particular circumstances to render greater yields, more flavourful foods, fewer pests and diseases and — not least — a more attractive garden.

Seedy Characters

"Well, Ultra Girl, he's quite the lad!"
"Yes, Big Boy, but whatever shall we name him?"

By Jennifer Bennett

For die-hard aficionados of seed nomenclature — those who can appreciate the rattle of an abused adjective, who rise to the allure of alliterative excess, and who savour a mistimed superlative — this is undoubtedly the finest time of year.

The new seed catalogues are out and one of mankind's less precise but most colourful arts is again on display. Connoisseurs of seed names are eager to pounce on such insipid appellations as *Early Hybrid*, an eggplant, but ready to relish titles like *Imperial Black Beauty* or *Dusky* — better eggplants, at least from the sounds of things.

Keen observers are keeping an especially close eye on the tomato sections of seedhouse catalogues for any clue that might answer two pressing horticultural questions: Where can the tomato go from here and what, after *Ultra Girl*, can seedsmen possibly call it?

It is not a matter to be taken lightly. While plant breeders are painstakingly cross-pollinating and hoping to find that elusive but fleshier, redder, tastier, bigger and earlier tomato, marketers are already scratching heads and wrinkling brows over potential names. A new vegetable variety's success often depends as much on the seed purveyor's ability with figures of speech as on the breeder's dexterity with plant genes.

This lesson was recently borne home to seed merchandisers at Northrup-King, a Minneapolis seed-producing giant. The company grows seed to fill the custom orders of many other seed houses and a number of years ago was commissioned by Stokes Seeds to produce a green bean of specified parentage.

Stokes' tests of the new hybrid showed that the cultivar (today's accepted term for a cultivated variety) stayed dark green when frozen or canned and that it sold very well at roadside stands.

Taking a certain amount of pride in the new snap bean, Stokes President John Gale dubbed it *Speculator* — a catchy name that Gale personally liked and that reflected the speculation involved in developing the new cultivar.

Speculator became the darling of many market gardeners, so much so that Northrup-King decided to market the bean under a different name — *Green Isle*. The title had neither the panache nor the meaning of *Speculator* and gardeners will be hard-pressed to find *Green Isle* in any seed catalogue today.

Too, there is evidence that many growers persist in buying outdated varieties merely because some seedsman of yore struck upon a zesty name. *Beefsteak* and *Bonny Best* tomatoes disappeared from agriculturalists' recommended lists a quarter of a century ago (Canada's prolific plant breeder E. A. Kerr points out that *Canadian Viceroy* is a far better beefsteak-type tomato), but their familiar names help keep them popular.

Those who dare slur such venerable sweet corn varieties as *Butter and Sugar* or *Golden Bantam* (introduced in 1902) will be labelled heretics by gardeners who have made a tradition of sticking to the first corn they really liked.

The profusion of names can be withering even for experienced gardeners. *Peaches and Cream* (corn), *French Breakfast* (radish), *Peter Pan* (peas), *Straight Eight* (cucumber), *Greenback* (cabbage), *Red Whoppa* (tomato) — just who are these seedy characters?

Facing page after page of seed names can prove as mind-boggling as deciphering the cata-

Bill Mayer/Graphics Group

logues' cleverly phrased descriptions and choosing between the too-perfect photographs.

In fact, the gardener who is able to crack the code can find in seed names a mini-history of vegetable gardening, a revelation of horticulture's most exciting moments, a storehouse of hidden seed breeders' names and a layman's guide to marketing strategy — information that can, at times, be turned to good advantage when making a final selection.

Much can be learned, for instance, by pondering the history of that plethora of *Boys* and *Girls* (be they *Big, Better, Early, Wonder* or even *Ultra)*, in the tomato family.

This confusing jumble can all be blamed on a toddler who was the son of David Burpee. Dr. Oved Shifriss — then a Burpee breeder — had an affection for the little tyke and addressed him with the pet name Big Boy. The name struck Shifriss as a natural one for his most recent tomato development, a variety which quickly became one of Burpee's tomato flagships, and which inspired the host of *Boy* and *Girl* spin-offs.

Not all seed namers have Shifriss' way with words, and regrettably some of the most outlandish names emanate from Canadian agricultural stations. The federal station at Vineland, Ontario has the devious habit of sneaking a "V" into the name of all of its cultivars. With names like *Vendor* (tomatoes) or *Viking* (asparagus), Vineland origins often slip past undetected. But with tomatoes like *Veemore, Veeset, Veebright, Veeroma, Veepick, Veepro* and *Basket Vee* the christener's verbal flat-footedness not only becomes irksome but smacks of bureaucratese.

"It's Twenties thinking," complains Stokes' John Gale, who has to sell many of these misnomers. Despite their handicapped titles, many are often excellent varieties geared to cold climate conditions, according to Gale. "You can't market something that doesn't have a good name on it."

CREATIVITY CURBED

The less-than-memorable *Veeroma*, Gale points out, is one of the best paste tomatoes to be developed in the last 20 years, yet the cultivar languishes in obscurity with a limp name.

Veeroma is the work of Dr. E. A. Kerr, whose other accomplishments at the highly respected Vineland station include *Veepick, Veeset, Veegan* (all tomatoes); *Sunnyvee, Tastevee, Northernvee, Polarvee* and *Buttervee* (all corn varieties).

No less stifled by government nomenclature rules is Walter Nuttal, another federal plant breeder. At one time, Nuttal worked in a position that gave him room to vent his naming creativity, one notable result being the favoured *Butter King* lettuce. Later transferred to a breeding station at Morden, Manitoba, his work all ended up with the syllable "mor" incorporated in the name: *Morden Yellow* tomatoes, *Morgold* sweet peppers.

Still hampered, Nuttal finds himself at the experiment station at Harrow, Ontario and — you guessed it — he has to work "har" into every name. *Harliton* seedless cucumbers get their label from the mandatory syllable and a combination derived from Little Hampton, England, home of the cultivar's parent stock.

New York's Robson Seed Farms leave their mark on seed varieties with the easily identified "Seneca" followed by a word with pioneer or Indian overtones: *Seneca Chief, Seneca Arrow, Seneca Explorer, Seneca Star* (all corn).

Inspired amateurs and private breeders often seize upon the act of christening as a chance to display paternal enthusiasm or to give themselves a pat on the back.

Proud fathers such as Thomas Laxton, a mid-nineteenth century horticulturist, earned immortality by naming their cultivars after themselves. *Thomas Laxton* peas are still planted by gardeners — a century after the breeder managed to cross *Ringleader* and *Maple* peas.

Borrowing a custom popular among racehorse owners, many seedsmen name new cultivars after parent plants: *Tendercrop* beans are a cross between *Topcrop* and *Tenderpod, Alton* is a product of *Alaska* and *Thomas Laxton* peas, and *Valnorth* comes from *Farthest North* and *Valiant* tomatoes.

This handy way of naming plants can have drawbacks: the medicinal ring to the name *Laxal* can be attributed to the variety's parents, *Thomas Laxton* and *Alton* (peas). (One wonders, too, about the future of the *Pacemaker* beet in our coronary-conscious society.)

Although we have yet to encounter *Eureka* zucchini or *Look-Ma-I-Did-It* peas, proud breeders have filled seed catalogue pages with *Bravo* broccoli, *Triumph* cucumbers, *Champion* radishes, *Perfection* cantaloupe and *Unique* leeks.

Nonetheless, these self-administered pats on the back are not always undeserved. In 1870 Ohio seedsman A.W. Livingston confidently dubbed his just-developed tomato *Paragon*. Now forgotten, *Paragon* was the first tomato variety that could be depended upon to produce abundant, smooth and uniform fruit, and it played a key role in breaking down North America's revulsion for tomatoes.

Livingston notes in his subsequent book, *Livingston and the Tomato*, that *Paragon* was born at the end of 20 years of "the most scrupulous care and labour." The first 15 years were discouraging, for "I did not then understand such stock seed would reproduce every trace of its ancestry, *viz* thin-fleshed, rough and undesirable fruits."

MISUNDERSTOOD TOMATO

Finally Livingston hit upon his "new method": selecting particular plants rather than individual tomatoes. He noticed a plant with heavy foliage and quantities of small, attractive tomatoes. He saved these seeds "with painstaking care," and, "by good cultivation and wise selec-

CHICORY BARBE DE CAPUCIN.

tion from season to season (the tomato) took on flesh, size and improved qualities. I then put it on the market. This was 1870. I called it the *Paragon* tomato."

Before *Paragon*, tomatoes were called many things by North Americans — few of them pleasant. *Mala insana* (the unwholesome fruit) was at best regarded as *pomme d'amour* — an aphrodisiac — by early nineteenth-century North Americans. At worst, as Dalechamps mentions in his *Historie des Plantes:* "These apples, as also the whole plant, chill the body. . . wherefore it is dangerous to make use of them." Even as late as 1832, a Connecticut botanist reported that the state agricultural station grew tomatoes chiefly as curios.

Although the tomato at its *Paragon* stage was still a far cry from *Better Boy*, *Glamour* or *Big Girl*, it has made remarkable strides. In its natural state, the tomato grows wild on the Andean slopes of Peru, Ecuador and Bolivia, and its currant-sized fruits range in colour from green with purple stripes to ruby red and have a texture that varies from hairy to marble smooth.

But even flashy *Ultra Girl* must pay homage to the tomato family's humble beginnings. In the "have your cake and eat it too" game that's called plant breeding, scientists are busy trying to breed the furry little berry's resistance to such diseases as verticillium and fusarium wilt back into the plate-sized fruits we grow today.

Tomatoes are easily North America's most popular garden crop — rare is the summertime backyard that doesn't support a half-dozen plants. And plant breeders have responded to consumers' demands for tomatoes — over 300 varieties (each one presumably better than the one before) have been named in the last 40 years.

The tomato family is not alone in its burgeoning numbers. Consider that the 1978 Stokes catalogue offers 51 cabbage varieties — everything from *Houston Evergreen* and *Copenhagen Market* to *Stonehead*. Burpee lists 20 cabbage varieties and Joseph Harris, a respectable 15.

Faced with *Perfection Drumhead*, *Crispy Choy*, *Red Acre*, *Savoy King*, *King Cole*, *Pee*

Wee and their cabbage brethren, the average gardener would be tempted to throw up his trowel in despair. He would certainly envy the classical Greek gardener. Theophrastus, a student of Aristotle, lists only three cabbage varieties in a third-century B.C. seed round-up. During the time of Christ, Roman gardeners could choose from six cabbage varieties. William Shakespeare would have had a more complicated task in selecting the best cabbage for his Stratford plot — by Elizabethan times the cabbage's ranks had swelled to 20 varieties. In the 19th century there were 30 cabbage varieties.

Not all horticulturists were pleased at this influx of bizarre new varieties and names. In the early 18th century, Englishman Richard Bradley was tearing his hair while contemplating the vast array of names of "Pease" available, and noted in *New Improvements of Planting and Gardening Both Philosophical & Practical:*

"I have often wondered at the Indescretion of some People who take delight in giving cramp names to plants and make it their business to multiply Species without Reason, as if a Fruit would be better for a Name that could not be understood, or that works of Nature were not already numerous enough for us to contemplate."

HORTICULTURAL HOODWINKS

Suspicions mounted and "Pease" were eventually brought to the attention of that august body, the Royal Horticultural Society, which immediately dispatched George Gordon, a member, to investigate the 43 varieties of "Pease" that had found their way into seed catalogues by the early 19th century.

Gordon, to the astonishment of the honourable Society, discovered that there were only 11 distinct varieties of "Pease." He then set about investigating other vegetables "to reduce the discordant nomenclature of the seed shops to something like order to enable the gardener to know the quality of the sorts he is unaccustomed to cultivate, and above all, to prevent his buying the same kind under different names."

Twentieth-century perusers of seed catalogues who find themselves confronted by lists of *Liberty Bells*, *Pennbells*, *Lincoln Bells*, *Golden Bells*, *Early Canada Bells* and *Bell Boys* (all sweet peppers) have good reason to wonder if they are sometimes buying the same kind under different names.

But the chances of this happening today are growing slimmer. The Plant Variety Act of 1970 safeguards, within guidelines, all registered seeds except those of celery, peppers, tomatoes, carrots and cucumbers.

Neither are hybrids under this umbrella, but seed companies keep their cross-breeding schemes well under wraps. (Ask a Burpee spokesman what goes into making a *Better Boy* and he'll give you a most bemused smile.) This genetic secrecy combined with honour among the better seedsmen makes it ever more probable that each name represents a truly different cultivar.

Bedding plant buyers, however, should exercise caution. Part of the reason that *Bonny Best* remains popular is that wily greenhouse operators occasionally sell similar but better varieties like *Glamour* or *Veebrite* under the faithful and beloved old name.

Before a vegetable is officially christened, its name must conform to the International Code of Nomenclature, and in North America it must be approved and catalogued by the American Seed Trade Association — a 600-member organization based in Washington, D.C.

International standards dictate that seed names must not exceed three words, including sets of numbers, nor should superlatives like earliest, biggest or best be used. The code, in its longsightedness, notes that names like Best Possible Beans "may become inaccurate through the introduction of new cultivars or other circumstances." It is a testimony to *Bonny Best's* age that the variety slipped its superlative through before the 25-year-old guidelines came into effect.

Names should be distinctive: not simply *Curled* endive, but *Curly Snowman* endive. Too, names must not be easily confused with those of other varieties of the same species.

INTERNATIONAL NAMES

Plant varieties regularly migrate from one country to another, and when immigration poses a language barrier the normal policy is that the name should be translated directly: *Ironhead* cabbage becomes *Eisenkopf* in Germany, *Tete de Fer* in France and *Jarhuvud* in Sweden. But this policy could lead to difficulties; we won't speculate on how sales of Joseph Harris' *Pink Lady* tomatoes would go in lands where pink lady translates into "scarlet woman."

Some varieties (*Sunnyvee* corn, *King Cole* cabbage) retain their original names as they move from nation to nation, and in other cases marketers attempt difficult linguistic hybridizations. Franglais, for example, saw one of its

finest hours when a seedsman at Quebec's J. Labonté & Fils coined *Hybride Big Boy de Burpee*.

William Dam (a popular Ontario seedhouse with Dutch roots) frequently dispenses with translation, and scanning Dam's catalogue is much like strolling through a European market: *Noordster* (beans), *Flakkee* (carrots), *Swart Duit* (carrots), *Rode Kogel* (cabbage) and *Langedijker* (cabbage).

But there will be no tongue-twisted *Langedijkers* for Stokes' customers. John Gale promptly re-christened the popular Dutch cabbages with names like *Ultra Green* and *April Green* — names that leave no question about the varieties' ability to remain attractive during long storage periods.

COMPUTER FUTURE?

As the seedsmen's battle for superlatives rages — hitting previously unattainable peaks with each year's influx of catalogues — new and better weapons are being called into the fray.

Today's seedsman no longer has to rely on a variety's natural colour or his own poetic sense to produce the *Detroit Dark Red* or the *Bibb* of the 1980's. Chances are good that the top seedsman has the latest in computer-age hardware at his elbow ready to tell him whether the name he is considering has already been used, and in some particularly advanced cases, able to offer him advice on what names will appeal to consumers.

Thus armed, today's seed marketers remain confident despite the immensity of the christening challenge. With an eye on the *Wonders*, *Bigs*, *Earlies* and *Betters* that have already been affixed to the *Boys* and *Girls* of the tomato family, we recently buttonholed John Gale — the man who actually gave the world *Ultra Girl* — and asked him point-blank: "Whither tomato nomenclature after *Ultra Girl*?"

"Well," said an unperturbed Gale, "guess we'll have to start working on the cousins."

Garlic: Gypsy King Of The Herb Garden

Bothered by werewolves, plagued by insomnia, tired of bland Anglo-Saxon dishes? Take a lesson from the Kootenay Garlic Man

By Edmund Haag

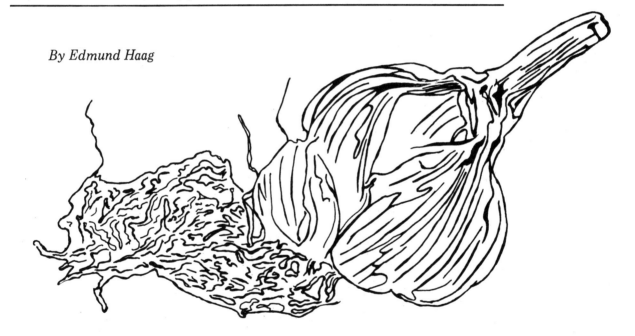

CANADIANS — hold your breath. We are in the midst of an invasion. Garlic The Terrible has arrived — what was once a social taboo is rapidly becoming high culture. That magnificent herb has struck our formerly bland-loving palate with all the pungency that has endeared it to more earthy cultures for centuries. Canada's finest restaurants now proudly boast of garlic-drenched escargots, garlic-laced Casear salads, and, for the true garlicophile, *Poulet Bearnaise*, whose recipe calls for two pounds of garlic per chicken.

What is this mania that has struck normally conservative Canadians — the same English-speaking stock who for generations snobbishly shunned the tasty herb? In actuality, North America is just beginning to catch up with the rest of the world in its appreciation of *Allium sativum*. Ancient Chinese writings make reference to the widespread use of garlic, both as a food and as a drug possessing magical properties. Petrified cloves of garlic have been found in Egyptian tombs, and an inscription in the Great Pyramid of Cheops records the garlic consumption of its labourers in the third century B.C.

The Roman legions marched on garlic — their equivalent of C-Rations being moulded balls of meal with cloves of garlic at the centre. Europeans in the Middle Ages considered it a great cleanser of the blood and preventative of plague.

The Anglo-Saxons, however, give us a clue to the origins of prejudice against garlic. "We absolutely forbid its entrance to our sailets," wrote John Evelyn in 1699. "Tis not for ladies palats, nor those who court them.'' Shakespeare associated garlic with beggars and brown bread (also out of favour), and even the eighteenth century fashionable French looked askance upon garlic.

The Russians, however, used it extensively and as late as World War II as an antibiotic, with extraordinary results. With such a glorious and savoury past, and with current interest both in ethnic foods and natural remedies running high, it is not surprising that even today there are people who have dedicated their lives to the plant.

GARLIC MAN

John Herman lives in Hills, British Columbia, a tiny hamlet nestled in the upper tip of the Slocan Valley. Originally from a large urban centre, he has lived in the valley for almost a decade, rebuilding and working an old homestead he and his wife purchased soon after their arrival.

To his neighbours, John Herman is known as the Kootenay Garlic Man — grower of the finest garlic in the area.

The quality of his plump, fresh buds has allowed John to barter them for three times their commercial value. He has, for example, exchanged $3 worth of garlic for $8 in handspun wool skeins. Herman began to have guilt feelings about such trades, but his customers assured him that his Number One Primo is worth the price.

Present orders for his Kootenay garlic are backlogged for more than a year, and he plans to put in a full acre of the herb to meet more of the demand. He has no plans to become a large commercial grower, however, choosing to maintain his simple life style.

"I am a poor man, and I intend to remain that way," he says.

Although he can't keep up with the demand for his product, many grocers in the same region of B.C. complain that the garlic they purchase from commercial sources goes bad before they can sell it. "I don't move more than $10 worth of the stuff a year," laments one large produce seller in a nearby town.

Herman feels that the reason some merchants have difficulties selling garlic is because the quality is poor. "If you've tasted good garlic and you are really into it, then there is no way you're going to buy the packaged stuff. It just isn't the same. People are better off growing their own."

JOHN HERMAN'S STEP BY STEP GROWING METHOD

1. The best time to plant garlic is in the fall, allowing the bulbs to lie dormant over the winter and get off to the earliest possible start in the spring. Garlic needs 100 days to reach maturity, and gardeners in cold northern areas may have to let the plants go two years before harvesting.

Soil should be properly mulched, with the addition of crushed oyster shells, wood ashes or limestone if it is too acidic. Garlic is not a particularly difficult plant to cultivate and no extensive pre-planting preparations are needed.

British researchers in 1972 concluded at the end of eight years of garlic investigation at the Doubleday Research Association that chemical fertilizers produced normal looking bulbs, but which lacked the bactericidal and pesticidal properties of garlic's essential oils.

Although many gardeners successfully use store-bought garlic for their planting stock, better sprouting percentages can be expected from seedhouse garlic, among which can be found cloves with such suggestive names as *Elephant* and *Red Snapper*.

If you have missed the fall planting, early spring is also a good time to start garlic, which, with optimal growing conditions, may be ready by fall.

2. The individual cloves (separated from the large bud) are planted about two inches deep, from four to six inches apart with at least one foot between rows. (If you are separating the cloves from a bud, take the largest outside sections.)

To plant in large quantities, a simple tool can be made by studding a long stick with six-inch spikes. The stick is aligned in the planting row and the spikes driven into the loose soil and wiggled around, producing a row of evenly spaced holes. Cloves are planted blunt end down and covered with a mulch.

3. Garlic is like the crocus. It is one of the first to come up in the spring, even beating most garden weeds. If the fall mulch was properly applied, the garlic will get a three week head start before weeding is necessary. John Herman mulches again after the first and second weeding. He says the garlic and mulch will deter most weeds. Don't worry about other garden pests — they hate garlic.

4. Most garlic planted in the fall reaches maturity the following September. A traditional way of determining maturity is by watching the stalks. Once they fall over, the garlic can be harvested. If some stubborn stalks dry up but refuse to fall on their own, kick them over and allow them to lie for several days before uprooting. This can speed the maturation process without affecting quality.

5. Weave garlic stalks into braids and store in a cool, dry place.

THE HEALER

Its gustatory qualities aside, garlic is an herb shrouded in mystique, a werewolf repellant, a curative of everything from arthritis to tuberculosis, in some people's minds.

Like other members of the genus *Allium*, (chives, onions, shallots, leeks), garlic contains allicin (more properly known as S-allyl-L cysteine sulfoxide). The volatile oil of garlic consists of diallyl disulfide, responsible for its pungent taste and odour, and allyl thiosulfonate, the active ingredient whose powers are again being investigated by science.

Garlic has long been known to be effective against certain bacteria, fungi and parasites, but its full powers remain a question to the medical world. Dr. Paul Sacco of Xavier University in New Orleans is presently involved in a federally funded project designed to test garlic's effectiveness as an antibiotic. (A 1944 U.S. test found garlic allicin to have only one per cent the activity of penicillin.)

"I have made for him a protective charm against thee, consisting of evil-smelling herbs, of garlic, which is harmful to thee. . . ."
—Magician of ancient Egypt
speaking to demons
attacking a child.

In some societies and by the faithful, garlic's medicinal value isn't questioned. In the mid-1960's a Russian team of scientists visited a remote mountain village known for the long life span of its citizens. When one centenarian was asked to describe his secret for a long life, he replied, "eat lots of garlic and work every day."

Garlic has been a traditional treatment for intestinal worms and parasites. One friend who uses herbal medicines worms his old tomcat monthly with garlic. Usually he does this by jamming a crushed clove down the unsuspecting feline's throat before it has the opportunity to protest. The animal is eight years old and still healthy in spite of a diet of tapeworm-infested rats and mice.

Recent medical studies done in both Great Britain and the United States indicate that moderate-to-heavy consumption of garlic by older patients reduces the likelihood of high blood pressure and hypertension.

TERRITORIAL IMPERATIVE AND GARLIC

Although garlic is growing in popularity, it has yet to make it with the puka shell and disco crowd. For those who want to hit it big on the social scene, it can still be a liability.

A garlic addict who is forced by circumstances to relate to a number of individuals each day — many of whom are abstainers — might find himself ostracized. One Vancouver dentist attributes his four-day work week to garlic. "I love garlic — if I work four extra-long days, then I can eat garlic three nights a week without wiping out my patients."

Pliny, who recommended garlic as a second century cure for tuberculosis, had an answer for early Roman garlic breath:

"If a man would not have his breath stink with eating of garlic, let him do no more but take a beet root roasted in the embers, and eat it after. It shall extinguish that hot and strong flavour."

John Herman's favourite neighbour pacifier is parsley fresh from the garden, chewed immediately after the meal. Others are convinced that the only thing that can successfully mask garlic

breath is two tablespoons of peanut butter. (Hands stained with the oils of garlic after cooking can be de-scented by washing with salt and cold water.)

CASH CROP POTENTIAL

As John Herman has found, the growing acceptance of garlic gives it natural possibilities as a cash crop for the homesteader or small farmer. Rough calculations indicate that one acre could produce up to $3,000 worth of garlic.

Thomas Horton, of a major herb supply house in Toronto, says that the bulk of the garlic used in Canada today comes from California. (China, Bulgaria, Egypt and Hungary are all garlic exporters, but their product faces much more rigorous import regulations imposed by North American departments of agriculture.) Horton says his sales of garlic in all forms have "easily doubled" in the past 10 years and he suspects that there would be a healthy market for home-grown garlic in this country.

Recently, while in Vancouver, I talked with the produce manager of a large supermarket chain which carries the California garlic and asked him if he would consider handling Canadian garlic. His answer was quick and straightforward: "If the price is competitive, why not?"

Despite its relative ease of culture and its tolerance of winters that drop to —40 degrees Fahrenheit, however, very little garlic is grown in Canada or the northern U.S. In such garlic-loving centres as Quebec City and Montreal, for example, home-grown fresh garlic accounted for between 8,000 and 19,000 pounds of the total consumed, while imports averaged between 500,000 and 750,000 pounds.

According to the Quebec Department of Agriculture, Canadian-grown garlic does not even meet five per cent of the demand.

Charles Lapointe, of the Agronome Crop Products Division, recommends that a marketing scheme be devised before anyone goes into commercial garlic production. This can be done by joining others to form a bulk supply or in direct sales arrangements with restaurants, supermarkets or other retail outlets.

Agriculture Canada makes the following suggestions:

Garlic will grow on any soil suited to onions, but thrives best on fertile, well-drained, medium-light soils. This crop requires a good supply of organic matter; therefore it is a good practice to prepare the soil the previous year by ploughing in a green manure or cover crop. . . .

More vigorous plants and larger bulbs will result if planting is done in the fall at about the same time as tulip bulbs. Where the land is well-drained, garlic can be planted in rows 18 to 36 inches apart depending upon the method of cultivation. The bulbs are separated into cloves and are planted four to six inches apart in the row. The clove is pressed into the soil with the fingers, the base pointing downward as with onion sets. The usual depth is one to two inches. To plant an acre, about 400 pounds of garlic will be required with rows spaced 18 inches apart.

As soon as the bulbs are mature, that is, when the tops are yellowed and dry, they are loosened and then pulled and placed in small bunches in windrows to dry.

After about one week they can be gathered into piles and then cleaned. This consists of removing the outer loose portions of skin and trimming the roots off close to the bulb. Then they may be either plaited or stored in open mesh bags for market. If plaited, the tops are left on and plaited together so as to hold the bulbs to the outside. These braids are hung in a dry, airy place to cure, after which several plaits may be fastened together in a long bunch. If not plaited, the tops are cut off to about one inch. The bulbs are allowed to cure for 10 to 14 days in the field, and are then graded and placed in mesh bags or slatted trays for storage.

Through subsistence farming and other work in a tree-planting cooperative, Herman and his wife find garlic growing supplies the small amount of extra cash and bartering power for their necessities and simple luxuries. Beyond that, John clearly delights in hefting a braid of his Number One Primo and is his own best customer. "Eating garlic gives me a feeling of well-being — besides, it tastes good."

Haag

GOURMET'S DELIGHT

As already mentioned, man's culinary love affair with this pungent herb has lasted some four thousand years or more. Although the ruling classes have had a traditional disdain for garlic, referring to it as peasant food and "the causer of an unholy stink," it has managed to work its way into some highly respected recipes:

Continental Garlic Soup
1 qt. soup stock
10 cloves of fresh garden garlic
2 tsp. lemon juice
5 oz. butter
1 tsp. brandy
½ tsp. thyme
½ bay leaf
6 oz. grated Gruyère cheese
6 slices French bread
Crush 5 garlic cloves, chop another 5 cloves, then sauté in butter until golden. Pour in the stock, add herbs, lemon juice and brandy, then simmer for 2 hours.

Remove bay leaf and pour contents into 6 heat-resistant soup servers. Float the bread slices and top with cheese. Bake in oven until cheese is melted.

Breath-taking Spaghetti Sauce
4 cloves garlic
2 medium onions, chopped finely
5 tomatoes
3 cups tomato juice
5 tsp. butter
½ cup red table wine
2 green peppers
1 can artichoke hearts (water packed)
½ tsp. oregano
1 cup sliced mushrooms
Crush garlic, chop onions and sauté in butter until golden and transparent. Add artichokes, green peppers, mushrooms, oregano and dry red wine. Simmer for 15 minutes and add tomatoes and tomato juice, salt and pepper.

Cover and simmer over very low heat at least 3 hours. Add more tomato juice if sauce thickens too quickly. Serve just less than steaming hot on firm spaghetti noodles and top with Parmesan cheese.

24-Hour Garlic-Dulse Dressing
½ cup olive oil
¼ cup vinegar
1 tsp. brown sugar (or substitute)
½ tsp. Worcestershire sauce
3 cloves garlic
½ tsp. ground black pepper
1 tsp. dry mustard
4 Tbsp. finely shredded dulse (a seaweed esteemed by many cooks)
Crush garlic in bowl. Shred dulse into pieces no larger than ¼ inch, and add mustard, pepper, sugar and Worcestershire sauce. Slowly blend in vinegar and oil. Mix well and allow to sit for 12 to 24 hours.

Haag

"Dressings are richer if prepared with garlic, then left until the next day," says John the Garlic Man, who looks upon the herb as a "flavour solvent." For an extra-rich dressing, blend with cream just before serving. This dressing goes best with fresh summer salads and greens.

Cream of Garlic Soup
4 cloves garlic
1 cup milk
cream to taste
This is a simple soup, but one which some garlic lovers regard as a lusty, warming tonic, and a fine treatment for insomniacs at bedtime.

Simply crush the cloves of garlic and simmer for 5 - 10 minutes in the creamy milk.

The rules for cooking with garlic are exceedingly simple. The more garlic flavour you want, the more cloves you use. They can be squeezed, chopped, crushed, pricked or used whole. They can be sliced and used to rub the inside of a cooking pot or salad bowl for subtle flavourings.

Garlic is a blender of flavours, especially if it is slowly simmered with other ingredients in a soup, sauce, stew or pickle recipe. The notorious reputation of garlic comes more from eating it raw or only partially cooked (as in many garlic breads); thorough cooking tames the odours somewhat.

New Roots

Forging a link to the past
— and the future —
with a grove
of black walnuts

By Samuel and Dorothy Wesley

When the American Revolution suddenly uprooted thousands of U.S. colonists and sent them fleeing north, it brought to Canada settlers already well-acclimated to the rigours of North American life.

These were hardy farmers of Dutch, German and Huguenot stock and they came seeking fertile land on which to establish new homesteads. Historian D.E. Reaman says that the settling of Upper Canada by these United Empire Loyalists followed a curious but highly pragmatic pattern: "The pioneers followed the black walnut trail, that is, they sought the land on which grew the black walnut — the limestone soil."

What these shrewd newcomers had discovered was that the native walnut required a deep, rich, well-drained loam to thrive. The black walnut was usually found in mixed virgin forest of red oak, white ash, sugar maple and beech. It was land upon which the settlers would survive — and prosper.

Today, of course, most of the continent's native black walnut has disappeared, much of it sold long ago to European craftsmen who prized it above all other woods. The demand continues stronger than ever as witnessed by the veneer company that recently paid $39,000 for a single black walnut tree growing near Johnson City, Iowa.

Samuel Wesley

Thus, shortly after purchasing our 10 acres of retirement property, we were delighted to discover that a huge black walnut still grew on a neighbouring farm. Our interest in this species — *Juglans nigra* — had been kindled by reading Reaman's book *The Trail of the Black Walnut* and one of our long-range goals was to replant our small pasture farm with trees. We were particularly interested in restoring some of our lost heritage by growing something of lasting value for future generations. The neighbour's old black walnut would give us the foundation stock.

SEED POACHERS

Our first task was to gather the nuts, and, as the tree's owner had no objections, one weekend

early in the month of September we filled a basket with the plump green walnuts.

We had not yet moved onto our new farm and the nuts were left outdoors awaiting our return. Arriving the next weekend with great plans and full of seed-planting zeal, we found that, in our urban naïvety, we had failed to reckon with the popularity of the walnut. Squirrels had carried off every single seed.

We diligently collected another basketful and began again. Walnuts are not among the fastest growing trees, but 10 years later, we have a small grove of healthy trees in what had been a grazed-over pasture and expect to collect our first walnuts this fall.

Perhaps the most important step in establishing a small plantation of black walnuts — or even

The Wesleys' 10-year-old grove of black walnuts, some of them already having reached 16 feet in height.

a single tree in the yard — is to secure planting material that is acclimated to your area. The best possible source is a healthy tree growing within your own climatic region — a mature walnut on your own land, from a nearby farmyard or growing wild in a woodlot or forest.

Identifying the black walnut tree is not particularly difficult, the nut itself being a useful guide. It can be spotted most easily at the end of the summer, when black walnut leaves drop earlier than most other species in Canada and the northern United States. For about two weeks the tree stands out stark and leafless amongst other species that are still in full foliage or turn-

ing colour. The tall dome of the branches will be studded with big round nuts, black against the sky.

In summer, the walnut's multiple leaves stand apart on long limbs allowing the sun to filter through and throw patterns of sunlight on the ground, unlike the dense shadows cast by other species of trees. The light green foliage contrasts sharply with the black bark of the long branches, and the leaf itself is distinctive. Twelve to 14 inches long, about six inches wide, it is composed of 15 to 23 leaflets arranged opposite one another along each side of the stout, hairy stem. The individual leaflets are sharp-pointed and deep-toothed, measuring approximately three to three-and-one-half inches in length.

The butternut tree is often confused with the black walnut, and, though they are related, the walnut is known by its rounded fruit, whereas the butternut is ellipsoid. Too, the bark of the mature walnut is rough, while that of the butternut is smooth.

Having obtained a supply of nuts (being sure to include extras to fill in for those which fail to germinate), a seedbed must be carefully prepared. A prime consideration is to keep the nuts from ground hogs and squirrels, which have an uncanny ability to ferret out walnuts buried by human hands.

PLANTING METHOD

We happen to be surrounded by these friendly marauders, so one year decided to try an experiment. One seedbed was covered with fine-mesh chicken wire and another left unprotected. Both beds were a complete success, and, made bold by this test, we left all beds unprotected the next year. Disaster — the squirrels struck again and we lost a full year of growing time. Dismayed, we resolved to cover future seedbeds at all costs.

We prepared the seedbeds by spading the soil to the depth of a shovel blade and then raking to remove lumps. Rows are made running the length of the bed and are spaced about two feet apart. The nuts are placed in these furrows, spaced six inches apart, pressed down lightly and covered with two inches of soil. Deeper planting prevents some nuts from sprouting.

OVERWINTERING

A screen of half-inch diameter chicken wire is then placed on top of the seedbed, widely overlapping the edges of the bed. The perimeter must be firmly weighted down with old fence rails, heavy boards or a border of stone.

Walnut seed must remain underground over the winter to allow sprouting in the spring. The authors plant directly into a seedbed of rich soil and cover to discourage nut-hungry squirrels and groundhogs. Above right, seedling at three years of age. Bottom left, at six years the trees have reached about 26 inches and, bottom right, at about eight years this tree is seven feet tall. Plantings in more temperate areas will often show faster growth.

We have also made larger plantings by ploughing a long single furrow, seeding as described above, and covering the whole length of the planting with a single run of chicken wire. Such a bed can be located alongside a fence or other boundary and thus not intrude upon a garden area.

Planted in the early fall, the seed goes through its natural maturing process during the winter and the impressive protective shell softens to allow a sprout to emerge in the spring.

Some growers have equal success with stratifying the seed over the winter: that is, they bury the nuts in layers of moist sand in outdoor pits and unearth them for planting in prepared beds in the spring.

We remove the wire mesh as soon as the walnut seedlings appear and the young trees are left in the original seedbed until they are about one foot tall and at least one-quarter inch in diameter at a point one inch above the root collar. This takes one to two years, depending on the soil and growing conditions.

Walnuts require fertile, well-drained soil and open sun — they should not be planted closer than 20 feet from the edge of other forest trees. Leave at least 10 feet between the walnut seedlings when transplanting them. However, as walnuts are subject to winterkill when high winds chill the terminal (growing) buds, it helps to locate a new walnut grove in an area protected from heavy, direct wind flows. Our own first seedlings went into an open field adjacent to a large group of poplar trees and we resisted the temptation to plant on the wind-swept hilltop.

Each tree site should be cleared of all grass and weeds before transplanting and a hole dug large enough to accommodate the roots without bending them severely.

In removing the young black walnuts from the seedbed, use care not to break or cut the long tap root. The transplants should immediately be put into a bucket of water or wrapped in wet burlap to keep the delicate roots and root hairs moist until planting.

SOAK & MULCH

Locate each transplant in the ground one inch deeper than its position in the seedbed and soak each planting with water to ensure that the roots are well-firmed with soil and that no pockets of air are trapped around them. The transplants must be watered in times of drought.

Damage from mice or other rodents can be prevented by giving each tree a 10-inch collar of wire mesh or hardware cloth. A five-foot-high stake positioned on the south side of each tree gives support (tie with old nylon stockings) and provides a bit of shade to help prevent scalding by the winter sun.

We find that transplanting in the fall gives the walnuts a head start in growth, but those in areas with frost heaving problems might be well-advised to transplant in the spring.

We surround the base of each transplant with

a mulch of old hay, straw, leaves or well-rotted manure to conserve moisture, cool the soil and add some organic nutrients. We add fresh mulch each year, but late in the fall pull it back from the tree trunks to eliminate the possibility of mice nesting near the trees.

As the trees grow in height, some tend to form multiple-branched crowns. Prune these competing lateral buds to encourage strong growth in the main trunk and to assure a well-shaped tree. However, if the terminal bud dies, it is sometimes possible to train one of the top-most lateral branches to grow vertically and become the main trunk.

"BABY THEM"

In southern Canada and the northern States, black walnuts can reach a height of 25 feet in 10 years, at which age they start producing nuts in appreciable quantities.

One man who has seen the value of establishing black walnut groves on small farms is Ingersol Arnold, chief of the New Hampshire State Forestry Service. Well-known to aboriculturists throughout the United States and Canada, Mr. Arnold has for the past 10 years tried to develop walnuts hardy to northern climates. He firmly believes that walnuts planted now will bring important monetary returns to small farmers in years to come.

"You see the lumber companies buying trees right out of people's front yards," says Mr. Arnold. "I know of one tree in northwestern Ohio that sold for $30,000 — of course, it was 57 feet from the base to the first branch.

"Some people think they can plant a few of these trees, forget them and then make a fortune. It doesn't work that way. You've got to baby them. They don't do well in poor soil. In fact, some of the best trees I've seen have been planted in abandoned barnyards."

Mr. Arnold says that it will take 40 years for a black walnut to reach the size when it can be sold for lumber. At age 62, he looks at his 10-year-old experimental grove and says with a laugh, "I'll never see that day. I'll be in the grave, maybe pushing one of them up."

Nevertheless, we take great pleasure in seeing our walnuts grow and have the continuing satisfaction of passing along some trees to young friends just establishing their first home and to others with space for a walnut tree on their property. Our pleasure in these gifts increases as we follow their growth and see the enjoyment they give to others.

We also take hope from the story of the late Sir William Mulock who, some years ago, (at the age of 80) planted several acres of walnut trees on his property a few miles north of Toronto. The old gentleman lived well beyond the century mark and saw the reality of his vision in an impressive stand of 20-year-old black walnut trees.

Black Walnut Seed And Tree Sources

Remedies For The Old Homestead Apple

Restoring neglected trees to fruitfulness

By Allan & Ellen Bonwill

"The planting of an orchard, which is a matter of great importance to the future comfort of the settler's family, is often delayed year after year, and that is done last which should have been attended to at the outset...

I cannot too forcibly impress upon the emigrant the advantage he will derive from thus securing to his household the comforts, I might almost say the blessing, of an orchard..."
— Catherine Parr Traill, 1855
The Canadian Settler's Guide

For those whose apple trees seem to date from the days of Catherine Parr Traill or the original wanderings of Jonathan Chapman (alias Johnny Appleseed), the blessings of an orchard or even a single tree may be well-disguised.

Neglected old apple trees seem to abound both in the countryside and backyard settings — trees suffering from gross untamed growth, insect damage and producing middling crops of undersized fruit.

Never, however, underestimate the value of an apple tree or its ability to respond to a bit of attention and judicious pruning. In fact, restoring such an old tree to production takes less patience than planting a new one. The results are often spectacular and far quicker than waiting for a young tree to come into bearing.

One noteworthy example in our orchard is a late-bearing pale yellow apple of unknown age and parentage that was producing only a half-bushel or so of inferior fruit before we gave it the rejuvenation treatment. The year after pruning was a foliage year in its cycle and it produced only a few handsful of fruit, but it now provides us with bushels of apples that make exceptional applesauce.

LEADERS & SCAFFOLDS

Many people seem to be intimidated by the thought of pruning, but it is actually a simple practice for small-scale farmers and gardeners. Pruning serves both to invigorate and dwarf a tree, removing unwanted and poorly placed limbs and keeping the tree manageable for the picking season.

In nature, a tree actually prunes itself to some extent, as branches are choked off and die back when over-shaded. To make best use of a domestic tree, however, the process should be speeded along with annual care.

The first step in bringing an older tree under control is to head back its upper growth. We like to keep our trees not more than 10 feet tall for ease of harvesting.

Most orchardists today train their trees in what is known as a "modified central leader", in which a main shoot is developed with three to five main branches radiating from it. (The tree's "leader" is its predominant upper branch, the one which seems to be leading the growth of the tree.) You can cut and prune an older tree into this formation (see accompanying illustration) by heading back its topmost growth and creating a leader which runs somewhat parallel to the ground (unless your farmstead is equipped with very long ladders or skyhooks, these heaven-bound branches will probably remain unpicked).

With the older tree it may not be possible to train the modified central leader, in which case the centre of the tree should be opened up (hence, the "open-centre type") so that a pigeon could fly right through it, in any direction. This admits sunlight to help ripen the fruit.

An old tree often shows snarled areas (sometimes known as "mare's nests"), and these should be cleaned out. Cut away crossed branches, prune out all water sprouts (adventitious, succulent, counter-productive growth), and remove down-hanging branches that are shaded heavily from above.

Try to leave main branches well-spaced around the trunk in a sort of scaffold arrangement and repeat at a higher level. This results in a branch pattern suitable for a tree-house and is sometimes called that type of growth.

Orchardists also remove any bad crotches — places where the tree forks into two branches of equal length and diameter from a common point. One of the two branches is generally removed,

APPLE TREE REJUVENATION

A. Head back the top, to make the tree lower.
B. Open up the crown to light.
C. Prune out crossed branches.
D. Prune off underbranches.
E. Fill cuts and cavities with pitch.
F. Cut away water sprouts.
G. Cut away root suckers.
H. Dig or punch holes 12 inches deep, 18 inches apart, in a circle at the edge of the foliage line. Fill holes with rich natural fertilizer (manure or compost).
I. Place a thick layer of mulch around the tree, leaving an 18-inch-radius clear space surrounding the trunk (discourages mice from gathering at trunk and chewing the bark). With young trees, mulch out past the foliage line.

Judith Goodwin

or at least cut back severely so that it becomes a lateral branch from its mate. At the same time, suckers at the base of the tree should be pulled out and any severe cuts or cavities in the trunk filled with pitch to prevent further rot.

Pruning is best done in late winter or early spring, when the tree is still dormant. Although pruning *may* continue up to blossom time, it should be remembered that the later the pruning the greater the negative effect on the tree's

94

vigour. On the other hand, frozen wood should not be pruned, and too-early pruning leaves the tree more vulnerable to infection and drying of tissues.

A SHOT OF MULCH

The old long-forgotten homestead tree is also very likely in need of feeding. Proper fertilization will not only boost growth and fruit yield, but will improve the size, quality and storage life of its apples.

A nitrogenous mulch is recommended for the spring to stimulate leaf growth, while a phosphate-rich tree food should be applied in late summer to encourage fruit bearing (ground phosphate rock, bone meal). Manures are best applied in the fall after harvest.

A Dutchman once told us that the place to plant a fruit tree was the spot where the outhouse used to stand. He claimed that after the roots got down to the right point, you would never need to fertilize the tree again. We question the wisdom of this. Once we planted lespedeza (a nitrogen-fixing legume) around a young apple tree, and after 15 years, it is still not bearing. As with other plants, too much nitrogen leads to heavy foliage growth, but almost no fruit. Phosphate is needed for fruit bearing.

To be sure, the fruit produced from many older trees may not be as versatile as some modern apple varieties. Grafting new stock onto a long-established tree is not a solution and often a waste of time, as the young top and the old roots are not matched in size. This leads to too-prolific top growth which is not particularly hardy and easily frosted.

Nevertheless, we manage to make use of large quantities of apples from restored trees, some that are better used for cider or sauce than fresh eating. We use cider fresh, we freeze it, and we make apple jelly from it, finding it far richer than jelly made according to the usual recipes.

We substitute apple juice for water in many recipes — jellied salads, meat dishes, salad dressings, some desserts. Apples themselves can be frozen, after being washed, cored and sliced (we do not even pare them). In this form they are fine for apple pies, applesauce or fried apples — great for breakfast on a winter morning with hot sausages and toast.

Whenever we speak of apples, we remember the quotation that says, "Anybody can count the seeds in an apple, but who can count the apples in a seed?" Trees are truly the engines of production, far outreaching any other plant and extending their usefulness through generations. A farmer we once knew planted a new peach orchard when he was in his eighties. When queried as to whether he expected to live to enjoy the fruit, he said, "Well, I've eaten fruit from trees that others have planted — why shouldn't later generations enjoy the fruit from the orchard I'm planting?"

We care for our old trees and — like Johnny Appleseed — we just keep on planting new ones.

Pruning Recommendations

The pruning of every tree, trained or not, calls for sufficient trimming to keep it open and to encourage moderate new growth every year. The following suggestions will serve as a guide.

1. Cut out broken, dead or diseased branches.

2. Where two branches closely parallel or one overhangs the other, remove the less desirable, taking into account horizontal and vertical spacing.

3. Prune on the horizontal plane. Leave those laterals growing horizontally or nearly so on the main branches, and remove those that hang down or grow upward. This cannot always be done but where possible it should be followed.

4. All varieties should be thinned out enough to permit exposure to sunlight and air.

5. Where it is desirable to reduce the height of tall trees, cut the leader branches back moderately to a well-developed horizontal lateral.

6. The lower branches of broad-headed or drooping varieties should be pruned to ascending laterals.

7. Varieties tending to produce numerous twiggy, lateral growths should have some of these removed to prevent overcrowding.

8. Make close, clean cuts. Stubs encourage decay and canker, thus providing a source of injury to the parent branch or trunk. Cover large wounds (over 1½ inches in diameter) with a suitable tree dressing.

9. Prune moderately. Very heavy pruning is likely to upset the balance between wood growth and fruitfulness, and generally should be avoided.

10. Prune regularly. Trees that are given some attention each year are more easily kept in good condition than trees that are pruned irregularly.

11. Prune the part of the tree where more growth is required. This is particularly important in older trees. New growth will be stimulated only in those parts that were pruned. Reduce pruning to an absolute minimum where growth is already excessive.

12. Do not remove a branch unless there is a very good reason for doing so. Remember that the leaves of a tree are the food-manufacturing organs. If the leaf area is reduced unnecessarily, the tree will be reduced in growth or fruitfulness or both.

From the fine booklet, Pruning and Training Fruit Trees, *available free from Information Branch, Agriculture Canada, Ottawa. Ask for Publication Number 1513.*

One Potato, Two Potato, Three Potato, Four,
Five Potato, Six Potato, Seven Potato, More.

He Says Potato, She Says Potahto

To others it is simply the Spud,
and no garden is complete without it

By Jennifer Bennett

It may sound like a nutritional believe-it-or-not, but a steady diet of seven pounds of potatoes and a pint of milk a day is said to provide all the essential nutrients needed by an adult male. The rural Irish knew this, of course — though the average adult male in Ireland might consider the diet lacking by an essential dram or two — and they put it this way:

> *"Be eating one potato, peeling*
> *a second, have a third in your*
> *fist and your eye on a fourth."*

No foodstuff is as humble as the potato. Scrooge called Marley's ghost "an underdone bit of potato." Van Gogh, possibly in a fit of depression, painted an abject group of potato-coloured peasants and called them "The Potato Eaters." And yet, what could be more aristocratic than *Pommes de Terre Chantilly* or *Potatoes Romanoff?* European botanists who first studied the newly discovered potato called it *taratoufli* — little truffle — for it was thought to resemble that delicacy in flavour. The Germans still call their spuds *kartoffel.*

Today, this enigmatic tuber is the fourth largest food crop in the world, behind corn, wheat and rice. It is fantastically productive -- cut a seed potato into three pieces, plant them in the spring, and by fall you will have at least 12, and perhaps upwards of 24, new tubers for your efforts. The potato produces more protein per acre than any other crop, with the exception of soybeans.

It wasn't always so, and even today visitors to remote Indian markets in the high Andes of South America can see the not-so-productive ancestors of the modern potato. The Inca tribes had been cultivating this rather lumpy, worm-shaped tuber for thousands of years — occasion-

ally sacrificing some of their more unfortunate citizens to the *Patata* God — when the Spanish conquistadores arrived. Hot on the trail of gold, these early sixteenth-century plunderers saw possibilities for the strange vegetable, and quantities of these Andigenum series potatoes accompanied the loads of *oro* home to Spain.

They were initially touted as a cure for impotence and sold for exorbitant prices, said to be as high as $1,000 per pound. This soon met with a decidedly flaccid reception, and the Spanish snake-oil dealers quickly changed tack and began to promote the vegetable as an exotic foodstuff.

This, too, proved a difficult selling job. The potato is related to the nightshade family (which numbers among its members the tomato, eggplant and tobacco) and is a member of the Solanaceae family. As such, its green parts contain solanine, a poison, and Spaniards who tasted its leaves or fruit doubtlessly suffered. In addition to the notoriety of its family tree, the potato is a lover of temperate climate and cool soil. Having just arrived from the mountain slopes of South America, it undoubtedly grew poorly in Spain.

In spite of attempts to popularize the potato, Europeans continued to prefer the familiar salsify, turnips and beets. In 1619, potatoes were banned in Burgundy because it was thought that they caused leprosy. The prudent Scots prohibited their culture in 1728, citing the fact that they were an unholy nightshade and not mentioned in the Bible.

The French reluctantly obeyed Louis XVI's orders to plant potatoes, never imagining the wildly popular "frites" their progeny would one day devour. The English took almost 200 years to accept the new vegetable, not really trusting it until the late 18th century. The Irish, however, too poor to be discriminating, ate them and thrived.

Fed by a balanced diet of milk and potatoes, the Irish peasantry of the 18th and 19th centuries increased in numbers as never before. It is said that the daily intake of the tubers of *Solanum tuberosum* amounted to 10 pounds or more per person in this period. No wonder, then, that young Irish colleens sang of a popular potato pancake:

> *"Boxty on the griddle,*
> *Boxty in the pan,*
> *If ye can't make Boxty*
> *Ye'll never get a man."*

The utter dependence of the Irish upon the potato brought disaster when a fungus disease known as late blight completely destroyed the crops of 1845 and 1846. Famine and wholesale emigration resulted, sending some 500,000 Irish people to early graves and another million to other countries — most notably Canada and the United States.

RARE FIND

The emigrants arrived to discover that the potato had preceeded them to North America. Just how it came to be there is open to conjecture, but one version states that Sir Walter Raleigh encountered potatoes in the Virginia colony — perhaps introduced there by the Spanish — and then carried them back to the unsuspecting Irish.

The potato had achieved popularity in North America by the mid-1800's, although this fact was not pleasing to everyone. In 1819, William Cobbett complained in *A Year's Residence in the United States of America*:

> *". . . I now dismiss the Potatoe with the hope, that I shall never again have to write the word, or see the thing."*

Cobbett's opinion notwithstanding, the potato was here to stay. New varieties occurred in great profusion. but very few showed much staying power. Burpee's, for example, offered the *Goldflesh* in their 1888 catalogue, and they claimed "the skin is reddish, and the flesh a pure, rich golden yellow." This was not the last attempt to foist the vitamin-A-rich, yellow-fleshed potato on North America. *Marygold* and *Calrose* were introduced in the United States in the 1940's and were promptly rejected. Although new moves are afoot to try again, the yellow-fleshed potato, popular in Europe, may have gone the way of the purple carrot in the land of "whiter than white."

Another ill-fated early variety was the *White Elephant* or *Late Beauty of Hebron*. In the book *Lark Rise to Candleford*, Flora Thompson explains:

> *"Everybody knew that the* Elephant *was an unsatisfactory potato, that it was awkward to handle when paring, and that it boiled down to a white pulp in cooking; but it produced tubers of such astonishing size that none of the men could resist the temptation to plant it."*

The fact is that the potato at this time was neither as tasty nor as reliable in its growth habits as most varieties we now know. The breeding stock had gradually deteriorated, due to years of entirely asexual reproduction (from tubers). It remained for Luther Burbank to rejuvenate the potato, and for fate to smile upon Burbank.

The American horticulturist was blessed by the good fortune of finding an extremely rare seed pod on an *Early Rose* variety plant. It was something he would never see again, though he offered a standing reward for one. From this fortuitous find, he managed to propagate two seedlings. "It was from the potatoes of these two plants," he wrote, "carefully raised, carefully dug, jealously guarded and painstakingly planted the next year, that I built the Burbank potato."

Burbank was never one to play down his own exploits, but his new variety was almost identical -- except in skin colour — to the *Russet Burbank*, the most popular potato in North America today, 100 years later. This wildly successful cultivar can be found under a confusing array of names (there is no Plant Patents Act covering potatoes). It is known as *Idaho*, *Idaho Russet*, *Golden Russet*, *Idaho Gem*, *Alberta Gem* and, perhaps most commonly in Canada, as *Netted Gem*.

It is the one Walter Nuttall, chairman of the Ontario Regional Potato Committee, says he would grow if he had a garden. "They're the ones wrapped in gold foil in the grocery store." It is a large, evenly-shaped potato, a good baker, and excellent in storage.

McCain's Foods Ltd. of Canada is the largest producer of frozen French fries in the world, and "450 New Brunswick farmers supply us with all the top quality *Netted Gem* potatoes they can grow." McCain's is now expanding its acreage in Manitoba, where a gigantic processing plant is being constructed.

The *Netted Gem* is a late variety, that is, it is harvested late in the season and meant for winter storage. The other mainstay varieties in the Maritimes — which supply seed potatoes to the world — are *Kennebec* and *Sebago*. Most potato acreage in the United States is planted in *Netted Gem*, *Katahdin*, *Kennebec* and *Red Pontiac*. The latter is an early variety, as are *Superior*, *Onoway* and *Irish Cobbler* — grown for harvest as early as July.

GARDEN CULTURE

Generally, the best (or at least the most popular) potatoes for your area will be those stocked by local nurseries and country general stores. Our source is a nearby marketeria, which stocks *Irish Cobbler*, *Kennebec*, *Red Bison*, *Sebago* and *Netted Gem*.

Irish peasantry seizing crop of an evicted tenant near Tralee, Kerry, during the Great Potato Famine of 1845-46.

The seed potatoes will be available in the early spring, and some varieties may sell out quickly, so it is wise to decide early what type and quantity you wish to grow. If you have no storage facilities at all, you may want to plant an early variety alone, or may choose to put in limited quantities of both an early and a late variety.

Ten pounds of seed potatoes, at one plant per foot, should be sufficient for roughly 100 feet of row. From this you can expect a minimum yield of 75 to 100 pounds; some gardeners report harvests as high as 500 pounds from this same space. The average North American adult eats about 150 pounds per year, but as your potatoes will be unlikely to keep until you start picking the next crop, you may wish to plant considerably less.

Potatoes prefer a fertile, slightly acid soil, and will be happiest in light, sandy soil in the 4.8 to 6.5 pH range. If you don't have a soil test kit, but have observed that tomatoes, eggplants and watermelons thrive in your garden, the soil is probably acid. If, on the other hand, the beets, cabbage and canteloupes usually take over, and you end up with scabby potatoes, you have a basic soil.

Lime or wood ashes will make garden soil more basic, so don't apply them to the potato area of your garden. Instead, give them compost or well-rotted manure from the barn. The soil should be friable·to a depth of one foot — the new potatoes are going to grow above your seed potato, but the roots, of course, must grow down. In any case, potatoes, like tobacco and peanuts, will grow in very poor soil. A gardening neighbour who gets a good potato crop swears that the more spuds are fertilized, the less flavour they have.

The generally accepted recommendation is to plant your potatoes one foot apart. Those who wish to conserve space may plant intensively — as close as eight inches between plants. The exception is *Netted Gem*, which requires at least

12 to 15 inches between plants to allow development of well-shaped potatoes.

Soon after the ground can be worked in the spring, you will be ready to prepare seed for planting. If you make the mistake of trying to use table potatoes from the grocery store, you may be unpleasantly surprised to discover that they won't grow, as many are treated with a sprout-inhibiting chemical to allow commercial storage. If they do sprout, table potatoes will generally produce less than certified seed, and may be a variety unsuited to your needs and climate.

Canada has one of the best seed certification programmes in the world, and it makes good sense to use new seed potatoes each year — the certification gives assurance that you won't introduce a new disease to your soil.

The "eyes" in the seed potatoes will become your new potato plants, so it is essential that at least one eye or sprout — preferably two — be present on each piece of seed potato. The potatoes should be cut into egg-sized pieces (1 to 2 ounces), or may be planted whole, if they are small. After cutting, the pieces should be left for at least one day, and even as long as a week, in a well-ventilated, well-lighted place. This allows them to heat and to "gather their strength" after cutting. (The formation of a dry, corky layer on the freshly-cut surfaces helps prevent the entry of decay organisms.)

The seed pieces are planted three to four inches deep, with from one to three feet between rows. The former will result in an intensive garden, which will be difficult to cultivate and really only suitable for growing under a mulch.

MULCH CULTURE

A great deal has been said and written about growing potatoes under mulch, and some gardeners occasionally go so far as to recommend placing seed potatoes directly on the soil and

Plant breeder Luther Burbank, originator of today's most popular potato strain.

then covering with mulch. This method is fraught with risk, because mice may finish off the potatoes after they've sprouted, or a dry spell may wither the seed.

The eighteenth-century Irish used to "lay them on the sward and cover them with six inches of mold and so hill them up as they grow." Although you may not have a great deal of "mold" available in your sward, the ideal method seems to be to plant the seed potatoes so that they are just covered with soil, and then cover them with six inches of leaves, grass clippings or straw. This system quite reliably increases production by about 40 per cent.

This increase can be attributed to the mulch keeping the soil cool and moist, which potatoes like, and adding nutrients as the mulch decomposes. Too, this method dispenses with the need to weed or cultivate, and it saves garden space while discouraging insects, which seldom venture far into a mulched plot.

Lacking a large quantity of mulching material, you should plant the seed potatoes in a trench about four inches deep and cover them with soil. As soon as the green plants appear, hoe them over with earth. Continue this hilling process as they grow until you have a ridge eight to 10 inches high. Remember, the potatoes grow above the original seed and you don't want the tubers to become exposed to sunlight. Now, relax.

By July you should be able to reach into the soft earth around one of your plants and feel an egg-sized spud. If you carefully remove a few of these, you'll have potatoes for dinner and will not have harmed your plant. This is known as "grabbing" and is a fine trick to know if you are — justifiably — tempted by the thought of delicate new potatoes.

From this point on, the plants will take care of themselves, if you remember to water when necessary (they need an inch per week) and to control the Colorado potato beetles. These fat, striped beetles and their repulsive orange larvae can best be controlled by physically removing the first ones to appear and dropping them in a pail of hot water, or, if you prefer, a small quantity of gasoline. Severe attacks can be quelled with dustings of rotenone.

POORE DUTCH MEN

There are other enemies awaiting your spuds — cutworms, slugs and many diseases, including the bane of the tomato, verticillium wilt, and, of course, the ubiquitous late blight. In areas where late blight is a problem, gardeners should plant resistant varieties (see accompanying article). In general, however, good seed potatoes and healthy soil will produce healthy plants that can resist attacks by most enemies.

Two or three weeks after the tops of the plants have died in the fall, you will be able to harvest the potatoes. You will now be open to description by the disdainful comment one Englishman had for the Dutch, who discovered the delights of tuber crops before the British: "The poore Dutch men, like swine, digge up the rootes."

Before you "digge all your rootes," test one by rubbing the potato with your thumb. If you cannot rub the skin off, the potatoes are ready to be dug. The drier the soil and the drier the day you harvest the crop, the better they will store. Choose a sunny day and dig carefully — any bruises will soon lead to rot in storage. Leave them out in the sunshine for the rest of the day to cure, then pack into burlap sacks, bags, boxes or baskets, sorting out the smallest for early use.

It is important that they be stored in total darkness, so if there is any light in your storage area, be sure they are completely covered. Light encourages the production of chlorophyll, which in potatoes produces the poisonous solanine. They are best kept at a constant temperature of 35 to 45 degrees (F) and at a humidity of 80 to 90 per cent. If you are fortunate, this spot will be your root cellar, or that of a neighbour. Otherwise, the potatoes may go into the basement or a room in the house that can be kept cool.

Do not attempt to freeze your potatoes. Because they have such a high water content (80 per cent), large crystals will form in freezing. Commercial food processors flash freeze their potatoes at an extremely low temperature, thus by-passing the crystal stage.

Gardeners in severe climates may be the only ones able to try a technique used by the Indians in the high Andes. They left slices of their potatoes out at night to freeze, then let the frozen water evaporate in the sunlight. The result, *chuno*, was able to be stored for long periods.

The potato contains about 2.5 per cent protein, and is a good nutritional source of calcium, phosphorous, potassium, iron and vitamins B and C. And, despite its reputation, a medium-sized potato, boiled or baked, supplies merely 100 calories. (Potatoes deserve their notoriety only because of the dressings they attract these days. Drenched in butter, gravy, sour cream or mayonnaise, the potato must be viewed in a new light.)

While many backyard gardeners are under the impression that it is hardly worth the space to grow potatoes, this notion is fast becoming outdated. In the first place, with home-grown potatoes, you won't be ingesting the common sprout inhibitors maleic hydrazide or chloro-IPC. You will also save money.

Twenty-five pounds of seed potatoes should cost $3 to $4, and will yield 250 pounds or more. This amount would cost you $25 to $60 in the store, depending on market conditions that year. Savings will be increased if your harvested potatoes are disease-free and you decide to save some for next year's planting.

When you prepare spuds for the table, whether they be *au gratin*, deep-fried, scalloped or in a salad, remember to remove sprouts and all green parts first. And when you sit down to dine, be thankful for the well-travelled potato. Were it not for its maiden voyage across the Atlantic and its trip back again to New England, you might be eating your hamburger with an order of deep-fried salsify, your T-bone with a nice baked turnip.

John Power Pettee, a nineteenth-century writer, composed a suitable grace for your meal:

> *"Pray for peace and grace and*
> *spiritual food,*
> *Pray for wisdom, and guidance,*
> *for all these are good,*
> *... But don't forget the potatoes!"*

Sources

The following companies offer seed potatoes by mail.

LINDENBERG SEEDS LIMITED
803 Princess Avenue
Brandon, Manitoba R7A 0P5
Plan to have the following available, as of May 1. Write for price list. (Sold in 10 and 35 kg. bags. Ten kg. plants 100 to 125 hills.)
Main Crop: *Norchip, Pontiac* (red), *Norland* (red), *Kennebec, Netted Gem, Burbank Russet*
Early: *Warba* (red or white), *Waseca* (red), *Early Ohio* (pink).

GEO. W. PARK SEED CO.
Greenwood South Carolina 29647
Idaho or *Red Russet:* $2.95 per bag (plants 25-foot row); 3 bags for $7.95. Postpaid.

Seed Potato Guide

Don't know a chipper from a masher? Read on ...

By R. G. Rowberry

Ninety years ago this spring, the W. Atlee Burpee & Co. seed catalogue could be used to order not only packets of pansy and carrot seed, but such weighty items as "thoroughbred" Cotswold breeder sheep, crushed oyster shells in 300-pound quantities and 21 varieties of potato — by the peck, bushel or barrel.

Today, seed potatoes are only rarely handled by mail-order seed houses, and the various attributes and drawbacks of different varieties are known to most gardeners only by word of mouth or from personal experimentation. By its exclusion from the colourful and descriptive pages of the seed brochures, the bulky potato now seems as dowdy and drab as the fat kid who stayed home from the party.

Although there are vast differences between varieties, many prospective potato growers have little idea whether a variety is early or late, mealy or soggy, whether it stores well, or if it is good for boiling, baking or chip-making.

Some varieties have been selected for use in certain small geographical areas (the highly specialized *Nipigon*, for instance, was released in 1976 just for the Thunder Bay area of northern Ontario). Others are particularly resistant to certain pests and viruses, while some potatoes are especially suited to filling the incredible maw of the fast-food industry, led by the French-fry hungry McDonald's empire.

There are 42 varieties of potato licensed in Canada and many more in the United States and Europe, but for brevity's sake I will describe those most readily available and likely to be of interest to the home gardener, moving from the earliest to the latest-maturing.

VIKING is red-skinned and sets four or five slightly rough tubers. These are scab-resistant and will grow to a good size even in dry conditions, giving a reliable over-all yield. *Viking* is very tasty as a boiled "new" potato but is not satisfactory for baking, chipping or for making French fries. Do not try to store *Viking* for any length of time, as the eating quality deteriorates after several months or so. A purple-skinned *Viking* is smoother and has better cooking quality but is not yet available in commercial quantities.

SUPERIOR has become a very popular early potato in the last few years. It is white-skinned, fairly smooth and scab-resistant. *Superior* is a good "new" boiling potato, and when dug in mid-season makes excellent chips and French fries. Care must be taken when using cut seed if the ground is very wet and very cold or hot, as the seed pieces will probably rot. *Superior* has a good yield and keeps well until about Christmas, after which the quality declines.

NORLAND is an attractive, deep-pink, very smooth potato which sets 10 or 12 tubers per hill. However, if it does not get regular and adequate moisture (25 mm — 1 inch — a week), these tubers will remain small and the yield will be low. *Norland* is scab-resistant and is very good as a "new" boiling potato, but again should not be stored for any length of time. It is very susceptible to the silver scurf fungus which kills the skin and causes the flesh to shrivel. A more serious weakness of *Norland* is that it is highly prone to ozone damage and you should not try to grow it where air pollution is a problem.

IRISH COBBLER has been the standard early variety for over 100 years. It has a good yield under almost any conditions, but the tubers are rough, deep-eyed and susceptible to scab. The quality is good and many people prefer the texture of *Cobbler*, which is more mealy than the other earlies. It can also be used for chips and French fries, but it is seldom stored once the smoother, late varieties are available.

KESWICK is a good mid-season potato if you live in the northern parts of the country. It sets about six scab-resistant tubers per hill, which may be slightly rough, but the quality is high, and *Keswick* may be boiled, chipped or French-fried. The foliage is light green and is resistant to both early and late blight. *Keswick* keeps well in well-ventilated storage if the temperature is kept at 3 to 5 degrees (C) with high humidity (about 90 per cent).

CHIEFTAIN. We are now into the mid-season or early main crop varieties. *Chieftain* has a bright red skin and a high yield. It is fairly smooth and is resistant to scab and late blight. There will be seven or eight good-sized, fairly smooth tubers per hill. *Chieftain* has excellent boiling qualities but is very poor for chipping, baking or French-frying. *Chieftain* is also sus-

ceptible to silver scurf, but if your crop is free from this disease it will keep well into the winter.

TOBIQUE. A new Canadian variety. It has a slightly different appearance in that it has a white skin with red patches, particularly at the "eye" end, which makes it easy to recognize. There will be at least a half-dozen fairly smooth tubers per hill, which are very good for boiling, chipping and baking but not for French-frying. *Tobique* is somewhat susceptible to scab and late blight, so you should avoid ground which you know to be infested with the scab fungus and should be prepared either to lose the crop or to mount a spray programme if late blight spreads into your area. This variety stores well at 3 to 4 degrees (C).

KENNEBEC is the highest yielding variety around and is widely used commercially for chips and French fries. It can be mid-season or early main crop, and sets about six large white-skinned tubers, so the in-row spacing should be kept at 9 inches (23 cm) or less. At wider spacings you are likely to get oversized tubers which may be hollow. *Kennebec* has large, vigorous vines which are resistant to early and late blight and the tubers are scab-resistant. It is a good potato for home use *if the tubers have not been exposed to much light*; if they have, they will turn yellowish-green or green very quickly and will become bitter. (A steady diet of bitter potatoes can lead to severe digestive upsets, even though the bitterness may be disguised by butter or gravy.) *Kennebec* keeps well in storage provided that the tubers are sound and have been kept in low-light or dark conditions.

BELLEISLE is another new Canadian variety from the same breeding programme as *Tobique*. It sets seven or eight fairly smooth tubers per hill but is quite late in maturing. The tubers adhere strongly to the vines even when the latter are almost dead, which makes for some rough handling in harvesting; fortunately, however, the tubers are quite resistant to bruising. *Belleisle* is also resistant to scab and late blight and is good for boiling and French-frying, although in my opinion it is not as good for baking as *Tobique*. *Belleisle* keeps well under normal storage conditions.

SEBAGO was the standard main crop potato for many years but is being superceded because of its lateness in maturing. It has a high yield of smooth, good-sized tubers which are fine for boiling and French-frying. *Sebago* is resistant to scab and late blight but is very susceptible to blackleg, which can reduce the yield considerably. This variety stores well up to Christmas, but will begin to sprout soon after if the temperature is not kept down to 3 or 4 degrees (C).

NETTED GEM (RUSSET BURBANK in the U.S.). This is the one you should try to grow if you can. You may know it as the "Idaho Baker" and perhaps have bought it wrapped in foil at three for 69 cents or more. The foil is a sales gimmick but also serves a very useful purpose in preventing greening, as most retailers display them in the best-lit part of their produce counters. *Netted Gem* has the highest quality in North America for boiling, baking, chipping and French-frying or for making scalloped (and instant) potatoes. It is very late and has long, heavily-netted russet-skinned tubers, which can become very misshapen if growing conditions are not optimum. For a good yield of reasonably smooth tubers it needs at least 35 cm spacing in 90 cm rows and *must* have a fertile soil and a *regular* supply of moisture (25 mm per week) throughout the growing season. *Gem* is scab-resistant but susceptible to late blight. It is important that aphids be kept off the vines, as they (the aphids) carry the leaf-roll virus which, in *Netted Gem*, causes the condition known as "net necrosis" in the tubers. This consists of a network of dark streaks throughout the flesh; it is quite harmless but looks most unpleasant. *Gem* is a very good storage potato.

The breeding programmes at Fredericton, New Brunswick and the University of Guelph have several promising new entries which will soon be released, and some readers will be interested to know that a number of them have yellow flesh. Many European varieties have yellow flesh and are considered to be superior in flavour and texture to white-fleshed potatoes. This, of course, is a matter of individual taste, but gardeners will soon have the chance to find out for themselves.

Having selected the variety or varieties you wish to grow, the next problem is to find suitable seed. The office of the Seed Certification Section, Plant Quarantine Division, Agriculture Canada in your area will have a list of seed growers from whom you may be able to buy seed, assuming that it is not available at your local feed store or garden centre. These offices are situated at Burnaby, British Columbia; Calgary and Edmonton, Alberta; Regina, Saskatchewan; Winnipeg, Manitoba; Barrie, London and Ottawa, Ontario; La Pocatière, Quebec; Fredericton, New Brunswick; Charlottetown, Prince Edward Island; Kentville, Nova Scotia and St. John's, Newfoundland.

Illustration from 1868 Burpee Seed Catalogue

Horticultural Heirlooms

Bringing out the Luther Burbank in you

By Robert Mariner

"This was the goal of the leaf and root.
For this did the blossom burn its hour.
This little grain is the ultimate fruit.
This is the awesome vessel of power."
— *Georgie Starbuck Galbraith*

Many are the poets and writers who have waxed horticulturally over the miracle of the seed, but, living as we do with the worst postal system in the civilized world, it sometimes seems that the greatest miracle of the seed is its safe arrival in the spring mail.

This first became dismayingly clear several years ago when a postal strike shut down the mails just when the heaviest flow of mail-order seeds normally moves around the countryside. What should have been an obvious truth suddenly became agonizingly clear to me: You cannot grow a garden without garden seeds.

I was reduced to digging through packets on supermarket display racks, searching in vain for my favourite varieties. Part of my prejudice against buying off-the-rack seeds is, I will admit, pure snobbishness (the best of the true seed aficionados wouldn't be caught dead with a packet of drugstore seeds).

Nevertheless, there is more than snobbishness involved here, for in all my searching I failed to find many favourite varieties among the mass market seeds.

As it turned out, the strike that year was settled in time for my seed orders to arrive for spring planting. Even these were missing some varieties that I had been counting on, either because of crop failures the year before or because they were sold out. By this time I had become determined to grow at least some of my own seed.

CALIFORNIA SEEDBED

To see what would happen, I planted some carrots (which produce seed in their second year of growth) from our vegetable cellar, as well as several celery plants that had wintered over in the greenhouse and which were now sending flower stalks five feet in the air, their blossoms swaying in a vegetable *Rite of Spring*.

The results of these first experimental efforts were mixed. The carrots produced seed, but because of my inexperience, it turned out to be only moderately viable. The celery plants, on the other hand, burst forth with such quantities of excellent seed that they even went toward replenishing the neighbours' spice jars. Many other seeds fell to the ground from the parent plants, producing, on the ground, a rich green carpet of little celery plants, and, in me, feelings of self-sufficiency.

When just a young man of 21, Luther Burbank purchased a 17-acre farm near Lunenberg, Massachusetts. Inspired by the works of Charles Darwin (most notably *The Variation of Animals and Plants Under Domestication*), Burbank set out to be a plant breeder.

Success came almost overnight, with the development of his *Burbank Potato* and, with a shrewdness that marked his outlook on life, Burbank immediately sold the plant rights in exchange for passage to California.

There he settled in Santa Rosa and began to establish the breeding farms that would make him famous. (He developed a total of 800 new strains and varieties.) Perhaps partly because of this move in 1870, the greatest concentration of seed production for North America remains in sunny southern California.

This is the heartland of vegetable agribusiness, and it is understandable that the seed growers

104

are most concerned with meeting the needs of their largest customers — the mega-farmers. In many ways, the needs of gardeners coincide with those of large agricultural interests.

Seed is so inexpensive, for example, that with very few exceptions (such as shell beans and other large-seeded crops) the amateur would be hard-put to save money on the home production of seed, considering the time involved and the space given over to growing seed instead of food.

Too, plant breeders have also provided marked improvement in many characteristics of vegetables: increased yields, earlier maturity and better resistance to such foes as disease, insects and frost.

Bill Milliken

Bolting lettuce: useless to the salad gardener, but a seed collector's prize.

The needs of home gardeners and small farmers do not, however, *always* coincide with those of agribusiness. The gardener, for example, will probably look for lettuce with taste and tenderness, rather than the bred-in, high-cellulose content that helps commercial lettuce heads withstand rough handling in picking and shipping.

Read any of the current research reports on tomato breeding and it will quickly become obvious that the qualities most sought in new varieties are toughness, uniform shape and colour and ease of ripening off the vine. Taste and nutrition, it seems, are forgotten.

One research director for a U.S. seed company stated frankly in 1967 that "not a single variety" of vegetable had been deliberately bred in that country for home garden use during the preceding two decades. Any new introduction which met the special needs of gardeners was purely accidental.

This trend goes back to at least the turn of the century, when *Danish Ballhead* cabbage was introduced with a proud claim: Its head was so firm and solid that, even if kicked all the way to market, it would still arrive in good condition.

There are exceptions, of course, and work of interest to gardeners is occasionally done at universities; for example the development of the outstandingly high vitamin A tomato (*Caro-Rich*) at Purdue University.

Another exception is Johnny's Selected Seeds, an unusual seed company that is dedicated to bringing back old varieties of garden vegetables.

Located in Maine, Johnny's specializes in vegetables for areas with short growing seasons (the first frost at Johnny's comes as early as August 20) and tolerance to cold. He is helping to bring back such venerable corn varieties as *Country Gentleman*, *Howling Mob* and *Stowell's Evergreen*, this last variety found in catalogues predating the U.S. Civil War.

Rob Johnston, Jr. is "Johnny" and he encourages his customers to grow their own seed and will exchange seed he has grown for varieties that gardeners have preserved over the years.

"A feeling of completeness — of closing the cycle in the life of the plant — is really experienced when you plant a row of your own homegrown seeds," says Johnston. "You can add a whole new dimension to the satisfaction of gardening."

LUTHER BURBANKISM

Though the uninitiated would find it strange to consider seeds to have heirloom qualities, there is a definite interest today in seeking out and growing plant varieties considered extinct.

For example, I was given an invaluable gift this spring — a dozen or so sugar pea (edible podded) seeds that had been passed down through the family of a gardening friend, Gladys Crooks, for at least 51 years. A seed company would be quick to christen these *Crook's Delight*, but they are of an unknown variety that has thrived year after year, suffering drought and the soggy, mouldy, fog-bound summers that are not-unheard-of here in Nova Scotia.

"We always just called them sugar peas," says Gladys, who has tried several other varieties that are commercially available. When asked if these proved as flavourful or productive, she says "No!" with the certainty of a no-nonsense horticulturist whose large garden has helped maintain a family of six.

Saving seed can be as simple or complicated as you wish to make it. Even gardeners with little experience will find particular varieties that taste better and that can be counted on as staples. Though it doesn't occur to many people that seed can be saved without expert help, there is no reason not to try your hand at plant breeding. (Burbank himself had little more than high school education and no formal training in genetics.)

106

In the wild state, a plant unable to withstand the full range of nature's vagaries in a particular environmental niche will be crowded out by plants that can. This is the essence of natural selection. In our gardens, however, we judge plants by their flavour, frost resistance, yield per foot of row, and so forth — not just whether they will survive the season. Plants bearing the qualities you want most are selected for seed production. Or, looking at it another way, any specimens which are clearly off-type are *rogued*: that is, removed from seed production.

Deciding just what characteristics you want in your vegetables will require some thought. The home gardeners' cabbages, for example, don't have to stand up to shipping; they generally travel no farther than from the row to the table, perhaps with a rest period in the root cellar.

For me, the first consideration for cabbage is whether it will store well in my extra damp root cellar. Winter cabbage is important to my family's economy, so, in this case, high productivity and reliability are more important than taste. Further, my decision to grow my own cabbage seed adds the requirement that the roots stay alive and frisky long enough that they can be planted in the garden the next spring and grow good seed.

A neighbour may rave about a huge squash, but will it keep well? Even if it does, perhaps

SOME COMMON MEMBERS OF FOUR CUCURBITA SPECIES (THE SQUASH/PUMPKIN GROUP)
1. C. MAXIMA Buttercup Delicious Hubbard Mammoth
2. C. MIXTA Cushaw
3. C. MOSCHATA Butternut Crookneck Kentucky Field
4. C. PEPO Acorn Connecticut Field Cocozelle Delicata Jack-O-Lantern Lady Godiva Scallop Small Sugar Straightneck Vegetable Marrow Vegetable Spaghetti Zucchini

NOTE: *Members of each species will pollinate each other. Some crosses might occur between species: 4 with 2 and 3; 1 with 3.*

SELF-POLLINATING	
Beans Broad Beans Chicory Endive Lettuce Peas Tomato	*All are annuals or perennials treated as annuals in temperate zones*

CROSS-POLLINATING VEGETABLES
(Including groups of vegetables which will pollinate one another)

	GROWTH HABIT	POLLEN DISTRIBUTION
ASPARAGUS	Perennial	Insect
BEET GROUP		
Garden beet	Biennial	Wind
Mangel	Biennial	Wind
Sugar beet	Biennial	Wind
Swiss chard	Biennial	Wind
BRASSICA OLERACEA GROUP *		
Broccoli	Annual	Insect
Brussels sprouts	Biennial	Insect
Cabbage	Biennial	Insect
Cauliflower	Biennial	Insect
Collard	Biennial	Insect
Kale	Biennial	Insect
Kohlrabi	Biennial	Insect
BRASSICA, OTHER SPECIES *		
Chinese cabbage	Annual	Insect
Mustard	Annual	Insect
Radish	Annual	Insect
Rutabaga	Biennial	Insect
Turnip	Annual	Insect
CARROT	Biennial	Insect
CELERY	Biennial	Insect
CORN	Annual	Wind
CUCUMBER	Annual	Insect
EGGPLANT	Annual	Insect
MELON GROUP **		
Muskmelon	Annual	Insect
ONION	Biennial	Insect
PARSLEY	Biennial	Insect
PARSNIP	Biennial	Insect
PEPPER, SWEET AND HOT	Annual	Insect
PUMPKIN/SQUASH See accompanying table	Annual	Insect
SPINACH	Annual	Wind
WATERMELON	Annual	Insect

* *Members of one Brassica group will not pollinate members of the other Brassica group.*

** *Other members of the muskmelon group are not generally grown in the north; e.g., Casaba and Honey Dew.*

you prefer several meal-sized squashes for your family. A tomato variety which ripens its fruit all at once is fine if you want a batch for home canning, but it's not so wonderful for prolonging the salad season. Such features are brought into sharp focus as you stand in front of a plant, recalling its performance through the entire season, debating if it should be a parent for your seeds.

In any row of vegetables one or more plants are likely to appear healthier and larger than the rest. These can be marked in some way, most easily by poking a marker stick next to the best specimens.

If, on the other hand, you are growing a field of something like beans just to propagate the seed, be sure to walk down the row and rogue out any inferior plants so that only the best will mature to produce seed.

It is essential that the seed grower know the reproductive cycle of any vegetable whose seed he hopes to save.

This may seem very basic to some gardeners, but if you expect to harvest beet seed, for example, the first year will be exceptionally frustrating unless you know that no seed will appear until the following season.

Annuals (see chart) grow from seed and produce new seed in one growing season and then die.

Biennials live for two years, producing seed during the second year.

Perennials grow and produce seed for more than two years.

Asparagus and rhubarb are two examples of northern perennials, while corn is a perfect example of an annual. In tropical climates, tomatoes, peppers and Lima beans are perennials, but we treat them as annuals because they produce seed in the first season and rarely make it through our winters.

The sturdy parsnip is a biennial that will survive in the ground throughout the winter in all but the coldest areas. In late fall or very early spring, the parsnips can be dug, the roots inspected, and perhaps 10 selected for propagation. These 10 best will either go back in the ground or be stored in a root cellar for spring planting. The treatment is essentially the same for carrots, beets, cabbage, celery, parsley, onions, cauliflower and Brussels sprouts.

With the cole crops, you can have your seed and eat the cabbage (or sprouts or cauliflower), too. I put cabbage plants, for instance, into the root cellar with heads and roots intact. Those that keep best through the winter are used last — we cut the heads off and save the detached roots for propagation.

Another method of producing seed from biennials calls for late summer or early fall sowing. Commercial cabbage seed is sometimes grown this way, because the young rosette stage of the cabbage is winter hardy (the head is not) and may therefore be left in the field.

The next year the cabbages never really form heads, but bolt to seed. This method allows for

Skeleton in the family closet? Showing its humble roots, this flowering parsnip in author Mariner's garden clearly resembles its relative, Queen Anne's Lace. Carrots are also closely related, and seed collectors must avoid cross-pollination between cultivated plants and their wild ancestors.

no selection (there were no heads to judge by) and the seed produced is called market seed. It is usually grown under contract for sale in large lots. This method is dependent on the use of stock seed, which is the best available. Stock seed is produced under the watchful eye of the plant breeder, who goes to extremes to isolate and rogue the crop.

When growing your own seeds, you are the breeder, and you must be equally assiduous in selecting and maintaining the purity of your strains.

Some annual vegetables produce seeds with very little effort required of the gardener; these include peas, beans and tomatoes. However, some of the hardy annuals pose more of a challenge for northern seed savers.

In the case of leaf lettuce, for example, you should always select plants that are the last to bolt (shoot to seed) when hot weather arrives, since this is the kind of behaviour you expect from properly bred leaf lettuce. Unfortunately, with the cool summer weather in my area, the lettuce bolts, but not in time to mature its seed that season.

This year, I treated my lettuce as if it were a biennial by fall planting *Black Seeded Simpson* lettuce in my greenhouse, where it wintered over, then burst into vegetative growth early in the spring. It produced well, then bolted. When hot weather came, I transplanted some of the potted plants into the garden. I have my eye on one plant which bolted three days later than the others, and I plan to save its seed separately for further comparisons.

ISOLATION

The prevention of unwanted cross-fertilization in your seed supply is essential, though it is really not very difficult unless your neighbours are also growing seed (and even then most problems can be solved).

The easiest and most frequently saved seeds are peas and beans. You can grow your favourite variety of peas among 15 other varieties of peas and you won't have to worry about whom your peas are out with. This is because they are self-pollinating. Cross-fertilization is extremely rare.

I also grow, in close proximity, four or five varieties of shell beans that were given me and which have been perfectly well-behaved in mixed company for decades. Peas and beans isolate themselves by having flowers which cover their reproductive organs, so fertilization occurs in the privacy of their own blooms.

Tomatoes are also self-pollinating, but bees will visit tomatoes and occasionally produce crosses of two varieties, so some professionals recommend that varieties be spaced about 50 feet apart for market seed and further apart for stock seed, perhaps with a stand of sunflowers or some other tall crop between varieties. I grow several favourite varieties and local bumblebees hum around the plants, so I cover an unopened flower or two on each variety with a small piece of cloth until the fruit has set.

Habitually cross-pollinated vegetables require considerable attention for satisfactory isolation. They are so anxious to play the field that they have devised methods for preventing self-fertilization. They may be sterile to their own pollen; the pollen of a particular flower may be discharged while the female portion of the flower isn't receptive; or male and female flowers may be separate, even growing on different plants. Not only will one variety of cabbage, for example, cross readily with another variety of cabbage, but matings can occur with and between all other members of the species: broccoli, Brussels sprouts, cauliflower, collard, kale and kohlrabi. This wild group is *Brassica oleracea*.

The simplest procedure to prevent a too-wild family reunion of that bunch is to stagger seed production, growing one variety of cabbage the first year, kale the next and so forth, returning to cabbage in four years and starting the cycle once again. This is possible because *Brassica oleracea* seeds will easily maintain viability for five years. Fortunately, long-lived seed (four or five years) is characteristic of all large groups of potentially intercrossing vegetables, so the same method of isolation may be used with them. These groups include another set of interbreeding *Brassicas* (see chart), the beet family, melons and pumpkin/squashes.

Staggering your seed production also means that your garden won't be overtaken by seeding plants. When leaf and root vegetables bolt, they often require a lot of space. Even when staked, tied, and wrapped round and round with string in an attempt to restrain them, rutabaga and Swiss chard seed-stems will sprawl over pathways or recline on a nearby bed of hot peppers. Many stems may be cut from such groping plants in order to save space and direct the plants' energy into ripening the remaining, still numerous seeds.

The pumpkin/squash group is made up of four species of the genus *Cucurbita*. Some of the more common varieties are listed on previous page. Most winter squashes are in *C. maxima*, but summer squashes, pumpkins and some winter squashes are interspersed among the other species. Varieties will cross with members in the same species. Some references say crosses also *might* occur *between* some species. However, Johnny's Selected Seeds reports that they've never experienced crosses between species in their fields.

That debate is academic for many home gardeners, who must isolate any variety of squash, because squash seed is likely to be grown by most of the neighbours. For isolation, paper bags or cheesecloth may be used to cover male and female flowers on the plants, allowing you to play the bee by transferring pollen with an artist's brush to initiate fertilization. Both male and female flowers should be covered before they open and each female flower — which has the miniature ovary, or fruit, at its base — should remain covered for several days after pollination.

If absolutely necessary, the bagging method may also be used to isolate your corn variety from the wind-borne pollen of a neighbour's variety (corn pollen may occasionally travel up to a mile). Each parent ear of corn should have pollen shaken on it from tassels of other specimens. The silk should be bagged and sealed tightly before and after pollination.

Beet pollen is very lightweight and may ride air currents for several miles. It is indeed fortunate, therefore, that beets are biennial — you needn't worry that your neighbour has a year-and-a-half old beet threatening to cross with yours. . . unless your neighbour also happens to belong to the *Savum seedum* variety of *Homo sapiens*.

There is one garden crop which will readily cross with a common non-cultivated plant, so, if you want to keep your favourite carrots pure, you must keep them away from Queen Anne's Lace — a totally untamed plant.

Inbreeding will occur if only a couple of ears of corn are saved for seed each year and used to grow parents for the next year. Too-close inbreeding of any normally cross-pollinating vegetable will result in degeneration of the species (except for squashes and pumpkins, which, though widely cross-pollinating if given the chance, seem to withstand inbreeding, at least for quite a few generations). I have found no formal research on the minimum number of parents required to prevent inbreeding degeneration in different vegetables. The results of some informal research indicate that 10 parents (e.g. taking seed from 10 ears of corn, not just one or two) should prevent too-close inbreeding.

The danger of inbreeding degeneration only applies to normally cross-pollinating vegetables. Those vegetables which habitually self-pollinate clearly do not degenerate from inbreeding. With peas, then, you may save seed from one good plant each year if you want. You might, however, consider saving seed from several plants, maintaining separate lines in the hope that one strain might be resistant should a new disease come along.

HYBRID ROULETTE

Before deciding that you've bought your last packet of premium priced *Burpee Ultra Girl* tomato seed, the whole question of hybrids must be considered.

For the non-geneticist, a hybrid is a plant produced from seed resulting from a cross of two different varieties of vegetable. For example, a breeder might take a pure variety known for its resistance to frost and cross it with another variety that yields extra-large fruit.

The next generation may show both of these desirable traits, as well as hybrid vigour (extra vitality seen in F1 — first generation crosses). Hybrid seed also tends to be much more uniform than that of non-hybrids (open-pollinated varieties).

The catch is that you are playing genetic roulette if you save the seed from your *Ultra Girl* hybrid tomatoes, for the next generation will revert to highly variable and generally inferior types of tomatoes. Furthermore, some hybrids do not even produce fertile seed.

There are good reasons for the premium price on hybrids, because the breeders must maintain pure strains of the non-hybrid parent stocks and assure that the correct cross takes place.

Anyone dreaming of becoming a contemporary plant wizard should be forewarned. First, successful experimentation with crossbreeding generally requires a considerable knowledge of genetics, including the ability to maintain genealogical records as extensive as Alex Haley's, if you are to keep track of the parents of *your* roots.

Second, a large area is needed to grow many offspring from a cross. Third most major vegetable varieties readily available to the gardener have been worked over thoroughly by the professionals, so a startling discovery, while not impossible, is extremely unlikely. Any dramatic development nowadays usually results from the discovery of a previously unknown strain, often in a distant land, growing wild or under cultivation in an isolated community. This introduces new genetic material.

The complexity and scope of experimental crossbreeding and hybridization does not mean that a gardener cannot develop a new variety through intentional crossing or by working with seed from an unusual plant that springs up somewhere. Anyone determined to develop a new vegetable variety might succeed with plants that are not generally used for food, just as the modern carrot was derived in the 1870's from Queen Anne's Lace. Remember, however, that what we might think of as unusual items may be popular vegetables in other parts of the world (*Witloof* chicory in Europe, Chinese cabbage in Asia) with a long history of both informal and formal selection and development.

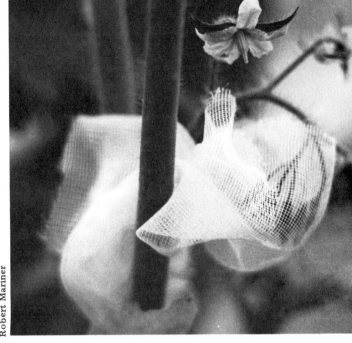

Robert Mariner

Sometimes promiscuous, the author's tomato flowers are covered to prevent uncontrolled pollination by errant bees.

It's interesting, of course, if you have the space, to grow squash seeds from crossed parents and maybe wind up with a large, hard, grey-and-yellow striped gourd-like thing laced with hints of a banana squash, or to try to preserve a discovered *sport* (genetic deviant). No one should be discouraged from these just-for-fun projects by any lack of theoretical sophistication.

HARVESTING, CLEANING, AND STORING SEEDS

Seed should be left on the plant until ripe if possible. Since you planted seed to start the whole process, you'll recognize mature seed. You wouldn't for example, try to extract tender, light coloured seed from *Black Seeded Simpson* lettuce.

In general, plants which form seed pods (beans, cabbages, etc.) or heads (carrots, onions, etc.) should be left standing until the pods or heads are dry — usually a tan or brown colour. The whole plant may then be cut and laid in the garden to complete drying the seed (several days to two weeks) or moved under cover if damp weather is a problem.

Alternatively, since many vegetables ripen seed unevenly, you may want to cut, say, ripened cabbage seed branches into large paper bags or periodically shake dried lettuce flowers over a container. You'll also want to collect pea pods before they either shatter (seeds drop out), sprout in the pods during damp weather, or serve as a main course for blue jays.

When you have as much seed as you want — remembering to save some for your friends and a reserve should your subsequent seed crop fail — the plants should be removed to the compost heap. I left some Swiss chard parents to overwinter last year, and I'm now weeding thousands of volunteers from among my carrot rows. Even using that area as a chard patch would require extensive thinning.

Some vegetables which bear fleshy fruits ripen seed when the fruit is ripe, such as watermelon, cantaloupe, muskmelon, pumpkins and winter squash. Others should have fruit left on the vine until it is somewhat overripe: tomatoes (a bit mushy), cucumbers (large and yellowish), eggplant (a few days larger than edible size), and summer squash (large and hard).

The slippery jelly-like sheath around some seeds, such as tomato, is usually removed by fermentation. The seed and any adhering pulp is placed in a jar of warm water and held at approximately 70 degrees (F) for three days, with occasional stirring, until the seeds no longer feel slippery. Though 75 to 80 degree water will complete fermentation in two days, longer time has a side advantage of killing any canker bacteria that might otherwise be transmitted by tomato seeds. The good seeds go to the bottom, while pulp and useless seeds float. The good seed is washed and then dried.

Fermentation may also be used with muskmelon and cucumber, but should be avoided with eggplant, winter squash, and pumpkin. Unfermented seed of winter squash can be rubbed to remove pulp and then washed. Extra care should be taken to ensure that these large, fleshy seeds are thoroughly dry before storage.

I have found no enjoyable procedures for removing seeds from some plants. You may get bored rubbing seed heads of carrots or onions in your hands or on a screen, but there's really no better method for a small quantity of seed. Cabbage seed stalks can be put into a cloth bag or pillowcase and beaten. (Remember to label carefully all those look-alike *Brassica* seeds, stems or pods as you harvest them, or you won't know your kohlrabi from your rutabaga.) Flailing seeds on a floor or driveway only makes sense for a large crop of seeds.

To separate your seed from the worst of the chaff, you can try winnowing — pouring the mixture back and forth between containers as the wind blows — or screening it through an appropriate size mesh.

CLIMATIC LIMITATIONS

The seeds should be stored in a cool, dry place. Storage in sealed jars may extend the life span of thoroughly dry seeds. Jars of seeds may be placed in the freezer, maintaining viability even longer. A large jar can hold many seed packets. Small manilla envelopes about the size of commercial seed packages will not leak seeds or tear as easily as white stationery envelopes.

Your particular weather conditions may make it impossible for you to save some types of vegetable seeds. For example, green bean seed is grown commercially in western regions where the dry air controls disease spores. If you live along the seacoast, humid air may encourage the spread of disease spores from bean plant to bean plant and garden to garden. Seed-borne blight may be carried over from year to year, so you shouldn't save seed from bean plants — or any types of plant, for that matter — which show signs of disease.

All seeds — whether home-grown or left-over commercial seeds — should be germination-tested each year before your seed order goes in. I usually plant my seeds in the house in a dishpan of soil. Any procedure normally used to sprout seeds is satisfactory. The important part is to keep the seeds damp (not soggy) and warm (not hot).

While northern gardeners may find it difficult to grow mature Lima bean seed or seed of other hot-weather crops, some vegetables, such as cabbage, carrots and beets, need a period of cold weather before they'll run to seed; this causes problems in warm-winter areas. But anyone determined to save seed from these particular crops can usually find a way, whether through experiments with starting Limas early and transplanting them or keeping beets in the fridge for a while.

Most seed savers will want seed from reliable crops — old standbys. Any gardener who saves seed from these favourite vegetables, even for just one or two seasons, will have learned a potentially valuable skill and will have an even stronger attachment to these old friends.

Sources

Growing Garden Seeds,
Booklet by Robert Johnston, Jr.
Johnny's Selected Seeds
Albion, Maine 04910
$2.30 plus 25 cents postage and handling.

True Seed Exchange
Kent Whealy
RFD 2 (HS)
Princeton, Missouri 64673

Exchange heirloom seeds with other gardeners in the United States and Canada (some foreign). Very informal organization, but seems to be growing quickly. Annual newsletter $2.00.

Sources for such seeds as:

Poke Weed, *Bumble Bee* beans, *French Breakfast* radishes, *Missouri Persimmon* (fully hardy), *Bird Egg* beans, *Quill* melon, *Norfolk* spinach and *Large Cob White* corn (especially good for making corncob pipes).

Behold, The Naked Seeded Pumpkin

*Will there be a Streaker
or Lady Godiva in your garden this summer?*

By Kenneth Allan

Soybeans, stand aside — here comes the Naked Seeded Pumpkin.

Boasting trendy names like *Lady Godiva* and *The Streaker*, these freaks of the squash/pumpkin family are already showing signs of being more than just another curiosity or novelty vegetable. With seeds containing 37 to 40 per cent protein, they are among the richest sources of protein that can be grown in the home garden. (Soybeans contain about 35 per cent protein, raw beef about 27 per cent, for comparison.)

The important difference between traditional pumpkins and the naked seeded varieties is that the seeds of the latter come wrapped in a thin green tissue — instead of the standard tough, white seed case — and hence are, if not quite naked, very thinly veiled.

Thus they are edible without having to be hulled, a procedure which puts most pumpkin seeds at several dollars a pound in health food stores. The seeds of *Lady Godiva* and *Streaker*, however, are simply washed and dried at home and can be used in granola, roasted and salted

for snacks or even made into a peanut-butter-like spread.

Only in the past five years have these pumpkins become generally available in North America, and most seed houses do not yet list them. Nevertheless, they have been grown in Germany, Austria and the Balkans since the beginning of this century. Folklore there gives the seeds credit for maintaining male vigour, preventing prostate disease and fending off senility. Nutritional testing seems to be lending some credence to this: pumpkin seeds are rich in phosphorous, iron, unsaturated fatty acids, and various trace minerals, including zinc.

DELICATE GERMINATOR

The unusual pumpkins did not attract the attention of the scientific community until 1934, when the German scientist Dr. Erich Tschermak-Seysenegg published the first of several articles on them with the avowed intent of studying, developing and promoting the naked seeded variety as an oil crop (the seeds contain 45 per cent unsaturated fatty acids). The war intervened, and, though there has been sporadic research in Germany and the United States, this impressive source of oil and protein has not yet caught on as a commercial crop.

German scientists, investigating the nakedness of the seed, discovered that all the usual layers of the seed coat are there — they just don't thicken and harden in the usual way. The filmy covering is so thin that it is a negligible part of the weight of the whole seed and does not affect its eating qualities. The thin seed-coat tissues are responsible for the green colour of the seeds; the kernel underneath is the same as that of an ordinary pumpkin seed.

The skimpy seed covering does pose one major problem, especially for commercial growers. The seed kernel is poorly protected from soil bacteria, and in average growing conditions it is a toss-up as to whether the seed will germinate or rot. Fungicides increase the chance of germination, but they are not foolproof and will not protect the seed through a spell of cool, damp weather.

This is less of a problem to the home gardener. Seeds can be planted at the beginning of a warm spell and then, if something goes awry with the weather, the patch can be re-planted a week or two later. The best way, however, is to start the seeds indoors.

Find a spot that stays at 90 degrees (F), near a heat register or perhaps over the coil at the back of the refrigerator. Here the seeds sprout in two or three days if kept moist. (Place between layers of cloth on a paper towel, as in a germination test). They should be set out immediately — the pumpkin and squash family do not generally take well to transplanting, but there is scarcely any large-seeded plant that can not be moved a day or two after sprouting. If the weather should take a turn for the worse just as they are ready to go, they will have to be transferred to peat

Vine crops react poorly to transplanting, but northern gardeners who must get an indoor start may find success with peat pellets or pots.

pots so that, when they are set out, the roots will not be disturbed. Naked seeded pumpkins take about 110 days to mature from seed, so far-northern gardeners will want to start the plants indoors in peat pellets or pots for transplanting.

AXMAN'S HARVEST

A study by R.G. Robinson (reported in *Agronomy Journal*, 1975) comparing the seeds of *Lady Godiva* with several varieties of sunflowers gave the edge to *Lady Godiva* in protein content from eight to 14 per cent. Robinson estimated that 95 grams (just over one cup) of seeds would meet the daily minimum protein requirement of the average adult and that 400 grams of either pumpkin or sunflower seeds would be more than enough for both protein and mineral requirements.

But if you are considering growing a year's supply of pumpkin seed protein, your garden may need some expanding. Even at the reported U.S. test plot yields of 6,000 pounds per acre, it would take a 30-foot-by-30-foot plot to produce 100 pounds of seed. I can't confirm these figures from practical experience — though I have grown naked seeded pumpkins for three years. I just plant a few along the edge of the garden and let the rampant growing vines ramble off through the grass. After the first frost, 25 to 40 mottled orange and green pumpkins are gathered and stored.

113

The pumpkins are about eight inches in diameter and have a tough inedible shell — so tough that if I had large numbers to deal with, I would split them with an axe rather than a knife. The hard shell has one advantage — if stored in a cool, dry place the pumpkins will last most of the winter. This means that you can put off clawing the seeds out of the shell until some quiet day in midwinter.

The seeds should be rinsed well and then, if they are to be stored, spread out in a thin layer on a tray or screen and allowed to dry completely (this takes several days to a week). A few, at least, should be salted while still wet from rinsing (the moisture makes the salt stick), put into the oven at 350 degrees and roasted for 15 to 20 minutes. Salted peanuts pale by comparison.

Raw or roasted, the seeds have a delicate, nutty flavour due to the high protein and oil content. The seeds can be made into a spread which is said to be better than peanut butter.

The choice of varieties is limited but growing. The U.S. Department of Agriculture embarked on a breeding programme in 1965, crossing and selecting from the strains that were available at that time, and came up with *Lady Godiva* and *The Streaker*. The variety from Stokes Seeds that I have been growing is just called *Naked Seeded Pumpkin-Squash*, but John Gale, president of Stokes, said in a telephone interview that this is really *The Streaker*. This makes it the same variety as that listed this year by Dominion Seeds.

GENETIC BREAKTHROUGH?

Dominion, however, claims that the flesh of their *Streaker*, unlike that from Stokes, is edible. I know that pigs will eat them, but the pumpkins I have been growing would require a culinary genius to render the tough flesh edible in any ordinary sense of the word.

Lady Godiva has been available for about five years from companies in the United States such as Harris and Burpee, and is being carried in Canada this year by William Dam. *The Streaker* tends to be oblong, whereas *Lady Godiva* is more often the traditional round pumpkin shape. Both are small, as pumpkins go, but one vine will produce up to one-half dozen of them.

The most recent addition to the list of varieties is *Triple Treat* which is being introduced this year by Burpee. It has edible flesh comparable to *Small Sugar*, a favourite pie pumpkin. The seeds are apparently not quite as plump as *Lady Godiva*, but if the flesh is all they claim, *Triple Treat* is a really exciting breakthrough in pumpkin breeding.

The naked seeded pumpkin is a member of the species *Cucurbita pepo* which includes most pumpkins, acorn squashes and the summer squashes like zucchini, crookneck, and *Patty Pan*.

Crossbreeding is possible between any of the members of this group, which means that it is theoretically possible to breed the naked seed characteristics into any of them.

This is easier said than done because, when two varieties are crossed, instead of yielding a fruit part-way between the two, they tend to revert to the tough inedible gourd from which they are both descended. Nevertheless, it is possible — *Triple Treat* is proof of that. When we consider the food value of the seed and the number of varieties into which it might be bred, it appears likely that the day of the naked seed has just begun.

Sources

STREAKER NAKED SEEDED PUMPKIN

Stokes Seeds
Box 10, Stokes Building
St. Catharines, Ontario L2R 6R6

OR

Stokes Seeds
Box 548, Stokes Building
Buffalo, New York 14240

Dominion Seed Co.
Georgetown, Ontario
(No U.S. orders accepted.)

LADY GODIVA

Harris Seeds
Moreton Farm
Rochester, New York 14624

William Dam Seeds
West Flamboro, Ontario L0R 2K0

Centennial Gardens
Rare & Unusual Flowers and Vegetables
Box 4516 M.P.O.
Vancouver, British Columbia
(Also offer *Eat-All Squash*, another naked seeded variety.)

Burpee Seed Co.
300 Park Avenue
Warminster, Pennsylvania 18991
(Also offer *Triple Treat* pumpkins.)

Some Thoughts On The Subject Of
BEANS

*Is there a row of Hakuchos
in your garden's future?*

By Kenneth Allan

With the winter seed catalogue season fast upon us, the pleasantly vexatious problem of what varieties to try this year will soon present itself. For most amateur gardeners, the ubiquitous snap bean — both green and yellow — is sure to be high on the list. They are easily the most popular members of the bean family, but by no means the only ones worth growing, or even the best.

My favourite snap bean, the standard by which I judge all others, is not green or yellow but a rich dark velvety purple (it turns an irresistible deep green after cooking). Its name is *Royalty* and it is a dwarf version of a pole bean called *Blue Coco*. The latter has a reputation in Europe as the best tasting of all beans, but it is not easily found in either Canada or the United States, and I haven't been able to try it myself.

BACK TO POLES

Royalty is more tender and tasty than any bean we tried except, perhaps, some of the green beans like *Tendercrop* (itself an excellent bush bean). *Royalty* withstands extremes of temperature better than most beans and its crop ripens over a longer period of time. Northern gardeners also find it a reliable germinator in cool, wet soils which usually are the bane of early bean plantings. *Royalty* produces a second crop, slightly inferior to the first, but still very good.

It is supposed to be a bush type but it is so recently descended from the climbing family that it sometimes forgets itself and wanders out of the row. The only fault I've found with *Royalty* is its tendency to lose flavour and crispness the day after picking. This is not a major disadvan-

tage for the home gardener, but it could be significant for those marketing produce.

Pole beans are slightly out of fashion these days, probably because they require the gardener to erect some sort of support system. The vines grow in a corkscrew fashion, always counter-clockwise, around anything provided for them to climb.

One-inch-thick poles, free for the cutting in any brushy rural area, are perfect supports (leave the bark intact to give the bean tendrils easier footing). It isn't difficult to improvise bean supports — planting a bean in each hill of corn so it will grow up the cornstalks is a traditional practice. Support for a row of beans can be made by driving a six-foot stake at each end and stringing wire at the top and another about 12 inches above the ground. Heavy string or twine is then run between the wires vertically at about one-foot intervals. (Keep horizontal supports and wires to a minimum, as they tend to impede the upward growth of the beans, resulting in snarled bunches which do not produce well.)

SCARLET RUNNERS

The added work of putting in pole beans has its rewards. They produce continuously — an inconvenience for the commercial grower (who likes to harvest the whole crop in one mechanized swoop), but ideal for the home gardener. Many of them boast superior flavours, and their six-foot growth gives satisfaction in watching them that can't be matched by a 15-inch bush, no matter how healthy it looks.

Kentucky Wonder is one of the classic pole beans that produces pods very much like those of the green bush types, but it has a distinctive flavour and texture. *Romano* beans are a bit more unusual in that they have a flat pod which cooks in half the time required by most beans. They are relished by many who have discovered their flavour and by old-time gardeners who still know them as *Italian Pole Beans* (these now come in bush varieties also).

Zebra is another pole variety of very good quality; it is green with purple stripes which disappear in cooking. *Blue Lake* is much touted by some gardeners, as is the *Scarlet Runner* bean, which is so fast growing and lush as to be useful in camouflaging unsightly walls in summer. They should be picked early for best eating, and the bright red blooms are a bonus.

PROTESTANT ETHIC

Culture of the bean family is much the same for the whole group and fairly simple.

1. Don't crowd them. Leave at least six inches between plants in a row.

2. Be careful of their shallow root system when cultivating.

3. Don't use a fertilizer high in nitrogen — like all legumes they play host to a bacteria which enables them to use nitrogen out of the air. They are lazy, so if you feed them nitrogen they will take it, rather than making their own, but they are Protestants, hence will do better on the nitrogen that they have had to work for.

Timing is very important in the harvesting of bush and pole beans. They should always be picked while still immature. In general, the best time is when the pod is almost full length but the beans have not started to bulge out. Since it takes only two days for the beans to go through this ideal stage, they have to be watched closely and picked often.

There is something very exotic about tender young beans cooked *whole*. Cooking is critical if you want the ultimate in fresh beans. About 15 minutes from the beginning of cooking the bean loses its raw crispness but remains firm — the flavour is then at its best. This stage lasts for about five minutes. If cooked any longer the beans rapidly acquire the texture and flavour of spaghetti noodles.

We haven't always been pleased with the flavour in some frozen beans, but the two varieties that have done best for us are *Royalty* and *Greencrop*, a long, flat green bean, prized by those who French-cut their beans.

A type of bean that does freeze very well is the soybean, one vegetable which seems to lose nothing, either in texture or flavour.

Joseph Harris Seeds

Above, Bluecrop *bush beans from Joseph Harris Seeds and* **below** *Stokes Seeds' new introduction called* Royal Burgundy Bush *a hybrid of the purple podded* Royalty.

Stokes Seeds

Although as yet largely undiscovered by home gardeners, they are easy to grow — almost any soil will do — and northern gardeners now have early varieties from which to choose (*Hakucho*, from Japan, or *Early Green Bush*); *Kanrich* is a standard table variety that takes longer to mature (103 days).

Green soybeans have a delicate, nutty flavour vaguely reminiscent of peas. The one drawback is that they are tedious to harvest, with each plant producing many small pods which contain two or three beans. The pods are boiled, steamed or blanched for about five minutes (longer if necessary) and the beans then popped out by squeezing the pod. Most of our soybeans get frozen because in summer they must compete with vegetables that have more flavour and are easier to prepare. By midwinter, however, they are one of the best.

Another bean which, like the soybean, is usually dried but can be picked green, shelled and frozen, is the kidney bean. The frozen beans (out of their pods) can be used as a replacement for canned kidney beans in things like chili con carne. The taste is slightly different, and, I think, better. This summer I discovered that the immature pods make excellent green beans — at least when raw. I didn't want to pick enough to try cooking them. To be sure, the mature beans would be stringy, but this spring there will be an extra row of kidney beans in our garden to be tried as young green beans.

Lima beans are especially sensitive to frost, and are probably best not planted until your tomatoes are set out. Those in short season areas should probably stick with the bush Limas (*Fordhook Bush, Jackson Wonder*, and *Henderson Bush)* known for their earliness. In especially cold climates, beans can be germinated about a week ahead of planting, but must be carefully handled and planted in moist soil. These bush types are usually taken as "baby Limas." *King of the Garden* is perhaps the most popular pole Lima.

Fava or *English Broad Beans* are actually members of the vetch, not the bean, family. Similar to the Limas in shape and taste, they can be planted and grown in cool, moist areas where other beans do not thrive (for this reason they are commonly grown in Britain). Unlike the other beans, they can be planted very early and are frost hardy. Shelled green, they are eaten as Limas, or they may be allowed to mature and dry for storage and later cooking.

I haven't yet tried asparagus beans or chick-peas, but intend to. So when the seed catalogues arrive this winter and you start planning next year's garden, try to find room for a new variety or two from the bean family.

Sources

Johnny's Selected Seeds
Albion, Maine 04910
(Ed. note: this is an organic seed house that ships both to the U.S. and to Canada)
Jackson Wonder & Henderson Limas:
Packet 60 cents; 1 lb. $2.60
Beurre de Rocquencourt (yellow wax bean newly introduced from Vilmourin, France):
Packet 65 cents; ½ lb. $1.70
Royalty: Packet 65 cents; 1 lb. $2.75

Vermont Bean Seed Company
Bomoseen, Vermont
Wide variety of bean seeds available

Stokes Seeds Ltd.
Box 10,
St. Catharines, Ontario
OR
Stokes Building
Buffalo, New York
Early Green Bush (soybean):
¼ lb. $1.60; 1 lb. $3.85
Bush Romano: ¼ lb. 85 cents; 1 lb. $2.20
Royal Burgundy: ¼ lb. 95 cents; 1 lb. $2.25

Wm. Dam Seeds
West Flamboro, Ontario
Asparagus Pole Beans: Packet 50 cents
Romano Pole: Packet 45 cents; 250 g. $1.50
Hakucho Soybeans: Packet 60 cents
(Wm. Dam sells untreated seed, including many European bean varieties, to both Canada and the U.S.)

Vesey's Seeds Limited
York, Prince Edward Island
Soldier Baking Beans:
125 g. 60 cents; 500 g. $2.10

Dominion Seed House
Georgetown, Ontario L7G 4A2
Royalty: Packet 55 cents, 250 g. $1.30

Peas By The Patchful

Intensive cultivation makes best use of space and boosts your harvests

By Kenneth Allan

If you have ever had the chance to witness a string of trucks sitting in fluorescent light on an early summer night, brimming with tons of green peas waiting to enter a processing plant, you will know why home-grown peas can never be matched by the commercial variety. The drivers joke amongst themselves, the weighing and sampling and waiting go on, and, seemingly forgotten, the natural sugar in the peas silently continues its conversion to starch.

As with sweet corn and many other garden vegetables, it is this sugar that supplies most of the flavour, and within two hours of picking the transformation to untasty starch has begun. And now that the most modern harvesting equipment cuts the vines and de-pods the peas immediately, the conversion begins before the crop has ever left the field.

Thus, in peas, the home gardener has a chance to grow a vegetable that money can't buy. I have never split the pod of a fresh young pea and been disappointed — even the field varieties, if caught before the peas reach full size, are sweet and delicious. There are many vegetables, such as carrots and beans, that are a delight to eat out of hand while strolling through the garden, but none quite equals peas.

A vegetable of great antiquity, peas were grown by the Egyptians, Greeks and Romans more than two millenniums ago and have never dropped out of favour. They are grown now by most home gardeners, but not as intensively as they deserve. Visit a small or medium-sized garden and attempt to cadge a few pods and you will probably be pulled away from the pea patch by a gardener intent on getting enough to the table for a meal or two.

INTENSIVE CULTURE

If peas are planted in the traditional single row, two to three inches apart, a 30-foot row will be just barely sufficient to sustain the occasional garden visitor, with a few handsful left over for the kitchen. At that rate it is little wonder that peas are shunned by some small-scale gardeners or included only as a luxury item.

There is no need, however, for such a poor return from the space allotted; the same row could produce five to 10 times as many peas.

The single-space row, though desirable for many vegetables, ignores a basic truth of pea nature — they don't mind being crowded. In fact, peas thrive when surrounded by their own kind. So the logical, but not often taken, first step to increased pea production is simply to plant more seed in the same space. This can be done in several ways, and the method chosen will depend, in part, on the type of supports you intend to provide.

Some peas are of a bush type (such as *Little Marvel*) and do very little climbing. Others are advertised as free standing, but they don't do it very well, and neither type comes close to the production of the standard peas.

Several years ago, one of my pea rows got staked for only half its length. This was pure negligence when it happened, but by picking time it had been elevated to scientific experiment status.

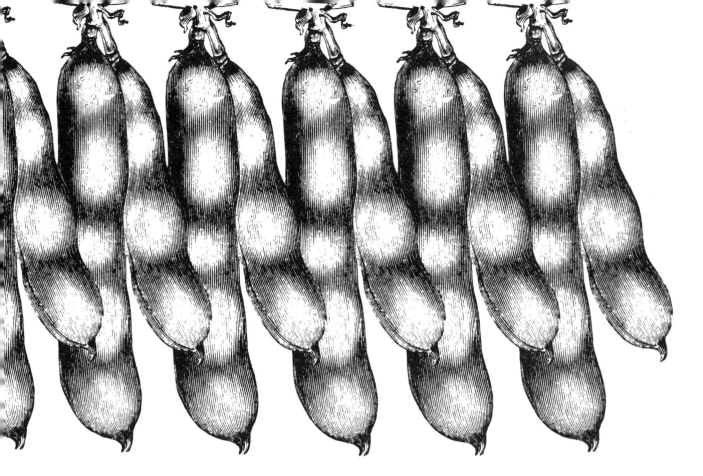

The staked half was twice as productive and easier to pick. The unstaked half was a tangled mess of mildewy vines with only the first few blossoms of each vine producing pods. From this experience I have concluded that, even though peas like to be crowded, they also need good ventilation.

Were it not for ventilation and picking problems, peas could be planted in square plots solidly packed with vines. The planting method I've developed is a compromise between a row and a solid patch.

The peas are planted in a grid pattern (see diagram) which is about 15 inches wide and runs the length of the row. The grid is a series of five or six rows, two-and-one-half to three inches apart, with the peas spaced from two-and-one-half to three inches apart in the row.

The peas get dropped on the ground roughly in this pattern (precision is not especially important), and are then pushed into the soil — one knuckle deep if it looks like rain or two knuckles deep if it is hot and dry.

WEEDS THWARTED

While the pattern is still visible in the fresh earth, the support system should be erected. We prefer a mini-hedge created by gathering dozens of untrimmed tree branches (saved from pruning or brought in from a field trip), each cut to about two-and-one-half-foot lengths. These branches are now stuck into the soil along the centre of the grid-row at about three-inch intervals. This is a bit time-consuming on planting day, but cheap and very effective. The vines that don't have a

twig within reach can grab a neighbour that has found some support.

The worst system is string or twine from stakes at either end of the row. The twine stretches and droops and then, in the first high wind, the interwoven mass of peas and string flops back and forth until something gives. In contrast, the brush hedge never moves, even when the twigs that compose it are fragile — the peas tie themselves together to make the whole network quite stable.

Another advantage of the grid system is that once the row is well established, the weed problem is minimal. Peas are strong growers and, if given a head start by weeding once or twice when they are small, they will choke out all but the hardiest of weeds.

Peas also produce a canopy to shade the ground and keep their feet cool. Some gardeners recommend planting in a trench which gets filled gradually as the peas grow. This buries the roots and keeps them cool — I am not convinced that this is important, but it pleases some to do it. Mulching would be another way of achieving the same end. The grid system, by producing its own shade, makes either expedient unnecessary.

Wire fencing of any sort makes a good support for peas and is easier to erect than the brush hedge. It is expensive if it must be bought new, but anyone who has some lying around should make use of it. (Small-opening poultry mesh works well, but can be a mess to clean at the end of the season.) Put up the fence first and then plant the seeds in a single or double row down *each* side of the fence, about three inches out from the base.

If planting two single rows, space the seeds one inch apart. I have not tried this myself, but apparently peas can even stand a spacing of slightly less than an inch. A friend, who just happened to have some fencing, planted two double rows with two-inch spacing (between seeds and rows) with excellent results.

Peas do best in rich soil but should not be given new manure or nitrogen-rich fertilizer — this produces lush foliage and empty pods. Peas, like beans and other legumes, secure their own nitrogen by playing host to a soil bacterium which converts atmospheric nitrogen to a form that the plants can use. Most soils contain small amounts of these bacteria which multiply rapidly when they find a legume root to grow on (forming the nodules that a healthy pea root will have in plenty).

Established gardens are usually rich in this bacterium (from previous crops of peas and beans), but new gardens may be deficient. To be sure that the peas get off to a good start they can be "innoculated" with the bacteria, available from most seed companies and garden centres — one trade name is *Nitragen*.

VARIETAL PREJUDICE

Peas are generally divided into two classes: those with smooth seeds and those whose seeds wrinkle upon drying. The wrinkling comes from a higher sugar content, and thus the wrinkled-seeded varieties are known for better taste and predominate in today's seed catalogues.

Smooth-seeded varieties tend to resist rotting in cold soils, and thus have an advantage in especially early gardens or in colder areas. *Alaska* is the earliest variety generally available and is sweet enough for table use and freezing. *Round Green* is another smooth-seeded pea, planted mainly for dry use, as in pea soup.

Actually, all peas taste fine if they are picked young enough, but they certainly don't all produce equally well. *Lincoln* is a consistent producer of large crops and is the one on which we

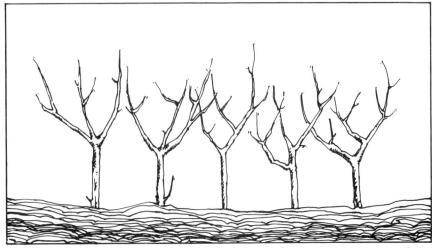

Above, *the author's inexpensive, ecological answer to pea staking, using untrimmed tree branches or prunings about 2½ feet high. The system allows for air circulation, and the canopy of vines which develops serves to shade out weed growth.*

12 in. to 15 in.

2½ in. to 3 in.

Above, *the grid system, in which peas are seeded in strips five or six across. Mr. Allan finds that only 2 to 3 inches should be left between seeds: "They thrive when surrounded by their own kind."*

rely for the main crop. It takes an average of 65 days to reach maturity, making it a late variety.

Last year I tried a row of *Laxton's Progress* which matured about a week earlier than the *Lincoln*. The peas were larger, slightly sweeter and juicier, but *Laxton's Progress* produced less than half as many peas as *Lincoln*.

Green Arrow is in the same, high-production league as *Lincoln*, with an average of nine to 11 peas per pod (thus taking the adventure out of the old notion that says finding nine or more peas in a pod entitles one to make a wish).

Wando is supposedly reliable under adverse conditions (especially hot weather), but I've found *Lincoln* equally hardy and more productive. *Tall Telephone* is very sweet and is big (three and one-half-foot vines) in every way except productivity (although I understand that, under ideal conditions, it does very well).

One pound of seed should be sufficient for a 50-to-60-foot row and should produce upwards of 25 quarts of peas, with the intensive system I've described.

The heavy yield from these wide rows makes it possible for an average-size garden to produce enough peas for fresh eating with a fair number for the freezer as well. Very hot weather may cut the yield a bit, but, on the other hand, a few years ago when the summer was cool and damp our *Lincolns* produced a second crop.

Growing the peas well is not enough, for the home gardener can encounter the same pitfalls as the agribusinessman. I don't think the lack of flavour of commercially frozen peas has anything to do with the processing, because professional blanching and fast-freezing facilities are much superior to anything available to the home gardener.

FINAL TOUCHES

The tastelessness comes mainly from the time elapsed between picking and freezing, and the lax gardening family can waste a fine harvest of peas simply by letting them wait overnight for processing.

It is always best to pick in the cool of the evening or early morning, or on an overcast day. If the peas are not to be frozen immediately, at least leave them in the pods and keep them cool.

Shucking the peas can be a chore, especially if done alone, or it can be an occasion for group activity. A number of mechanical shellers have appeared on the market — one of the most promising, a simple, hand-cranked device recommended by the Vermont Bean Seed Company.

However the shelling is done, no time should be wasted before the peas are blanched in steam to stop enzyme activity. We leave them in the steam for three minutes and then hold under cold running water until cool. Next roll them on a towel to remove excess water — this keeps them from freezing in a clump.

Peas are irresistible in the garden where the picking and splitting of each pod is rewarded immediately, but there is work involved in preparing enough for supper. Thus we eat a lot more beans than peas during the summer. But nothing (except soybeans, which are even more work) can match peas as a freezer vegetable. Corn is excellent in the fall, but its flavour fades quickly, so that, after Christmas, peas have the field virtually to themselves. It is then that the hours of podding in July become worth all the effort, and new resolutions are made for the next year.

Seed Sources

THOMPSON & MORGAN
132 James Avenue East
Winnipeg, Manitoba R3B 0N8
Thompson & Morgan (U.S.)
Box 24
Somerdale, New Jersey 08083

Self-confidently subtitled *"The Seedsmen"* and bearing appointments from both His Majesty King Gustav VI Adolf of Sweden and Her Majesty The Queen of Denmark, this British firm has the personality of a well-travelled eccentric Englishman. Like an explorer just back from the tropics, Thompson & Morgan seem always to have a touch of the exotic.

Year-round greenhouse gardeners will find many temptations here: banana plants, guavas, passion fruit, Mexican water chestnuts and a flashy selection of South American bromeliads.

JOHNNY'S SELECTED SEEDS
Albion, Maine 04910

Northern organic gardeners will find a kindred soul here — all seeds are untreated, many bred with an eye to chilly Maine summer weather and thus easily adaptable to Canadian conditions. Slender but useful catalogue with honest descriptions and an increasingly wide selection of varieties — many not found elsewhere (*Champagne* heirloom pole bean, *Black Mexican* sweet corn, and *Beagle* peas — for those who prize the earliest fresh peas of the spring). Prices of standard varieties not competitive with those of bigger seed companies. New customers should send 50 cents (refundable with order) for first catalogue.

STOKES SEEDS LTD.
Box 10
St. Catharines, Ontario L2R 6R6
Stokes Seeds (U.S.)
Box 548
Buffalo, New York 14240

If limited to a single source of seeds, many home gardeners and truck farmers would probably pick Stokes, which combines a selection of some 600 varieties of vegetables with excellent service. Canada's biggest mail-order seedsman puts out a no-nonsense catalogue with descriptions that are specific and conservative — to the point of not recommending some varieties.

W. H. PERRON & CO. LTD.
515 Labelle Blvd.
City of Laval, P.Q. H7V 2T3

Canada's third largest seed house annually issues a lavishly-illustrated, coffee table-type catalogue. Has the All-American selections and a commendable offering of both flowers and vegetables. Descriptions are reliable (but often scant) and prices fair.

VERMONT BEAN SEED COMPANY
Garden Lane
Bomoseen, Vermont 05732

Claiming the "largest bean & pea seed selection in the world," this company offers a useful, handsome catalogue with many hard-to-find varieties: *Wren's Egg Bean, Oregon Giant Paul Bunyan Bean, Jacobs Cattle Bean, Black Turtle Soup Bean* and *Pink Eye Purple Hull Cow Pea.* All seed is untreated with fungicides and the back-of-the-book listings include a new bacterial powder to kill Japanese bean beetles.

Also have some interesting garden products such as a pea and bean sheller for $13.95.

DOMINION SEED HOUSE
Georgetown, Ontario L7G 4A2

There is a sign in a Toronto record store that says, "Sam has everything . . . If you can find it. If you can't, ask Sam's clerks . . . If you can find Sam's clerks."

Dominion, which has Canada's second largest mail-order seed selection, makes Sam's system appear fastidious by comparison. Many good varieties are included, but the catalogue is a bewildering jumble of descriptions that are glowing to the point of being nearly useless. Dominion does not accept U.S. orders.

WILLIAM DAM SEEDS
West Flamboro, Ontario L0R 2K0

A small, low-key seed house with Dutch roots and an unusual selection of European vegetable varieties (*Kalbo Veense* pole beans, *Dunsel* leaf lettuce). Descriptions and growing instructions occasionally too terse, but Dam has a faithful following and a number of items not carried by others: garlic bulbs, horseradish roots, Jerusalem artichokes, *Everbearing* lettuce for chickens and rabbits and both rhubarb and asparagus plants.

LOWDEN'S BETTER PLANTS AND SEEDS
Box 10
Ancaster, Ontario L9G 3L3

Edward Lowden, Canada's grand old man of organic gardening and plant breeding, puts out a mimeographed flyer with an extensive list of raspberries and blackberries. Far northern growers will like his selection of very early tomato varieties.

BURPEE SEED CO.
300 Park Avenue
Warminster, Pennsylvania 18991

Burpee's lavish, all-colour catalogue seems increasingly to resemble a gardener's mail-order supermarket, with non-seed items ranging from the useful (sauerkraut shredders, soil test kits) to the inane (inflatable, vinyl "Farmer Fred" scarecrows). Nonetheless, the business end of the catalogue is rife with excellent varieties not found elsewhere.

GEORGE W. PARK SEED COMPANY
Box 31
Greenwood, South Carolina 29647

Big, slightly disjointed catalogue with lots of colour and a good number of off-beat flower, foliage and vegetable seeds (*Magnolia Black Eye Peas*, *Red Okra*).

ALBERTA NURSERIES & SEEDS LTD.
Bowden, Alberta T0M 0K0

The company's specialty is seed and planting stock for short-season, high-altitude growing areas, and they have a deserved following in the West. Has gooseberry, raspberry and currant plants, as well as asparagus and rhubarb roots.

HARRIS SEEDS
Moreton Farm
Rochester, New York 14624

Harris distinguishes itself with an attractive, not-at-all flamboyant catalogue that is careful to point out the strengths and weaknesses of each entry. Those who have had little success with cauliflower may be interested in their vigorous, fast-growing (50 days) hybrid cauliflower *Snow Crown*. Does *not* offer untreated seed.

VESEY'S SEEDS LIMITED
York, Prince Edward Island

The Vesey family limits its listings to varieties that will grow well in their own market gardens and that have proved suitable to Maritime and northern New England growing conditions.

OTTO RICHTER & SONS
Box 26
Goodwood, Ontario L0C 1A0

An extensive, 300-strong listing of herbs, including sacred plants used by Tibetan lamas to stimulate psychic powers, herbs used by other cultures for ritual suicides and an herbalist's black bag full of other rare offerings. Their stock in trade, however, is a fine selection of culinary herb seed. (The more common kitchen herb seeds are appreciably cheaper elsewhere.) Send 50 cents for catalogue.

GRACE'S GARDEN
Autumn Lane
Hackettstown, New Jersey 07840

A horticultural freak show — seed from 6½-pound tomatoes, 23-foot-tall sunflowers and 451-pound squash. Big is better in these pages.

REDWOOD CITY SEED CO.
Box 361
Redwood City, California 94064

Craig Dremann runs this mellow little company almost single-handedly, and the slim catalogue is most notable for its herbs and other rarities: *Siberian Ginseng, Buffalo Berry, Cardoon* (an artichoke-like vegetable), *Orach*, (a cultivated green from France). Those with patience may

want to try *Redwood's Giant Sequoia*. (Send 50 cents for catalogue.)

J. LABONTE & FILS
560 Chemin Chambly
Longueuil, Quebec J4H 3L8

This catalogue features everything from plastic yard swans to riding lawnmowers, including a 300-variety, French-only listing of seeds (75 cents for new customers).

CENTENNIAL GARDENS
Box 4516 M.P.O.
Vancouver, B.C. V6B 3Z8

Twelve-page catalogue with organic orientation and a few nice surprises (*Elephant Garlic, French Red Endive* seed, Jerusalem artichokes, Giant (20-foot-tall) *Peruvian Tree Corn*).

BISHOP FARM SEEDS
Box 338
Belleville, Ontario K8N 5A5

Specializes in bulk orders of forage seeds (alfalfa, clover, timothy, trefoil) and grains (oats, barley, buckwheat, field peas, corn). Will, however, sell in small quantities and has extremely reasonable prices on vegetable seed.

MCFAYDEN SEED CO. LTD.
Box 1600
30 - 9th Street
Brandon, Manitoba R7A 6A6

This Manitoba firm is the mail-order division of McKenzie Seeds, who cover the country with retail seed racks.

T&T SEEDS LIMITED
Box 1710
Winnipeg, Manitoba R3C 3P6

Short listing of high-quality seeds and a number of nursery items: *Saskatoon* berries, currants, apple trees.

SEMENCES LAVAL INC.
3505 Boul. St.-Martin
Laval, Quebec H7T 1A2

Their catalogue is free, colourful and lists more than 400 varieties, but it is *en français seulement*.

C. A. CRUICKSHANK LTD.
1015 Mount Pleasant Rd.
Toronto, Ontario M4P 2L9

Canada's best selection of flower, ornamental and other non-vegetable seed. Especially good source for bulbs of all sorts. (Cruickshank is the Canadian representative for the Royal Zwanenburg Nurseries of Haarlem, Holland).

TRUE SEED EXCHANGE
Kent Whealy
R.R.2
Princeton, Missouri 64673

Kent Whealy's idea for establishing a clearing-house for the exchange of unusual seeds appears to be catching on. The third annual newsletter has grown to 32 pages, mostly filled with the names and addresses of gardeners willing to trade their hard-to-find seed for your hard-to-find seed. (One member of the exchange has 448 varieties of beans.)

Annual Newsletter: $2/copy.

PIKE & CO. LTD.
10552 - 114 Street
Edmonton, Alberta T5H 3J7
A 200-item list geared to Prairie conditions, with some excellent prices.

CLYDE ROBIN SEED COMPANY INC.
Box 2855
Castro Valley, California 94546
A lovely little catalogue of wild flowers and tree seeds — natives of both U.S. and Canada.

BRIEFLY NOTED:

GAZE SEED COMPANY
Box 640
St. John's, Newfoundland A1C 5K8

TREGUNNO SEEDS LIMITED
126 St. Catherine Street N.
Hamilton, Ontario L8R 1J4

DEGIORGI COMPANY, INC.
P.O. Box 413
Council Bluffs, Iowa 51501

ONTARIO SEED COMPANY
Box 144
Waterloo, Ontario N2J 3Z9

CHAS. C. HART SEED COMPANY
Wethersfield, Conn. 05109

SEED CENTRE LIMITED
Box 3867, Station D
Edmonton, Alberta T5L 4K1

LINDENBERG SEEDS LTD.
803 Princess Avenue
Brandon, Manitoba R7A 0P5

D. V. BURRELL SEED GROWERS COMPANY
Rocky Ford, Colorado 81067

SHADES OF GREEN
Box 57
Ipswitch, Mass. 01938

JENKINS
P.O. Box 2424
London, Ontario N6A 4G3

NEIGHBOURHOOD MAILBOX (herbs)
1470 East 22nd Avenue
Vancouver, B.C. V5N 2N7

HERBS 'N HONEY NURSERY
Route 2, Box 205
Monmouth, Oregon 97361

NURSERIES

KEITH SOMERS TREES
10 Tillson Avenue
Tillsonburg, Ontario N4G 2Z6
Great organic source for fruit, nut and reforestation trees of all sizes, as well as a nice selection of native Canadian wild plants.

BOUNTIFUL RIDGE NURSERIES
Princess Anne, Maryland 21853
A graphically attractive catalogue with an outstanding choice of berry bushes, fruit and nut trees and grapes. They will ship to Canada.

SEARS-MCCONNELL NURSERY
Port Burwell, Ontario N0J 1T0
Canada's largest national nursery has now joined with the Simpsons-Sears department store chain and most plant material will now be shipped by Sears. Interesting selection of fruits, nuts, berries and grapes.

KELLY BROS. NURSERIES INC.
Dansville, New York 14437
Long-established family firm with wide-ranging selection including strawberries, blueberry bushes, nuts, grapes, wine grapes.

MUSSER FORESTS, INC.
Indiana, Pennsylvania 15701
Thorough listings of trees, shrubs and other nursery stock.

CAMERON NURSERY
R.R.2
Cameron, Ontario K0M 1G0
Simple, mimeographed catalogue (30 cents) specializing in herbs and wild plants (e.g. four kinds of garlic, French shallots, Sneezewort, Wild ginger). Seeds, bulbs and plants shipped to Canada and U.S.

ROSMARINUS HERBS
R.R.1
Souris, P.E.I. C0A 2B0
(902) 687-3460
Live herb plants shipped airmail May 15 to October 30. Ships within Canada only.

GAYBIRD NURSERIES
Box 42
Wawanesa, Manitoba R0K 2G0
To order from Gaybird is a real treat. The catalogue is small, but includes many unusual selections, such as western-hardy apricot trees. Ed Robinson runs Gaybird and his stock comes lovingly wrapped in recycled puffed wheat bags and bread sacks.

WILDLIFE NURSERIES
P.O. Box 2724
Oshkosh, Wisconsin 44903

DEAN FOSTER NURSERIES
Hartford, Michigan

Ah, Mother Nature, Bless Your Fickle Soul

Contingency plans for a northern garden

By Kathryn Sinclair

E. Estey

Yukon-grown radishes, a hardy crop guaranteed to make any gardener look good.

The British comedy team of Flanders & Swann has a singing routine in which the Sceptered Isle's weather is berated, month by month, ending with ". . . *Bleak December days and then . . . Bloody January again!*" Here in Alberta, I often feel a great kinship with Messrs. Flanders and Swann.

As the cold rain drips down the back of my neck, and the garden blooms with a dandy crop of mould, I find myself muttering, "Bloody August again!" This is not exactly the Arctic Circle — we live 20 miles due west of Edmonton, Alberta — but to grow a successful vegetable garden here is a character-building experience.

When we first moved here from a relatively warmer clime, I found myself with the uncomfortable suspicion, in mid-July, that we had somehow missed summer. Then the radio announcers confirmed my fears: "Well, we've had summer this year. Our records show it clearly. It was on July 5, from 2 to 4 p.m." That passes for a joke around here.

By now we have learned to bear with the short seasons and have acquired the masochistic tendency of all western gardeners: We feel we somehow deserve the wretchedly unpredictable weather, just for having had the brash thought that we might actually succeed.

Now, there is an absolute rule in the West, and that is that it is not safe to plant before Queen Victoria's Birthday. This is the beginning of the supposed 90-day frost-free period, theoretically between May 24 and September 1, when gardeners vehemently attack the soil, flail it into life with frantic haste, and then rip out the fruits of their labour with equally manic haste when frost threatens.

Still, I have seen it snow in June and in 1974 we had a killer frost on August 18th. The panorama was grim. *Scarlet Runner* and *Horticultural* beans hung like limp laundry from the fences;

wax and green beans lay flattened in their rows; purple-podded beans were decidedly brown. "If this is August, it must be Australia," I thought to myself.

That year was the turning point in my gardening philosophy. True, the tomatoes, tucked around the corner and in the lee of the huge spruce trees had survived. The broccoli, cabbage and cauliflower seemed refreshed by the chill and the peas were totally unconcerned. But I was angry and vowed then to devise some way to outfox the climate.

The first step was to realize that it is not enough to plant with the great western knee-jerk on May 24th. Each area must be treated individually, each garden viewed as a micro-climate. What worked for me might bring disaster to my neighbour who put his garden at the foot of the yard or for a friend four miles away who is out in the open and at the mercy of the prairie winds.

REFORMED HUNCHBACK

I recalled strolling in my garden on a warm day in April, turning-in the odd clump of half-done compost, with the temperature hot enough to bring out the summer clothes. And, I began to remember, many is the time I have had to shed a warm fall coat while cleaning up the last of the garden on a bright day in autumn. Why not put these mild days on either side of the theoretical gardening period to good use?

I came to the bold conclusion that I would no longer stagger around madly like the Hunchback of Notre Dame on May 24th, trying to put everything in at once. After all, you cannot freeze seeds once they are under the ground. Furthermore, it is much easier to cover tiny plants on a frosty spring night than to try to blanket their hulking adult forms in the fall.

Early in, early out is the rule that has evolved in my garden.

I start first in the beginning of spring with the hardier seeds — species that can take cool soil and some freezing — and work down to the most tender plants. This spreads the work out over several leisurely weeks and has one's garden looking lush and green, in one area at least, by the time the conservative neighbours are just coming out to plant their first seed. This method also tends to spread the harvest out at the end of the season and avoids the absolutely hectic time in fall when every square foot of kitchen floor is covered with boxes of squash, corn, beans, beets and Swiss chard. Traditionally this is the time when tomatoes line every window sill and are hidden under every bed, fugitives from the cold. This harvesting-in-a-crunch method brings on the urge to scream if one is presented with just one more bushel of beans to snap or basket of berries to pick over.

By my system, all garden vegetables are divided into four categories, with their planting times grouped accordingly.

CATEGORY ONE

These are seeds whose packages read: "Can be planted as early as the ground can be worked." These plants can usually take a bit of frost without keeling over and kicking their roots in the air.

As soon as the frost is out of the soil, turn it or rototill it immediately. It doesn't matter that the soil is still cold. These seeds don't mind. You will usually have the added advantage of good moisture in the soil and plants will have a head start if a dry spring follows. My Category One seeds go in anytime from April 12 to the first week of May. They include:

Peas	Turnips
Onion Seeds and Sets	Beets
Lettuce	Parsnips
Radishes	Leeks
Swiss Chard	Broad Beans
Spinach	Broccoli
Cauliflower	Brussels Sprouts
Cabbage	Savoy Cabbage

Surprised? This list includes a great deal of your garden. Many of them can be given a second planting later if you wish. (If you have started your cabbages and other *Brassicas* indoors, of course, they shouldn't be transplanted until warm weather has truly arrived.)

In planting extra early peas, there is a chance to be doubly crafty. Put in tall and thick pea rows (not dwarf varieties) across the garden in the path of the main frost flow to help protect delicate plants further up the way.

CATEGORY TWO

Two weeks or so before the last expected frost (around May 10 in our area) you can rouse yourself for more action. By now tiny peas and

radishes will be showing, and this is most encouraging. Putting Category Two seeds in at this time will raise eyebrows, and to be sure, there is some risk involved.

I feel it is a matter of weighing the chance of getting off a crop of some slow grower against the risk of losing it to a late frost. Every year it is a game of Russian roulette around here anyway, so plunge ahead bravely. Category Two plants include:

Potatoes	Squash
Corn	Soybeans
Beans	Carrots
Cucumbers	

These seeds will all germinate in coolish soil. The warmer, the better, of course, but if your sunshine is adequate the soil will be fine, even if the air is still nippy. Potatoes could be included in Category One, but I feel there is no rush with them as they grow quickly and love our cool climate.

You must be prepared to throw a cover over these tiny plants if a frosty night is predicted. Cucumber and squash are especially sensitive to freezing, so I hedge my bets by putting in seed early and also setting out plants later.

CATEGORY THREE

By now things are really coming to life in your garden. Neighbours are hauling out Rototillers and carts of fertilizer and looking over the fence at you harvesting radishes and lettuce. In Category Three I include all of the following:

Broccoli	Cauliflower
Cabbage	Brussels Sprouts

These are all *Brassicas* and, yes, I know that you planted all that seed earlier. Again, I plant these both times to double my chances of success. Remember, seed is the cheapest element in gardening (a single packet of Brussels sprouts seed will produce 200 seedlings and costs but 50 cents; simply start several dozen indoors or in your greenhouse six to eight weeks before transplanting time and save the rest of the seed for direct sowing).

If you have started plants indoors, harden them off gradually by taking them outdoors for lengthening periods of time before transplanting. A good nursery should sell you seedlings that have been hardened off, but if you buy from a greenhouse operator who hasn't done so, you may also want to harden off purchased plants before setting them out. Now, dash on to Category Four.

CATEGORY FOUR

Finally comes Victoria Day (or Memorial Day or whatever the standard date is where you live) and out go all the other young plants that you have started or bought in flats. Of course, watch them like a hawk. These are pampered beauties that blanch at the threat of frost.

Celery
Tomatoes
Green Peppers
Eggplant
Pumpkin
Muskmelon
Cucumber

I plant both squash and pumpkins in the corn patch, mainly because I can't spare room for them elsewhere. Last year my *Spirit* pumpkins from Stokes Seeds managed to climb the fence and festoon themselves from the spruce trees.

When it comes to harvesting in the fall, some species will survive more frost than others.

Category One is frost hardy. Don't worry unless the weather gets really awful. I am of the opinion that Brussels sprouts are improved by a frost.

Category Two: Corn and beans (if they aren't picked) will be ruined by freezing. Broad beans, however, will take a light frost and the underground crops won't be affected until the soil begins to feeeze.

Category Three plants are all frost-hardy.

Category Four plants are not! If you cover some of these plants, the first frost may be followed by two or three weeks of warm weather. If you have planted early however, you may already have a fine harvest when all the nearby gardens are being bundled nightly under quilts, blankets, sheets, tarps and newspapers.

I like to think I can approach the growing season now with a little less hostility toward Mother Nature. Forgetting the rigid calendar dates is the gardener's surest way to outfox the old lady in this northern climate.

Sources

The following seed companies offer a number of vegetable varieties I have found notably good in this climate.

DOMINION SEED HOUSE
Georgetown, Ontario L7G 4A2

Windressa is my overall favourite pea variety, especially for cold hardiness and early planting. It is a dwarf plant and the small peas and pods can drive one to distraction when shelling. Nonetheless, I will swear that the "peas-to-pod" ratio is higher than for any other type. My mother-in-law in Saskatchewan swears by *Green Arrow*, a large plant that tends to sprawl if not staked. *Dwarf Grey Sugar*, from Dominion and most others is said to be the earliest edible-podded pea.

She also has great success with the *Peaches and Cream* and *Illini* sweet corn, but the shorter growing season here seems to rule them out for me.

Tomatoes, admittedly, are a challenge to grow in this climate and I always suspect that the big clusters one sees at the fairs must have been grown by those lucky enough to have greenhouses.

Still, there are two tomatoes for the rest of us. Dominion Seeds and others offer the *Sub-Arctic* type, which actually has three varieties: *Sub-Arctic Delight, Sub-Arctic Plenty* and *Sub-Arctic Maxi*. They are all relatively small-fruited, but will take the most miserable weather, short of a killing frost, in stride. Last summer I picked all useable-sized fruits off in late August when we had a frost. The plants were damaged somewhat but went on to produce a good second crop in September from the tiny tomatoes I had left. The other good northern producer is *Rocket*, from Alberta Nurseries and Stokes Seeds.

STOKES SEEDS LIMITED
Box 10
St. Catharines, Ontario
(or Stokes Building,
Buffalo, New York.)

Stokes is especially valuable for their selection of early sweet corn varieties, including *Polar Vee, Butter Vee* and *Early Vee*, the latter considered the best overall by many growers.

Their *Wando* pea runs *Windressa* a close second, in my estimation, and the peas and pods are larger. I never thought Lima beans would grow here, but their *Limelight* was the hit of the garden last year.

ALBERTA NURSERIES AND SEEDS LTD.
Bowden, Alberta

Our very best success with sweet corn has come from planting *Amazing Early Alberta*. I have had corn that was already three inches high touched by frost without being set back. Even if corn loses a few leaves to nippy weather, it will usually surge on. In any case, if I were to wait until all danger of frost was past, the corn would never ripen here.

Alberta Nurseries' *Beautini* is my favourite zucchini squash. Stokes' *Kindred*, among the winter varieties of squash, is by far our most successful.

We also buy this company's mixed bag of loose-leaf lettuce varieties and find them to grow well all summer without bolting.

Northern Melons & Other Growing Phenomena

Clear plastic mulch shows promise
for cold climate gardeners

By Jeffrey C. Hautala

"It is not necessarily those lands which are the most fertile or most favoured in climate that seem to me the happiest, but those in which a long struggle of adaption between man and his environment has brought out the best qualities of both."

— T. S. Eliot

TS. Eliot never tried to garden in an area where frosts in both June and August are commonplace, but he doubtless would have been cheered by the efforts of two University of Wisconsin researchers to thwart the short growing seasons that plague northern gardeners.

By mulching both row and hill crops with clear plastic, Garit Tenpas and Dale Schlough have successfully grown watermelons on the chilly shores of Lake Superior, hundreds of miles north of their normal growing range.

The researchers back up their contentions with 10 years of experimental data which show that muskmelons, tomatoes, corn, cucumbers and squash can all be grown in areas formerly considered inhospitable to warm-weather crops.

Using clear plastic, it is now possible to mature winter squash in the far north, even the 100-day-plus Hubbard varieties, provided that they are started indoors or in cold frames four to six weeks before planting time.

The Wisconsin research, conducted at the same latitude as Sault Ste. Marie, Ontario, shows that the ripening date of some vegetables can be hastened by up to two weeks, while yields also increase substantially in most cases — dramatically in others. Early tomatoes, for example, showed a 119 per cent increase in total harvest when grown with clear plastic.

In effect creating a ground-hugging mini-greenhouse atmosphere, the clear plastic works to raise soil temperature, thus improving the immediate environment of seeds and tender young plants at the beginning of the growing season. Research has shown that soil under the clear plastic is often 20 degrees (F) warmer than uncovered ground nearby.

Plastic mulch, of course, is hardly a new idea. The clear plastic is an obvious spin-off from the use of black plastic mulch, which has gained general acceptance as a weed retardant and moisture retainer.

But the results in Wisconsin indicate that clear plastic has definite advantages over the widely used black material. Tenpas and Schlough say that black plastic absorbs heat, but that soil temperature increases only at the soil surface, just below the sheet of plastic.

If an air space develops between the plastic and the earth, the soil is barely heated at all. Even if it is put down properly, cool winds blowing over the surface quickly dissipate the heat that has accumulated.

Initial trials at the Ashland experiment station convinced the researchers that soil under clear film not only warmed more quickly, but that it retained its warmth through cool nights.

Elaine Macpherson

Most significant was the discovery that, in comparison to unmulched plots, *black plastic doesn't increase yield or hasten maturity.*

HARVEST BOOSTED

However, over a four-year period, sweet corn planted under clear plastic showed one-third more useable ears, with the weight of these ears from 39 to 51 per cent higher. Corn planted on May 11 matured 11 days sooner than that in control plantings.

The number of early tomatoes jumped by 76 per cent, with the average weight per tomato up 23 per cent. Late tomatoes showed production increases with clear plastic, although the improvement was not as spectacular as the early varieties.

Because of the warmth and moisture trapped in the plants' root zone, growth is stimulated, resulting in the earlier ripening and better yields observed. The mulch has been most effective in heavy wet soils.

Excessive weed growth between clear plastic has been the drawback most frequently cited by those who feel that it is impractical.

Dr. Herman Tiessen, professor of vegetable science at the University of Guelph has also been experimenting with plastic mulch for seven years. He feels that unless one is willing to undertake the not-very-organic application of commercial herbicides under the clear plastic, crops will be overrun by weeds which tend to grow profusely beneath the plastic, receiving the same benefits as the garden plants.

Tiessen believes that competition between weeds and plants for moisture and nutrients would counteract any benefits of plastic mulch.

"Small gardeners," he said, "could skirt this problem by lifting the plastic periodically, weeding by hand, and replacing it."

This sounds like enough of a handicap to make the whole idea unrealistic, but Schlough disagrees with Tiessen's contention that weeds pose a serious threat.

"We have used clear plastic for 10 years and have encountered no serious weed problems," he says. "Yes, weeds do grow beneath the plastic in a tangled, green mass, but if the plastic is anchored tightly to the earth, the weeds become spindly and unhealthy because of the excessive heat and moisture retained by the plastic.

"I would never recommend that a small-scale gardener use herbicides. Even with weeds present beneath the plastic, we have recorded 200 to 400 per cent increases in the yields of some crops."

Schlough says that the plastic may balloon out with weeds, but it can be removed after July first or once the full warmth of the summer is assured. The greatest benefit from the film has already been achieved by the time weeds become a problem, he says.

The major drawbacks for home gardeners are aesthetics and cost. Plastic is not the most natural of substances, but for those in northern areas

Brassicas, such as these Brussels sprouts can be given an early boost by clear plastic mulch.

the alternative will be to buy food grown on southern California mega-farms or to build a greenhouse.

Costs can range as high as $225 to cover a full-acre garden according to Tiessen. Smaller quantities of 4-mil plastic can be bought at hardware and builders' supply stores for 75 to 80 cents per linear yard in 10-foot widths.

Other drawbacks include the extra time required to lay the plastic and the increased possibility of frost damage early and late in the growing season. Schlough explains that plastic retains heat in the earth, and the warmth that normally radiates from unmulched soil and can save plants from being damaged by frost, is prevented from escaping.

FIRSTHAND EXPERIENCE

Using clear plastic in my own garden I have found the explanation for the experimenters' excellent results apparent. Since the plastic is clear, the sun's rays penetrate with almost the same intensity as if there weren't any barrier at all.

The plastic then acts as a trap, holding warmth in the soil. Increased temperatures promote germination and stimulate plant growth. End results in my home garden, as in the tests, are better yields and earlier maturation.

Three years of using plastic mulch have brought me award-winning green peppers and the only successful *Beefsteak* tomato crop in this area.

Plastic mulch can be adapted to either a hill or row planting system, and 4-mil material has been found the optimum thickness to work with, as this is the cheapest grade that can be used without undue tearing. Heavier plastic can be used, but don't count on stretching more than one year out of it, as the sun causes it to become brittle. Plastic, of course, will not break down in the soil. After it is used it has to be taken up and discarded.

A little careful scrounging will often turn up a free source of clear plastic. Last year we were

Protected from killer frosts, vegetable seedlings will thrive if planted in the greenhouse atmosphere of clear plastic stretched between earth mounds.

Bumper crops have been reported by University of Wisconsin researchers who stretched clear plastic tightly against the ground to warm soil in the growing area.

fortunate enough to find a cottage that had been covered for the winter with a double thickness of 8-mil plastic. The cottage owner let us have it all for the asking, and even helped remove it from the building.

Construction sites and other industrial areas often have plastic that will only be thrown away if you don't salvage it for your garden.

Because plastic traps evaporating moisture, a minimum of watering is required, provided that you plant when there is adequate moisture in the soil. Soon after applying the plastic, you will observe small droplets of water beginning to condense on the sheet, indicating that moisture which would have evaporated is being trapped close to plants — where it can be used.

Careful preparation of a three-foot-wide planting bed is important, and it is advisable to make a continuous hump up the middle of the row to assure a tight fit with the plastic. Any organic matter or compost should be worked into the soil before the plastic is laid.

Next, with a spade, make a trench several inches deep on each side of the mound, 18 inches

Above, left in the dust, productivity of unprotected cucumbers paled in comparison to that of cucumbers reared in the mini-greenhouse atmosphere of plastic mulch. Protected sweet corn, below, produced yields 51 per cent greater than unprotected corn and ripened 11 days earlier in the chilly summers of northern Wisconsin.

from the centre of the row. Short trenches should also be made at the ends of the row, and the soil reserved for refilling later.

Begin by placing the plastic in one of the end trenches and cover it with earth in order to hold it securely in place. Work toward the other end, covering about four to six inches of the edge of the plastic sheet in each trench. Make sure the plastic fits tightly — it often helps to use small rocks to hold down the plastic while it is being laid. Too, laying the plastic mulch is one job where many hands make light work (and fewer headlong dashes across the backyard in pursuit of a windblown sheet of plastic). If a friend or your spouse is nearby, enlist some help in the name of horticultural experimentation.

Planting is usually done through the plastic after it has been laid. Use a sharpened two-inch-diameter pipe or knife to cut through the plastic, making an X or T-shaped slice directly above the area where you are going to plant seeds or young plants. Use the same planting depths as recommended for unmulched seeding.

It's quite true that weeds do sometimes begin to crawl beneath the plastic as the growing season progresses, but I have found that if the plastic has been laid tightly against the surface of the soil, only a few of the hardiest weeds will appear. Even these are usually stringy and poorly developed.

Schlough reports that the following early varieties showed the best performance of any he has grown under plastic mulch:

Tomato — *Way Ahead, Spring Giant, Spring Set*
Cucumber — *Sweet Slice, Victory Hybrid*
Corn — *Early Sunglow, Jubilee*
Squash — *True Hubbard, Golden Hubbard, Buttercup, Table King*
Green Pepper — *Early Hybrid* (Stokes)

Cool weather crops that should not need plastic mulch in most climates are peas, beans, carrots and spinach. Subarctic gardeners, however, may find clear plastic a boon to even those vegetables.

Agriculture Canada has reported 16-degree (F) increases of soil temperature with clear plastic, which can also be stretched over small frames to cover the plant foliage as well.

Cornell University has tried, with success, creating a trench for such crops as corn and stretching clear plastic over the trench to give such vegetables a running jump on the growing season (they find the film must be removed when the weeds start to become a problem).

Schlough and Tenpas are now directing their experiments with clear plastic toward cold climate crops such as the *Brassicas*. "We have had ripe cauliflower and broccoli by the first of July," Schlough told us, "but we found that forcing the plants produces very loose heads."

My own abiding goal is to ripen a tomato by the first of July. Using clear plastic mulch and some new subarctic varieties I've recently come across, I'm convinced that my ambition may not be as outlandish as it seems.

Gardener, Know Thy Bugs

A good insect is not necessarily a dead insect

Harrowsmith Staff

Picture mile-long rows of carrots, not a weed in sight, acre upon acre without an insect, land lifeless but for the lone crop that has been singled out for maximum production.

Such sights are common in the current agribusiness publications, but they illustrate monoculture at its worst, the form of agriculture that most invites serious insect problems. These too-perfect fields of highly intensive crops are easily found by insect pests whose populations may burgeon overnight. Enter the pesticides.

Unfortunately, many gardeners and small farmers have been affected by the spillover of this monoculture philosophy. They can be seen at regular intervals spraying or dusting everything in sight, whether it has insect problems or not.

On the other hand, there are the idealists who believe that nature will take care of everything. This is exemplified by people whose apple trees fill with tent caterpillars and whose cabbages are ruined by hosts of worms and loopers. The argument most often heard is that the birds will eventually take care of the insects, that nature's balance must be kept.

Between the extremes is the gardener who has no intention of losing a crop to a sudden insect explosion, but who dislikes the approach that says, "The only good bug is a dead bug."

Although there are some 80,000 species of insects in the world, the average garden will probably see only one or two a season that seriously threaten the harvest. Ecologically aware gardeners will deal with these outbreaks swiftly — plants can be set back very quickly by such insects as the Colorado potato beetle or Mexican bean beetle — but without blanketing an entire plot with pesticides.

By recognizing the insect and attacking it, and it alone, we can minimize harm to the beneficial insects that usually keep pests in check. Even the mildest of chemical insecticides — such as *Sevin* — recommended by the majority of gar-

dening magazines are death to honeybees and many insect predators.

The most sensible progression is this: (1) Pick the offenders by hand, if possible. Small numbers of most insects are easily controlled and can often be tolerated. A blast of water may even be enough to dislodge certain caterpillars or aphids that will perish once separated from their food source.

(2) Repellant sprays *do* work, most notably those heavy with garlic. Controlled experiments have verified the effectiveness of this English decoction:

Chop or blend three ounces of garlic and let it soak in about two teaspoons of mineral oil for one day. Add one pint of water that contains one-quarter ounce of pure soap (not detergent). Stir or blend and then strain through several layers of cheesecloth.

This may be diluted at about one part to 20 parts of fresh water and sprayed on the vegetables or flowers to be protected. Store in a glass or plastic container (the spray will react with metal).

(3) Turn to *rotenone, pyrethrum* or *Bacillus thuringiensis* if necessary. All are derived from natural sources, they do not persist in the environment and are harmless to humans and pets.

Rotenone, from various South American roots, can be used against many garden insects, and many growers use nothing else. *Pyrethrum* comes from the pyrethrum daisy and has better knock-down power, but is harder on the beneficial insects.

Bacillus thuringiensis is a bacterial disease that stops most caterpillars (Lepidoptera larvae) from eating within 24 hours and then kills them. It comes in a powder (endospores) that is mixed with water and sprayed. Trade names are DIPEL, THURICIDE and BIOTROL.

But before setting out to decimate the ranks of an insect that has suddenly appeared in your tomato patch, be sure to know your adversary. There are many more beneficial insects than the

Dr. E. Ross

U.S.D.A.

Right, *Mexican Bean Beetle, from egg to destructive adult.*

Colorado Potato Beetle with larva.

Dr. E. Ross

One of 600 species of North American grasshoppers.

U.S.D.A.

Corn Earworm, a distasteful pest, is also known as Tomato Fruitworm.

Harlequin Bug.

Dr. D. Christman

Colourful but not usually found in significant numbers, is the Old Horseblanket.

Tobacco Hornworm.

Asparagus Beetle.

Dr. V. Tipton

Left, *two Sap Beetles, drawn to "honeydew" left by aphids.*

Cutworm pupae.

Dr. V. Tipton

Looper, destructive caterpillar that is unselective in its eating habits.

U.S.D.A.

Cutworm and its damage — others work below ground.

Celeryworm or Parsleyworm.

Dr. V. Tipton

Aphids come in a variety of colours and are prime prey for Ladybird Beetles.

137

Often unnoticed, Parasitic Wasps lay their eggs in the larvae and pupae of many garden pests.

Ladybird larva with aphids.

Moth caterpillar bearing dozens of tiny wasp cocoons. Such caterpillars should be left to die in the garden.

Praying Mantid.

Green Lacewing adult, an attractive beneficial whose larvae are called Aphid wolves.

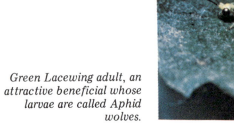

Predaceous Antlion, the curious larva of a dragonfly look-alike.

Lacewing larva with victim.

average gardener realizes, and they should be protected. Other bugs are more fearsome than destructive. The following guide shows some of the more common North American pests and beneficials, compiled with the aid of Dr. Mark Sears, Department of Environmental Biology, University of Guelph.

TOBACCO HORNWORM *Protoparce sexta.* It is likely that you will discover this huge caterpillar's handiwork before you find the actual insect. Evidence will include denuded branches on any of its solanaceous hosts (the nightshade group of tomatoes, peppers, eggplant and tobacco). Droppings approaching the size of mouse feces will be found under the plant.

The tobacco hornworm has seven diagonal stripes, and a reddish horn at the rear, while its near relative, the tomato hornworm, has eight, and its prong-like horn at the rear is green. Both attack the same plants. The adult form is the large, grey/brown sphinx moth, also known as the hawk, or hummingbird moth.
CONTROL. Hand-picking is usually enough to eradicate these impressive insects, as they often appear singly or in numbers of less than a dozen. Larger infestations may be cleaned out with the use of *Bacillus thuringiensis.*

CELERY WORM, PARSLEYWORM or **CARROTWORM,** *Papilio polyxenes.* A strikingly beautiful caterpillar, this is the larval stage of the black swallowtail butterfly. The celeryworm can be frightening to encounter unexpectedly, for it will extend a pair of black-tipped orange horns and release a sickeningly sweet defensive odour.

It favours dill, parsley, celery, parsnips and carrots, but is not often found in numbers nor is it frequently a serious pest.
CONTROL. Hand-picking is simple because of the insect's bright colouration. *Rotenone* may be used if necessary.

COLORADO POTATO BEETLE, *Leptinotarsa decemlineata.* Many gardeners find this the worst of summer's pests and one that is well able to strip an entire crop of potatoes of its foliage, or turn rows of tomato plants into barely visible stubs.

Curiously, the insect was almost unheard of until settlers brought the potato to North America. The Colorado potato beetle has lived in obscurity in the foothills of the Rocky Mountains, feeding mainly on sandbar weed, but found the potato to its liking and began moving eastward at a rate of 85 miles per year. It managed to reach even Europe, with World War I supply ships, despite frantic efforts to stop it at the Atlantic.
CONTROL. Small infestations may be curbed by hand-picking, but many gardeners will turn to regular applications of *rotenone* in bad years.

MEXICAN BEAN BEETLE, *Epilachna varivestis.* The worst species in the generally beneficial ladybird beetle family, this can be an especially destructive garden insect.

The Mexican bean beetle will attack stems and seed pods as well as the foliage of all beans, especially the Lima. It is slightly larger than the ladybird, with faded, yellowish wing-coverings bearing 16 dots. The photograph shows its development from egg to adult.
CONTROL. These beetles winter as adults in the garden, and cleaning and burning vegetative matter or composting after harvesting is a good preventative measure. *Rotenone* is effective, but must be repeated after 10 days.

Some gardeners report success with planting their beans and potatoes in alternating rows. It is said that green beans (but not Limas) are repellant to the Colorado potato beetle, while the potato plant can ward off Mexican bean beetles.

SPOTTED CUCUMBER BEETLE, *Diabrotica undecimpunctata.* Although small, these quarter-inch beetles attack many garden crops — beans, all of the cucurbits, asparagus, tomatoes, cabbage — and others — and they will infest flower blooms in the late summer.

Serves as a vector for cucumber wilt and other diseases carried as bacteria in its intestines.
CONTROL. *Rotenone* or *pyrethrum.* May also be repelled with botanical sprays (garlic, red pepper).

CORN EARWORM or **TOMATO FRUITWORM,** *Heliothus zea.* This pest is also called the tobacco budworm, but whatever crop it attacks, the results are often dismaying.

It is especially fond of sweet corn and will begin with the fresh silk — preventing effective pollination — and work down the cob. Tomatoes are burrowed into from the stem end and become unuseable. The corn earworm will also feed on beans, cabbage, lettuce and fruit.
CONTROL. If your corn patch is small, a half-dropper of mineral oil can be syringed into the tip of each ear. Do this after the silk wilts. The oil is tasteless to humans, but it will suffocate the earworm.

Marigolds are said to help repel these worms. Early planting in the north avoids some of the problem, but *Bacillus thuringiensis* can be relied upon to kill them — especially if applied as a dust rather than as a liquid.

CUTWORM, *Nephelodes emmedonia,* and others. One of the most insidious of common garden insects, the cutworm works only at night and may take down whole rows of tender seedlings in short order.

Some varieties work just below the surface of the soil, while others nip off the stalk just above the ground. Their adult form is a night-flying moth, while the pupa stage is a shelled case commonly seen when tilling or forking over garden soil (see photo).

Cutworms attack mainly cabbages, tomatoes and beans, but other crops are also vulnerable.
CONTROL. The above-surface cutworms can be spotted easily at night with a flashlight and hand-picked. Many gardeners put a handful of

J. Lawrence

Between chemical overkill and complete surrender to insect attack is the selective use of biological sprays.

wood ash around new transplants, while others stick a toothpick in the soil immediately next to the stem. The cutworm, which encircles the stem to eat it, will find this obstruction and, hopefully, move on.

ASPARAGUS BEETLE, *Crioceris asparagi.* Although but one-quarter inch long, these beetles can strip an asparagus patch of its foliage if left unchecked. The black-headed grey-bodied larvae will be seen clinging to the stems and are equally voracious.
CONTROL. Hens or ducks can safely be let into the asparagus patch following harvest, and they will eat large numbers of these insects. Ladybird beetles and chalcid flies help keep the asparagus beetle under control, but *rotenone* can be brought into play if the plants' survival is threatened.

CABBAGE LOOPER, *Trichoplusia ni.* Loopers attack almost everything, with an especial affinity for lettuce and cabbage. Their adult form is a mottled, brownish moth and they may go through three or more generations per year.
CONTROL. *Bacillus thuringiensis* is particularly effective in killing these, along with other caterpillars (such as the green cabbageworm, progeny of the common white cabbage butterfly) that attack lettuce and the cole crops. *Rotenone* dust may also be used.

SAP BEETLES, *Nitulidae.* Beetles comprise about 40 per cent of all insects, with 26,000 different species. These sap beetles are quite common, especially when fermentation is taking place. (They are attracted to sour fluids and exposed sap.)
   Sap beetles most commonly cause trouble for the gardener by infesting ears of corn or raspberry canes. They are small (3/16 inch), but can cause disheartening damage.
CONTROL. *Rotenone.* Remove stalks and vegetation in which they seek shelter for the winter.

HARLEQUIN BUG, *Murgantia histrionica.* This is a true bug, of the order Hemiptera, suborder Heteroptera, classified by the thickened section of the front wing where it attaches to the body.
   Harlequin bugs occasionally become numerous enough to pose a serious threat to cabbage, turnip or Brussels sprout crops.
CONTROL. Clean up the garden each fall, as this insect overwinters in old cabbage stalks and other hiding places. *Rotenone* is effective against most outbreaks.

OLD HORSEBLANKET or **SADDLEBACK CATERPILLAR,** *Sabine stimula.* An insect curiosity, this caterpillar can be found attacking the foliage of trees, flowers and rose bushes.
CONTROL. It is not often a problem, but can be controlled by hand-picking or *Bacillus thuringiensis.* (Old Horseblanket's sharp tufts are as formidable as they look — they will pierce bare skin and cause painful irritation.)

APHIDS, many species. Aphids or plant lice come in a variety of hues — black, red, yellow, brown, lavender and others — and rare is the garden that sees none of them in a season.
   Fortunately, aphid populations often subside before serious damage is done, although they can tap the strength of plants and affect flower and bud formation.
   They are sucking insects that exude "honeydew" as a waste product — this being digested plant sap which can attract ants and black fungus.
CONTROL. Ladybird beetles, lacewings and syrphid flies are all important enemies of the aphid. Gardeners can often throw aphids off balance with nothing more than a strong blast of water from a hose. *Rotenone* and *pyrethrum* are also effective.

140

LACEWING, order Neuroptera. The larva of this beneficial insect is aptly named the aphid wolf or aphid lion for its hearty appetite in the midst of aphid infestations.

In fact, the female lacewing lays her eggs on slender hair-like projections so that the cannibalistic larvae cannot reach each other when they hatch out.

Both green and brown lacewings are commonly seen. They are important in controlling mealy bugs, thrips, mites and cottony-cushion scales.

MANTIDS, PRAYING MANTIS, family Mantidae. There are 20 species of praying mantids in North America, both green and brown in colour.

The green Chinese mantid was introduced with a shipment of nursery stock from China in about 1895 and is one of the most hardy species.

Mantids are fearless, and have been known to strike against frogs and lizards. They eat aphids, bees, hornets, each other and any other insect that passes within reach of their grasping forelegs.

Cases of 100 to 200 eggs may be purchased through the mail, but, as with all beneficial insects, they move where the hunting is best.

BRACONID WASP, *Braconidae*. This small parasitic wasp is not often noticed by gardeners, but it is an extremely important beneficial insect.

The female wasp lays its eggs on the larvae and pupae of butterflies and moths, and on aphids, after which the host insect is slowly killed as its strength is sapped by the developing wasps. If you discover a caterpillar covered with tiny white spots or cocoons, do not remove it from the garden. The wasps will hatch out and continue their valuable work.

Similar beneficial insects are the TRICHO-GRAMMA WASPS, CHALCID FLIES and ICH-NEUMEN FLIES, some of which can be purchased through biological supply houses.

ANTLION, family Myrmeleonitdae. The antlion is the famous "doodlebug" of the southern United States, and is not an especially important insect at this time.

It does, however, have the fascinating habit of digging a pit about two inches in diameter by two inches deep, to trap passing ants and other insects.

LADYBIRD BEETLE, family Coccinellidae. The adult ladybird beetle or ladybug is familiar to everyone, but little known is its important larval form (see photo).

Both adults and larvae have insatiable appetites for aphids, and they also form an important natural control of scale insects and mealy bugs.

There are numerous forms within this family, with varying numbers of dots on their backs. the convergent lady beetle is the one most commonly pictured, and is the type purchased in quantity by organic gardeners from west coast suppliers. It has 12 spots.

141

A Classical Plot

Elementary, my good friend — a raised bed for dooryard gardens

By Barry Estabrook

If Confucius and Aristotle ever paused from their philosophical musings to plant a vegetable patch, chances are excellent that these ancient wise men would have opted for some sort of raised-bed plot.

With roots extending back 3,000 years to Classical Greece and China, raised-bed gardening is currently undergoing a noteworthy revival in (of all places) suburban North America.

But whether your gardening goals are a modest salad patch in the back corner of a cramped city lot or a small herb garden handy to the country kitchen, you might be wise to heed the example of our forefathers. Intensive, raised-bed gardens will mean greater yields and far less work than the squares of widely spaced rows the land-wealthy have come to feel are the only true gardens.

A city-bound friend of ours, finding himself temporarily forced to take refuge in a garish condominium for a year between permanent residences, discovered that horticultural urges did not subside with the alteration in environment.

When we visited him toward the end of August, a small sunny patch, between his front door and the sidewalk, burgeoned with lush tomatoes, cabbages, lettuce, carrots, radishes and scallions, while the sandy front-yards-to-be of his neighbours still produced nothing more than a few slender blades of grass scattered among shards of brick, metal siding and tar paper.

The humble keys to his astounding horticultural *tour de force* were a half-dozen weathered railroad ties, a pickup truckload of rich, river bottom loam, and a few wheelbarrows full of aged manure.

Intensive, raised-bed gardens use the natural mulching properties of plant leaves which shade the ground, retaining moisture and retarding weed growth. Plants are positioned so that their leaves just barely touch. They get ample light, yet there is not a square inch of wasted space.

For northern gardeners, the raised bed has other advantages. The higher your small plot is raised above surrounding soil, the faster it will warm up in the springtime. Because the beds are small, it is an easier job to cultivate thoroughly. This, combined with the increased possibility of a high humus content, creates excellent drainage and ventilation and further accelerates spring planting.

The first step to successful raised-bed gardening is deep cultivation with a spade. This task might be laborious the first time you undertake it, but you will reap dividends in future years when you will have to spend only a few moments turning your rich, non-compacted soil.

If your soil is poor and you want immediate results, you would do well to follow our friend's example of adding a layer of rich topsoil to the raised portion of the bed. Into this layer mix manure, compost and other organic fertilizers.

Be prepared to cast aside traditional visions of rows when the time comes to sow your seeds. A curious anomaly of seed packet planting instructions might already have dawned on you: Why, for instance, do radishes not object to growing only one inch from neighbours in the same row, when they demand a solid foot between rows?

The answer is that they don't. Whatever spacing is recommended between plants in the same row will prove sufficient for the distance between rows in your raised-bed plot.

Careful staggering of plantings and intercropping will make even the tiny garden produce yields that belie its size and the small amount of labour you have put into it. Sneak smaller, faster-maturing vegetables (lettuce, carrots, spinach, radishes, shallots, beets) among your tomatoes and peppers. Stake your pole beans and cucumbers. Look upon sowing as a weekly activity, not a once-a-year task.

Of the fast-growing species, plant only a few seeds at a time — your harvest will be extended throughout the season.

The final result of this intensive method of planting is that, although yields per plant are somewhat reduced, yields per square foot of garden space rise dramatically.

During the early stages of growth, ensure that the moisture content of the soil in your raised bed remains adequate. Because it is above the surrounding earth, the raised bed will be prone to drying out before the protective umbrella of plant foliage has been produced.

Also, during these early stages, you should hand-cultivate frequently, eliminating weeds before they get a good start.

Slugs and some garden insects are usually too lazy to undertake the climb over railroad ties to gain access to your plants, and weeds and grasses from the perimeters will not creep into your vegetables.

Too, there is an aesthetic pleasure to raised-bed gardens. Neat, squared edges, beautiful borders of insect-repelling marigolds blending with the variegated greens of carrots, lettuce and peppers, accented by the magenta of beet-green stems and the bright red of tomatoes — it's a sight that would bring grace to the most dismal setting, and still add fresh produce to the summertime dinner table.

S. D. Warren Company

A Swineherd's Notes

Bringing home the bacon—on the hoof

By Hank Reinink

Most farmers would have known better. The piglet was a runt, too small to amount to anything, too weak to get a fair shot at her mother's milk. Her dozen larger and more aggressive brothers and sisters had the ability to stay in the front row and kept pushing her away from the nipples. Instead of growing, she became weaker and weaker.

The traditional cure for a situation like this is to take the piglet out and give it a killing knock on the head. Having recently arrived from the Netherlands and having taken work on the farm with the intention of someday buying our own land, we were rookie farmers.

We asked and were allowed to take the piglet home, intent on nursing her back to health. With the aid of a Coke bottle equipped with a nursing nipple, she soon learned to drink a mixture of cow's milk, water and corn syrup. Her flat stomach and hollow sides filled out and became nicely rounded. Treated as a foster child, she responded physically and her spirits were soon on the rise, as well.

The piglet came to be called Susi (urbanites may find it curious, but many a farmer names his best animal after his wife), and she became a happy member of our small family. She had the run of the kitchen, coming when she was called, taking great enjoyment in being petted and spending many evening hours sleeping in a warm, soft lap.

Now a pig, it turns out, can be housebroken, but this particular young thing was never educated in the ways of piggy-litter. When she was just a little four-pounder, clicking around on tiny hooves, the small deposits were cleaned up easily enough. However, as her eating and drinking habits really began to blossom, her presence became, how to say it? — *noticeable.*

Susi was pronounced well and fit and promptly returned to the barn, where she lived to a ripe old age and where her descendants can still be seen today.

In later years, with a farm and pigs of our own, our swine herd grew as large as 500 head. Recently, with a shift in our interests, the herd has been reduced to about 100 animals. We would never do away with the pigs altogether — their public image notwithstanding, pigs are more than walking mounds of pork chop. They are intelligent, have characters of their own and are a constant joy to work around.

MORE THAN PORK CHOPS

A pig moves through a number of clearly noticeable changes in personality as it goes through life. When newborn, it is cuddly, sweet-smelling and its only interest in life is to suckle its moth-

er and find a warm, dry place to sleep. Not unlike human babies, they do little else for the first several days.

But when young pigs are a week old, they begin to stay awake longer than it takes to fill their stomachs. They look for other things to pass the time. Piglets invent all sorts of playful games, with mock battles and much racing around the pen. Pigs often appeared in entertainment acts in Merry Olde England, and their still-unrepressed curiosity and prankster instincts are really evident when they are young. They explore their quarters, scale the sides of their sleeping mother, frisk along her back, attacking her ears and tail. In general, all is well in hogdom at this stage.

When weaning time comes, at about six weeks of age, the young pigs will still play, but with less abandon. It seems to me, however, that they are at their most intelligent at this time of their lives. They will figure out how to undo the nuts and bolts that hold their feed trough in place, and if you get in the pen with them, they will swarm around your feet to tug at your clothes and untie your bootlaces.

By the time the pigs reach 150 pounds — perhaps in four or five months, depending on many factors — they concentrate mostly on making hogs of themselves.

These barrows (castrated males) and gilts (unbred young females) eat, drink and grow very quickly. At about 200 pounds, the most advantageous size for slaughter, there seems to be little in the way of personality left in the beast. And that, of course, is just as well, especially if you have violated the common-sense rule of never naming anything you intend to eat.

PIG LATIN

Of course, you will find that hogs raised to full maturity to serve as breeding stock come up with strong personalities. The mature sow that is kept by a loving farmer will be affectionate and pleasant in a quiet way. She will come when called and will enjoy being scratched on the back or behind the ears — two places a pig can't reach by itself. She will return, with interest, any amount of attention extended to her and you will have a friend as long as she remains in the herd.

Before going on to the meatier aspects of swine-rearing, it should be pointed out that once you've worked with pigs for a time, you will learn to understand their language.

When you walk into the barn a low roar from a sow may tell you she is in heat. A high-pitched, prolonged screaming from one of the weanlings possibly indicates that he has managed to get his head stuck trying to sneak out of the pen. Soft, short, low grunts say that all is well — the sow talking to her young as they suckle. An angry or alarmed hog will break into the porcine equivalent of a bark: *woof, woof.*

Perhaps the hardest pig sound to distinguish is the difference between piglets fighting over a favourite nipple and the cry of a tiny pig who has ended up beneath his mother when she has flopped down. (A 300-to-600-pound breeding sow can easily kill one of its own offspring in this manner, and in large commercial piggeries, even with special birthing pens, 10 per cent of all piglets may perish under their own mothers.) In each case the screaming is quite frantic, and even after many years around swine, I never trust that it is simply a fight, but rush over to check the situation. If it's a fight, let things proceed, but if there turns out to be a little pig caught under the sow, you must get the old lady up in a hurry if the young one is to have any chance at all.

HOME-GROWN BACON

Pigs are ideal livestock for even the smallest of farms. Their omnivorous habits make them easy to feed, and they require very little space. Two pigs are plenty for stocking the freezer of most families with a year's supply of pork. The average slaughter-age hog weighing 220 pounds will provide 135 to 155 pounds of dressed meat, not including head or hooves.

The most sensible plan for raising your own pork without actually starting a breeding herd of swine is to buy young stock in the spring, feed them over the summer and butcher when cold weather arrives. This system allows a maximum use of free food — culled vegetables from the garden, weeds, pea shells, corn husks, surplus skimmed milk or whey if you have a dairy animal.

I would strongly advise against buying a single hog. Pigs are herd animals and they do best with company. A lonesome pig will mope around and feel sorry about his plight. Even if you care not

J. D. Wilson

a whit for the pig's feelings, consider that he won't eat as well and consequently won't grow as well, either.

The best place to buy your pigs is at a neighbourhood farm where you can get some idea of the health of the herd. There are many swine diseases, most of them contagious, so the fewer strange pigs yours have been in contact with the better. Pigs bought in the livestock sales barn, for instance, have had a chance to pick up every disease in the book.

Your choice of an individual animal is more important than its breed considering that its destination is the freezer in a few months. As a producer of pork rather than as a hog breeder, you do not want to pay the premium price for a purebred animal. Crossbreeds grow faster, convert feed more efficiently and are hardier than pure stock. Too, the vast majority of North American pigs are crossbreeds and are less expensive to purchase.

For all that, you should know a bit about the foundation stock. The most popular breeds — Hampshire, Yorkshire, Landrace and, in the

146

U.S.A., Poland China — are all considered "bacon" types, which is what you want. They have been bred, particularly in recent years for meatiness — not lard — which in the days before vegetable oils was a valued and sought-after pork product.

When you buy your young porkers avoid the sentiment that could send you home with the runt of the litter. Pick big, young animals that are long in the body and that seem alert and lively. Your pigs should be old enough to have been weaned — hence the term "weaners" — and to have recovered from the shock of losing their mother's milk. They will have enough trauma to cope with when you take them to their new home.

They should weigh about 35 pounds or more and their bodies should be well-rounded without a sharp ridge of backbone showing through too thin a layer of flesh. The hair should be smooth against the body rather than appearing too long and standing on end. The skin should be free from crusts or eczema.

Chances are that you are looking for a young gilt, a female that has never given birth, or a barrow, a male that has been castrated. Although that job is reasonably simple, it is too critical to the health of the pig to take the chance if you have never done it before. Ask the farmer who supplied your weaners if you may watch the process at his farm or, failing that, pay the vet to do the job in the first place. Otherwise you will just end up paying him to cure problems that you caused.

BOAR PREVENTION

Pigs are usually castrated when they are very young — often within the first week of their lives — so by the time you bring home your young porkers the job should already have been done and the wound healed. If you are harbouring any notions that this is all barbaric and unnecessary, be advised that castrating is absolutely necessary if you plan to eat the pork.

Meat from a fully-equipped boar has a very strong and unsavoury smell that will drive you right out of the house while it roasts in the oven.

The price of weaner pigs depends on their weight, the present market price for pork and the state of the grain market. In this area, in early 1978, the average price for a 40-pound pig was $35; you will have to ask around in your own area to determine what the fair, going price is.

If you start with a pair of healthy pigs, there shouldn't be any need to medicate them, but practically every swine herd has some level of intestinal worm infestation and it is wise to give new pigs a dose of wormer a few days after they arrive on your farm. They shouldn't need to be wormed again, and if the farmer from whom you bought them happened to have just wormed the pigs you needn't bother at all.

In the heyday of the zany twenties, jazz chanteuse Josephine Baker reportedly set Paris abuzz by using dancing pigs in her act, walking them proudly along the boulevards and bringing them into her home to dine on champagne and cocks' combs.

Homestead pigs require no such treatment and their housing need not be at all fancy. A simple run outside any small building or barn where they have dry bedding and can escape the weather will do. Strong fences make happy pig farmers, and you will be well-advised to assure that the run is escape-proof. Bury the wire fencing a foot or more into the ground or they will root their way out. Another solution to the burrowing problem would be to run a single strand of barbed wire or electric fence (which gives a memorable but harmless jolt to a wet nose) around the bottom of the fence.

If you have spare land and a good fence you can let your pigs run out to pasture. They will make good use of the grass and other vegetation and improve your soil, but the danger of escape increases drastically. Commercial swine farmers calculate they can crowd 35 pigs to the acre, but that concentration is work-intensive — you have to supply extra feed — and it is not especially good either for the pigs or the pastureland.

A more contained and controlled approach involves the use of a portable run which is moved from time to time to furnish fresh pasture. In this case, the fencing must be rather light and the use of a strand of electrified fence (which runs off a 6-volt battery) is recommended. The pigs will do a fine job of clearing weeds, even stubborn twitch grass roots, and will leave the soil friable, well-fertilized and ready for planting next spring. In fact, pigs do an excellent job of clearing rough land for a garden site.

If you prefer a permanent pen but can't take advantage of an existing building, a very inexpensive alternative can be improvised (see diagram). Drive four sturdy stakes into the ground in a 10-by-10-foot square and wire or spike rough cedar logs against the posts. The higher front end of the pen should face away from the

Agriculture Canada

147

prevailing wind and should have an opening 30 inches wide to serve as a gateway. The back end need be no more than 30 inches high.

The walls may be fortified by banking the exterior with sod, earth or bales of straw. The roof may be of corrugated metal or painted plywood or, to be truly rustic, cover the roof with a layer of plastic sheeting, soil and a layer of growing sod.

A sod roof provides excellent insulation, an important factor as your weaners will appreciate a cool, shady den in the heat of the summer. Pigs can stand almost any amount of cold, but they cannot sweat, and suffer on hot days. Hence their love of a shallow pond or mudhole in which to loll. This desire to wallow will be alleviated if they can find a cool, breezy place in the shade.

In winter, any pig shed should stop drafts and keep the bedding dry. Heat is not required, except to keep the drinking water flowing. Scientists who carried out sperm counts on boar semen under winter conditions which dropped as low as 40 below zero (F), found that swine can scoff at the Canadian prairie winter. Even at that temperature sperm count was not affected.

A PIG'S DIET

Swine are the most easily fed of farm animals. They will eat almost anything and care not where it comes from. A pig will thrive on the wasted grain that passes, undigested, through the many stomachs of a cow, and if this feeding regimen is unappealing, consider that many a commercial pork producer draws his hog rations from the pavement of a beef feedlot.

You will, doubtless, want to start your weaner pigs on something slightly more appetizing. Feed stores sell several types of prepared swine feeds, each geared to the age of the pig.

Creep feed, for baby pigs, is high in protein (24 per cent), rich in glucose, and the piglets will start eating it when only a few days old. When each of them has eaten about five pounds of creep feed, they are gradually switched to *pig starter*, which is more economical and contains 18 per cent protein. Standard recommendations call for a total of 75 pounds of starter per pig before they are moved on to a 16 per cent protein ration called *hog grower*. It will take an average of 200 pounds of this feed to bring the live weight of your pigs to 120 pounds. And then it will require 300 to 350 pounds of 14 per cent grower ration to carry the pigs up to their 200 pound slaughter weight.

A quick call to your feed dealer and some elementary arithmetic will show that you are not producing exceptionally cheap pork if your pigs are raised largely on commercial rations.

There are many ways to supplement a pig's diet with extremely cheap — if not free — feeds. Swine relish all manner of kitchen scraps and garden wastes. They can be fed cull potatoes and most grocers daily throw out wilted vegetables and stale bread that would make exceptionally good pig feed.

If you grow field corn or grain, it will form the basis of a good hog ration. We feed our herd on a mixture of nine parts of our own ground grain to one part of a swine protein supplement that is purchased from a feed mill. Cooked soybeans make a high protein hog feed and the new varieties being developed for northern climates mean that many farmsteaders may be growing more of their own stock feed.

The dominant pig feed is corn — shelled, ground, mixed on the cob, wet or dry, but by itself, it won't keep a pig alive. Corn lacks most of the essential amino acids.

Pigs do very well on surplus milk and will eat with glee any extra summer eggs, even those that your ducks sat on, in vain, long after the hatching date has come and gone.

This habit of eating whatever is offered can get a pig into trouble. They should not be fed table scraps with fine bones (e.g. chicken) that can splinter and puncture the stomach. Pigs will eat mouldy grains and bread, but some moulds produce hormone-like substances that can cause serious problems ranging from abortion to death.

Unlike most farm animals, pigs are bright enough to know when to stop; they won't keep eating until they make themselves sick. You can gauge how much to give them by how much they eat. The young animals efficiently convert feed to pork, but by the time they are nearly mature, at about 150 pounds, you should impose a stricter diet. The pigs will eat all they get, using the excess to put on layers of fat. Since you want chops and roasts rather than lard you must reduce their intake to about five pounds of balanced ration per pig per day.

INEXPENSIVE SUMMER SHELTER *for a pair of feeder pigs, as suggested by Hank Reinink. The sod roof is optional but affords added heat protection for the pigs, which do not have the ability to perspire. The back or low end of the shelter should head into the prevailing winds.*

Water is actually the most critical requirement of your pigs. They will drink large quantities, especially in the warmer weather, and if you don't allow them enough, they may develop a condition known as "salt poisoning," a brain disorder. The pigs will stagger around and bump into things, and as the condition worsens, they will stop reacting to any stimulus and death is the only possible outcome.

The best solution to the water requirements of your pigs is a fountain they can operate them-

selves, and which you can have hooked up to your pressure system. There are "nipple drinkers," which cost about $10 each, which the pigs operate themselves by biting the valve, causing the water to squirt straight into their mouths. Float-operated bowls are available for a slightly higher price, but have the advantage that you can use them for any kind of livestock, once the pigs are in the freezer. But where the nipple drinker guarantees that the water is always clean (even if it has been sitting around in the water pipes for some time) the float-operated bowls have to be cleaned out periodically, especially in the float-box. Algae and bacteria gather there, making it necessary to scrub them once a week.

When buying a float-operated bowl, look for the type which allows you to open the box by hand for clean-outs. Some bowls are constructed in such a way that tools are required to undo a set of bolts just to get into the box. I favour the Fordham bowl (about $12.50 each), which has a hinged float-box cover equipped with a hole. By poking your finger into the hole you can open the box instantly, but pigs can't.

Your porkers will be ready for slaughter sometime after five months of age. Repeated studies have shown that feeding them past 200 pounds (90 kilos) is an inefficient use of feed and results in pork that most of us would judge too fatty.

Most rural communities have a slaughterhouse which will kill and custom-butcher your pork for surprisingly reasonable rates. Many people prefer, however, to complete the cycle themselves.

Sources

RAISING THE HOMESTEAD HOG
By Jerome Belanger
Rodale Press, 226 pages
Excellent guide to small-scale pork production with good information on minimal cost feeding.

LANGSIDE DISTRIBUTORS LIMITED
Box 369
Lucknow, Ontario N0G 2H0
Sell Fordham waterer, as well as nipple drinkers and other equipment.

K. G. JOHNSON LIVESTOCK
EQUIPMENT LIMITED
Box 201
Elmira, Ontario N3B 2Z6
Fordham drinkers, nipple waterers, feed troughs and other pig husbandry equipment.

BERRY HILL LIMITED
75 Burwell Road
St. Thomas, Ontario N5P 3R5
Many types of livestock equipment — tell them you are interested in hog waterers or feeders.

Bee Fever

Be forewarned: Beekeeping is fascinating, addictive and, yes, you will get stung

---

By Michael Shook

The other morning I was delivering a pail of clover honey to a neighbour and was just getting back into the pickup with a filler of eggs that was part of the bartered payment, when he came out with a question that I hear over and over at this time of year: "What is it you beekeepers do in the winter?"

Now there are a variety of answers to this question, and, depending on the mood of the day, I'll answer, "Sit with my feet as close as possible to the fire," or "Nail together new bee equipment."

What any honest beekeeper would answer is, "Think about bees." Certainly there is time to catch up on reading, sleep late some mornings, and putter with new frames and hive boxes. The bees have plenty of honey or syrup to see them through the winter, so this is the season for evaluating the past year's experiments and for plotting the new season's strategies.

For would-be apiarists, however, this is a crucial time — beekeeping, to be started right, should not be a spur of the moment decision. There is equipment to buy, bees to order and hives to prepare. Most important, a good basic knowledge of what you are about to get into must be acquired — well before the soft maples come into bloom and signal the start of the honey flow.

Be forewarned: those strange little beasts are addictive. Many a commercial beekeeper has started out by catching a swarm or buying a couple of colonies and contracted bee fever in the bargain. One needn't live on a homestead, surrounded by cubic miles of wilderness and farmland to keep bees. Hives kept on New York City rooftops have produced honey, and beekeepers can be found in nearly every city and township of southern Canada and the United States.

Those who keep bees tend to be of a rather mild, easygoing temperament, but not all of us were thus constituted before starting with bees. The sweet little honey bee will not tolerate manic behaviour and rough manipulation — bee fever has been known to slow some pretty frantic characters right down.

BEE PASTURE

In a good year, as old beekeepers are fond of saying, the bees will make honey from the pump handle. In the average year, however, there must be some appreciable "bee pasture" of floral nature nearby, whether it be cultivated alfalfa, buckwheat and clover or wild plants such as fireweed and basswood trees.

The type of flowers within a two-mile radius of a colony will determine the colour and flavour of the honey produced. Honeys range from heavy-tasting, purple-brown buckwheat to the light-flavoured, water-white clearness of alfalfa or fireweed. Not all flowers secrete nectar, the raw material from which honey is made, while some rather unlikely ones do.

In many cases, from the Maritimes of Canada to the southeastern United States, bees obtain their first appreciable nectar of the spring from soft maple trees, which few people ever notice to be in bloom. On the other hand, a colony in the midst of the watery-blue blossoms of flax plants could starve to death.

And not all honey plants will yield nectar every year. Even the best-yielding plants will dry up without sufficient ground moisture, while other flowers may have their nectar washed away by a light rain. There's nothing quite as

150

Dr. M. V. Smith

—Courtesy Diane Birch

Pollen-foraging honey bee, **above centre,** *hovers at a pear blossom raking and brushing the pollen—essential for colony fertility—from its body to its legs. An average trip from hive to field may take the bee to some 50 blooms, with 50,000 miles of flying necessary to make a single pound of honey.* **Right,** *the annual ritual of expelling drones before winter. Prevented from reaching the honey stores, the weakened drones are physically removed from the hive by the workers. A single colony may have upwards of 80,000 worker bees (whose task it is to gather nectar), a single queen and several hundred male drones.* **Above,** *beekeeper Diane Birch inspects a comb-filled frame. As with most beekeepers, she foregoes the use of bee gloves except in special situations.*

All photography, except where otherwise noted, by Dr. M.V. Smith, Department of Environmental Biology, University of Guelph.

Wild swarm, **bottom**, gathered on a small tree. Swarms at times become so heavy they will break branches on which they gather, but the mass of bees, in search of a new home, are in a holiday mood and easily managed. **Below**, the interior of a honey super in which the beekeeper has left out a number of frames, giving the bees room to build uncontrolled comb on the side of the hive.

—Alan B. Stone

Below, *the bane of North American bee-keepers, a disease called foulbrood in which bee larvae are seen as dying and coffee-coloured (in contrast to the healthy white larvae).* **Right,** *the worker bee, an imperfect female, is the smallest of the hive's inhabitants. The queen, the largest, is capable of laying up to 2,000 eggs per day to maintain and increase the strength of the hive. Drones have the sole purpose in life of fertilizing the queen.* **Bottom,** *the bee smoker causes the bees to engorge themselves on honey (perhaps an instinctive reaction in preparation for fleeing a fire), allowing the apiarist to work in the hive.*

frustrating as having an apiary beside a huge field of buckwheat on a dry year.

The available floral pasture, the weather, and the strength of each colony will determine how much honey will be produced per hive in a season. The goal of the beekeeper is to manage his bees in such a way that they reach their maximum strength in numbers just prior to the main expected honey flow, the time at which the most blooms of major honey producing flowers are expected in his area. It's an exciting and challenging job, but we can't control the weather, and disappointment has rewarded the hard work of many a beekeeper, especially when rains prevent his bees from flying every day of the anticipated main honey flow.

Every beginning beekeeper would like to know what magnitude of reward he will derive from his investment, risk, and labours, but it's quite impossible to quote a projected yield per colony. Obviously, honey crops will vary from colony to colony, from year to year, and, most dramatically, according to the area in which the bees are kept. In southern Ontario for instance production ranges from zero to over 100 pounds per colony, but our 50-pound long-term average looks pale in comparison to some of the crops on the prairies, where rapeseed, sunflower, and buckwheat fields stretch away to the horizon.

Another vital question for anyone considering taking up beekeeping as a hobby and to supply personal honey needs is what the costs will be in time and money. One could easily spend more than $300 to set up two colonies and purchase new protective and extracting gear. If one is in a position to pick up used equipment and established colonies, the cost could be cut to under $200.

With reasonable care and attention, the investment should be returned in three years. If the bees are neglected, as sometimes happens, the novice may have nothing more than a stack of weathered, empty boxes at the end of three years.

Proper management does not mean a lot of time spent in the beeyard. Indeed, many beginning beekeepers over-manage their bees — probing, exploring and disturbing them every few days. One inspection every two weeks will keep a colony in shape from April through August, unless a re-queening or splitting operation is undertaken, in which case, two visits in one week may be necessary. Otherwise, it is best to leave the bees alone. One could easily manage 20 or 30 colonies with one afternoon's involvement every two weeks. When the honey crop is taken off and processed in August or September, some additional labour will be necessary. With fairly rudimentary extracting equipment, the surplus honey from three or four colonies can be processed in a day by inexperienced hands. Between harvest time and the onset of winter, three or four visits should suffice to ensure that the bees are in shape to endure the rigours of cold weather and emerge in the spring ready to produce another crop of honey.

Thus if one has a bit of time and money to invest, and is willing to run the risk of contracting bee fever, the salient question is whether there is anything in the area from which the bees can make honey. Wild flowers provide good bee pasture in much of the wilderness area of North America, while gardens and ornamental plantings will support a few city-bound colonies. In agricultural country, it's a safe bet that a few farmers will have a field or two of clover or alfalfa.

If a commercial beekeeper has apiaries in the area, it's highly unlikely that he has them there entirely for the fun of it. Pay him a visit; most beekeepers are more than happy to talk about bees and will usually know of any colonies that are for sale nearby. Make that a winter visit, though, because come spring, a person who's running several hundred colonies becomes a very busy beekeeper indeed.

SEED STOCK

A full hive of bees has from about 80,000 to 100,000 workers, with 200 to 300 drones and one queen. The drones' enviable role is spent in the warm weather months roaming the skies looking for receptive queen bees, while the colony queen has the task of producing eggs — up to 2,000 in a single day.

The female worker bees, flying at 10 to 15 m.p.h. put in some 50,000 miles of travel between hive and nectar source to make a single pound of honey. Creating a good population of these worker bees is the goal of the beekeeper, and your first management problem will be to decide how many colonies to start the first year.

Most experienced beekeepers, I think, would advise *not* to start with a single hive. A one-colony apiary is in grave danger of extinction if that hive's queen dies or, for some reason, quits laying eggs. Another colony next door can help its stricken neighbour with either a queen cell or eggs from which to raise a new queen.

Two colonies is a good number for a start, and after a person is sure that he enjoys working with bees, the number is easily increased by catching swarms or splitting established colonies.

If you don't care to wait for a swarm to pass by and don't have an established hive to draw stock from, how are the first bees acquired? The most common way is to buy new equipment and install a two-or-three-pound package of bees and a queen. These are purchased from breeders in the southern part of the U.S. (Bees from these shippers cross the U.S.-Canadian border with no problem.)

A two-pound package contains about 10,000 bees. However, only during exceptional years are they able to establish themselves and produce surplus honey. Unless one lives on the prairies, where honey flows are much stronger, one really shouldn't expect to take any honey from a colony the year it is started with packaged bees. In my experience a two or three-pound package simply won't build up the required population in time to harvest a crop of honey.

If you plan to start with mail-order bees, start thinking about ordering early. The bee breeders need time to plan their spring production schedules, and are often unable to fill orders placed after late February.

Most of the beekeeping supply houses handle packaged bees and queens, but, of course, they must make a few dollars for their services of ordering and handling. Order directly from the breeders for the best prices. When ordering, you must state the date you wish the bees to arrive — six to 10 weeks before the main nectar flow in your area.

In contrast to a new hive of packaged bees, purchasing an established colony will usually assure that you can take from 50 to 100 pounds of surplus honey that first year.

In addition, an experienced beekeeper can pass on a wealth of useful information at the time of the purchase.

ESTABLISHED COLONIES

The packaged bees and associated equipment will cost about $90. The beekeeper should give a bit better deal, though a well-established colony housed in good equipment is worth much more than a package in a brand-new hive.

In most provinces and states, any colony of bees that is sold or moved must be reported to a state or provincial apiarist. Be sure this is done, and acquire any necessary permits before buying or moving bees. This is not just another government hassle; the purpose of the registration is to prevent the spread of bee diseases such as American foulbrood. Each spring, in most provinces and states, a bee inspector checks all colonies for disease. If he finds American foulbrood, he will probably be obligated by law to burn all infected colonies. Though this seems a pretty heavy rap, the disease is extremely contagious and can kill a colony in a matter of weeks. Better to burn one hive than to lose the whole apiary.

SWARM CHASING

Swarm chasing is a favourite pastime of some beekeepers, and early swarms can start or add to an apiary. An old beekeeping gentleman of my acquaintance always claimed that, "A swarm in May is worth a load of hay, a swarm is June is worth a silver spoon, but a swarm in July ain't worth a fly," which is folksy, but wise. An early swarm will have time to build up to full strength before the main honey flow, but a late emerging swarm may not even produce its own winter supplies.

If one has an extra hive body and frames on hand in the spring, it's a simple matter to bring home any swarms that come to your attention and simply dump them in.

An advertisement in any small town newspaper saying, "Will remove swarms of bees from trees, fences, etc." will probably bring in several productive calls each year. But it will also bring calls about bumblebees and yellowjackets in barns, and about honey bees that established

themselves, "way up there at the peak of the eaves about five or six years ago."

There are ways of getting bees out of homes without tearing the house down, but unless such "swarms" are easily accessible by tearing off a few siding boards, such bees will be more trouble than they're worth.

Of course, a house swarm or bee tree that has been established for a number of years may yield a couple hundred pounds of honey, but the job of removing it generally leaves the honey full of bees and the beekeeper and his tools covered with honey and stings. It's best to stick with newly emerged swarms which are hanging in a classic cluster from a branch, post or wall.

These swarms can be transported in just about any large container that won't smother or squash the bees, including the branch they're found on. My mother tells of driving a beekeeping neighbour home in a Model 'A' with a branch crawling with bees on the seat beside her. Though she had serious reservations about the trip, she received no stings. Swarming bees are in a holiday mood, filled with honey, with no home to defend. They will seldom sting unless one inadvertently squeezes them.

For those who don't care to truck with bees on such an intimate basis, a cardboard box makes an excellent "cage". It is light enough to hold up under the swarm with one hand while the other hand gives the branch or pole they're on a good sharp shake. If the bees have clustered on a wall or post, the box can be set under them, and they can be scooped into it with the hands or a stick. Once the queen is in the box, the bees will settle there too, and when the majority are inside, the lid can be closed and the bees trucked home.

Once in the apiary, the bees may be dumped into or in front of a hive body full of empty combs. Many beekeepers transfer a frame or two

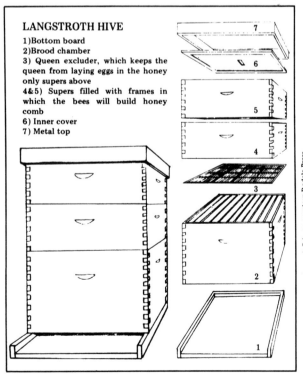

LANGSTROTH HIVE
1) Bottom board
2) Brood chamber
3) Queen excluder, which keeps the queen from laying eggs in the honey only supers above
4&5) Supers filled with frames in which the bees will build honey comb
6) Inner cover
7) Metal top

Illustration from Buzz—A Beekeeper's Primer, courtesy Rodale Press

of brood from a strong, well-established colony to ensure that the swarm will stay in its new home. If the transferred comb contains a few eggs, it also provides the swarm with the wherewithal to raise a new queen if the old one has been killed or injured during the move.

BEEKEEPING GEAR

Three categories of equipment are required by anyone planning to keep bees: personal and protective gear, the hive itself and some type of honey processing tools.

A beginner can get along with three items for personal use: a veil, a hive tool and a smoker. If you are paranoid about stings, you may want to wear gloves. If terrified of or allergic to stings, either buy a bee suit or take up maple sugaring to obtain your sweets.

A bee suit ($18—$27) is a pair of white coveralls with straps or elastic at the wrists and ankles. If kept free of holes, it will exclude bees from all parts of the body except hands, feet and face. Gloves, tall boots or long stockings, and a veil complete the protective costume.

Snow mitts or driving gloves tend to make the hands clumsy and probably won't prevent stings. Commercial bee gloves ($5—$9) also make the hands clumsy, but provide good protection. There are many types of veils available ($5—$12), all of which work well if worn properly.

If one is always in a position to work his bees on hot sunny afternoons, when they are gathering nectar at full pace, no protective gear is needed. On such a day it is possible (and delightful) to work the whole afternoon in a T-shirt or no shirt and never get stung. Work schedules being what they are, though, one eventually finds himself in the bee yard on less than ideal days when the bees have more time to be defensive. Then a veil and long sleeves are in order.

A hive tool is the best and only piece of equipment for taking a beehive apart, which one must do occasionally. It's a piece of flat steel bent 90 degrees at one end; with it a person can pry frames out of hive bodies, break apart hive bodies that the bees have glued together, scrape bottom boards, open bottles and repair trucks. It's a beekeeping necessity at $2.50.

A smoker is a metal can fitted with bellows and a nozzle for directing the smoke ($5—$9). A smudgy combustion of wood chips, burlap, bailer twine, etc. can be maintained in the can by pumping the bellows, and the smoke directed over the bees just prior to working a colony. Smoking causes bees to gorge themselves with honey, which keeps them busy and out of the way for awhile, and, as rumour has it, makes them more tractable.

The hive itself consists of (from the ground up): a hive stand, a bottom board, two hive bodies with frames known as the brood chamber, a queen excluder, honey supers, an inner cover and a telescoping outer cover (see diagram). All these components can be purchased unassembled for about $60. Anyone with access to woodworking tools can save money by making some items himself. Hive stands are extremely easy to make, as they generally consist of a couple of fence posts, two-by-fours, bricks, or whatever, placed under the hive to keep it off the ground.

Bottom boards, inner covers and telescoping covers are quite simple building projects, but buy one of each from a beekeepers' supply house and copy their dimensions.

Hive bodies and frames can be made on a table saw, but *all* inside dimensions are critical. If any piece is off size, the bees will glue frames and boxes together much tighter than usual. Unless one is set up to duplicate factory-made frames and boxes exactly, it is best to buy them.

Each hive will hold 10 frames, though many commercial beekeepers use only nine per box for ease of manipulation. Each frame must be wired and have a sheet of wax foundation set into it. The foundation is also wired for strength and is imprinted with the bees' hexagonal comb pattern, so that the bees will build their combs within the frames, allowing for the frames to be removed from the hive body for inspection and extraction.

The "queen excluder" is a flat sheet of zinc, plastic, or welded wires with holes through which the bees can enter the honey supers. The queen, with her wider body, cannot pass through the excluder, and so is confined to the lower two boxes or brood chambers. Thus her eggs, the developing bees, and the pollen and honey which they are fed are contained in the brood chambers, and only honey is carried up into the honey supers. Complex machines are required to produce both the queen excluder and foundation, so these must be purchased from a supply house.

LANGSTROTH EQUIPMENT

Whether one makes or buys equipment, it is wise to stay with standard Langstroth-size hives. An incredible number of shapes and designs for beehives have been used over the years, but nearly everyone has settled on a hive body that is 20" long by 16-5/8" wide by 9½" high. This is known as the Langstroth Full Depth Super, after its inventor, Rev. L.L. Langstroth, the father of modern beekeeping (circa 1850).

The problem with off-size equipment is that the frames will not fit into a Langstroth hive body or into any of the extractors that are readily available. Nor can replacement parts be obtained from any beekeepers' supply house. A beekeeper sometimes wishes to transfer frames from one colony to another or between the brood chamber and honey supers of the same colony. This is only possible if all hive parts are standardized. Non-standard equipment reduces the value of a colony of bees to nearly zero when buying or selling.

All the components described thus far comprise the colony's year-round furniture. Most beekeepers leave each colony with two brood chambers, bottom, covers and excluder in the apiary all winter. In the fall, the hive should contain at least 65 pounds of honey or sugar syrup to provide for the bees' winter and early spring needs. On the first spring visit to the apiary, usually in April, the bees will generally be found to occupy only the upper of the two brood chambers. As the season progresses, a healthy colony will build up to occupy both brood chambers, and by mid-May will probably require more space to store the year's crop of honey.

It is at this stage that the honey supers are put on, and the contents of those honey supers is what beekeeping is all about. Bees are such inveterate workers that they don't stop producing once they have secured sufficient honey to see them through the coming winter, they just keep piling up surplus as long as the flowers bloom and yield nectar. It's the beekeeper's job to provide storage space for this surplus and eventually reap the rewards of the year's work.

COMB SECTION HONEY

The beginning beekeeper must make a decision early in the game whether he wishes to produce comb honey or liquid (extracted) honey. This decision will determine what type of supers and honey processing equipment he will need to build or buy. In either case, one should have on hand honey supers with at least the storage capacity of two full-depth Langstroth supers per colony. A great deal of honey production and innumerable swarms have been lost because beekeepers failed to provide adequate storage space for surplus honey (crowded bees will swarm and head for more spacious parts).

Both comb and extracted honey supers are made in standard Langstroth width and length but in varying depths, from 5-1/8" to 7½". These are known as shallow supers, and are appropriate for anyone who can't handle the 70 to 90 pounds that a full-depth super full of honey can weigh. The main drawback to shallow supers is that the frames will not fit in a full-depth honey super or brood chamber. Also, it takes the same amount of crank turning to spin out an extractor load of shallow frames as full-depth frames, and the yield is substantially less.

The production of comb honey requires less equipment and less time than does extracted

Clockwise, from top left: *Wetting a package of bees for installation (a warm weather trick only); the queen, largest bee in the colony, with her attendants; typical section of worker brood; development of the honey bee.* **Bottom left,** *supercedure cells, from which the new queen will emerge;* **centre left,** *brood comb with oblong bee eggs.*

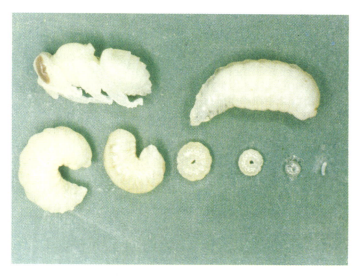

honey. The end product, however, is in a rather awkward form for storage and use in cooking, canning or baking.

Colonies used for comb honey production require more attention in early summer, as the bees are often more inclined to swarm than to move up into comb honey supers. There is less work at harvest time, though, since the sections of comb honey can simply be removed from the frames, wrapped and stored; no extractor or extracting process is required.

Two types of comb honey are produced, comb sections and cut comb honey. The former method uses a special type of frame which holds a number of small (4''x5''x1½'') wooden sections. When the bees have filled these open-ended boxes with honey and capped the cells with wax, the sections are quite an attractive and marketable item.

Cut comb honey is produced in ordinary extractor or brood chamber frames which have been fitted and capped by the bees, they can be removed and sections of "chunk honey" of varying sizes cut out with a sharp knife. The empty frames can be cleaned up and reused the next year.

Both section and cut comb honey are delicious treats, as the wax is eaten along with the honey and adds to the flavour and freshness. However, neither type of comb honey lends itself to cooking, canning or baking, where the wax is not always a welcome addition to the recipe. Pressing honey out of combs is at best a messy and wasteful operation. In addition, comb honey must be kept in a freezer if it is to be stored any length of time, to prevent the ravages of wax moth larvae. A hundred pounds of comb honey will occupy quite a volume of scarce and expensive freezer space.

While the production of liquid honey requires more equipment and effort, the resulting product is easier to store and use, and the frames used in honey supers can be interchanged with those in the brood chambers.

Some beekeepers involved in liquid honey production simply pile two or three honey supers on each colony in the spring and forget about the bees until harvest time. Though this approach will not generally result in the highest possible yield per colony, it does maximize production per man hour spent working bees. In any case, the use of colonies to produce extractable honey will reduce early summer management and increase the beekeeper's late summer or early fall labour. In fact, in a large commercial operation, the extracting process during August and September can approach pandemonium.

A beginning beekeeper will need two fairly major pieces of equipment to extract his honey: an uncapping knife and an extractor. Uncapping is the process of slicing the wax cappings from the face of a frame of honey. A kitchen knife or "cold" knife ($5) can be used for this purpose, but if one has any number of frames to uncap (realistically, more than one!) he'll want a heated uncapping knife ($22—$25). These knives are heated either by steam or electricity. The choice depends on the individual's access to hydro or steam under pressure.

Another uncapping tool is a "cappings scratcher" ($4), which punctures the cappings instead of slicing them off. Again, any amount of uncapping with a scratcher will prove extremely tiresome and will also leave a lot of fine wax particles in the extracted honey.

The function of the extractor is to spin the honey out of the uncapped frame by centrifugal force. They come in a variety of sizes that take from two to more than 100 frames per spinning. A two-or-four-frame, hand-cranked extractor ($70—$100) would be adequate for an operation of up to 20 to 30 colonies, though many small commercial beekeepers spin out huge crops (50 to 100 tons) with these little machines every year.

WINTER WORKS PROJECTS

So, if the decicion is, "Yes, let's get some bees," what do you do now? As I've said, winter is a time for reading and planning, so invest in one of the beekeeping books listed at the end of this article and perhaps subscribe to one of the magazines, and curl up by the fire.

The pace of living, and especially beekeeping, picks up rather quickly in the spring, so it is best to get all necessary equipment assembled, painted, and ready to go while snow and cold weather enforce some leisure time on us. Send for catalogues and price lists from several of the beekeepers' supply houses and write out your order.

Even if you arrange to buy established colonies from a beekeeper, honey supers are often not included in such deals, so you may have to buy and assemble these. If possible, pick up that first order personally. Some suppliers offer discounts for cash-and-carry, and most will pass on a few hints on assembling their frames and hive bodies.

Give all exposed surfaces of hive bodies, bottom boards and covers a coat or two of a good weather-resistant paint. By this time spring fever (and perhaps bee fever) will be setting in.

Sources

BEEKEEPER'S SUPPLY HOUSES

Hodgson Bee Supplies Ltd.
7925 - 13th Avenue
New Westminster, British Columbia

The Walter T. Kelley Co.
Clarkson, Kentucky

Lorh Co-Operative Industries
Box 1921
Tisdale, Saskatchewan

F.W. Jones & Son Ltd.
Box 1230
Bedford, Quebec
OR
68 Tycos Drive
Toronto, Ontario

Dadant & Sons Inc.
Hamilton, Illinois

Bee-Care Supplies Ltd.
22 Morton Avenue
Brantford, Ontario

Cooks Bee Supplies
91 Edward Street
Aurora, Ontario

Benson Bee Supplies
Metcalfe, Ontario

W.A. Chrysler & Son
1010 Richmond Street
Chatham, Ontario

Marioara Apiaries
129 Ossington Avenue
Toronto, Ontario

The A.I. Root Company
Box 706
Medina, Ohio

PERIODICALS

Canadian Beekeeping
Orono, Ontario
($5/year)

Gleanings in Bee Culture
A.I. Root Company
($5.50/year)

American Bee Journal
Dadant & Sons
($6.50/year)

The Speedy Bee
Route 1, Box G-27
Jessup, Georgia
($3/year)

QUEEN BREEDERS & SUPPLIERS OF LIVE PACKAGED BEES

Beemaster's
Route 1, Box 502-E
Arroyo Grande, California

C.E.M. Apiaries
Box 303
Grass Valley, California

The Stover Apiaries Inc.
Mayhew, Mississippi

Weaver Apiaries
Route 1, Box 111
Navasota, Texas

York Bee Company
Box 307
Jessup, Georgia

BEEKEEPING BOOKS

The ABC and XYZ Of Bee Culture
A.I. Root Company — $10.00

The Hive and the Honey Bee
Dadant & Sons — $10.00

The Complete Guide to Beekeeping
from The Speedy Bee — $8.00

First Lessons in Beekeeping
Dadant & Sons — $1.00

How To Keep Bees and Sell Honey
Walter T. Kelley — $1.50

Honey Plants Manual
Bee-Care Supplies Ltd. — $1.60

Practical Beekeeping
available from Harrowsmith Books — $5.95

A note on the books and periodicals: *The ABC and XYZ Of Bee Culture* is generally considered to be the beekeeper's bible. It gives encyclopaedic coverage of the subject and is well worth the money.

The other texts and handbooks listed are written by recognized authorities in the field and are also good value for the money. *Gleanings In Bee Culture* carries more articles of interest to the hobbyist than the other periodicals, which cater to the commercial beekeeper to a greater extent. All periodicals carry a great deal of advertising, which can put the beginner in touch with additional resources. Most periodicals will send a sample copy on request.

And now for the bad news ...

By Diane Birch

Thinking back to our first hive of bees, I often wonder how we decided to take up beekeeping for a living. That first hive was a terror. We wanted to raise most of our own food, and a hive of bees seemed like a good idea. Of course, we knew that the little buggers stung, but we thought we would just get all dressed up, and everything would be fine.

We talked about this to the beekeeper from whom we bought our first hive. He just laughed and said,"Oh, you can't keep bees without getting stung." We should have listened.

Bee gloves are almost useless. If they are thick enough that the bees can't sting through them, they're too thick to wear without making your hands very clumsy. This means you crush a lot of bees moving the frames around, including, perhaps, your queen.

This, of course, makes the bees angry and much more likely to sting. Also, the gloves themselves will anger the bees. As more and more bees sting them, the gloves get saturated with bee venom, and the smell of this venom causes bees to identify you as an enemy and try to sting you. A lot of bee gloves are sold, but most end up seeing only occasional use.

Protective clothing also has its drawbacks. It has to be thick to prevent stings — two layers are better than one. But we now find that working the bees on a hot summer day with that much clothing is really more uncomfortable than a few stings.

I remember a day when we had had the bees for about a month. We had looked through the hives two or three times and were starting to feel courageous. But that day one enterprising bee — perhaps two or three, I couldn't be sure — found its way under my bee veil. This was extremely disconcerting, especially since the veil prevented me from being able to get at the bee. And if one or two bees could get in, why not more?

In blind panic, I ran away from the hive, taking off the bee veil as I ran. As I removed it, however, all the bees who had been waiting outside just wishing they could get at me now saw their opportunity. Eight or nine of them zapped me all at the same time. Dave and I spent the rest of the summer arguing about whose turn it was to work on the bees.

However, at some point in the long winter we must have decided that it wasn't as bad as all that. The next spring we bought 20 more hives, and now five years later, we have 130.

Beekeeping, of course, has become quite popular lately, and not everyone who tries finds that it suits him. They have a saying at the Apiculture Department of the University of Guelph, which goes something like, "There'll be a lot of bee hives sitting on the fence posts in a few years. . ." Generally people who sell their hives say they find the stings too bothersome, although there may be other problems, such as bears, disease, or irate neighbours. A friend of mine recently confessed that the only time anyone looked inside his hives in the year that he owned them was one day when I came over and went through them for him.

The best defence against investing in honey bees only to give them up after a season is to talk to other beekeepers, and to read everything you can get your hands on about apiculture. You'll find that bees are fascinating creatures and, even if you never get around to starting a hive, your time will have been well-spent. So, with the next big snowstorm, plan to settle down in a nice warm chair and take up a book about bees. While the winds howl outside you can dream about warm, lazy fields of yellow dandelions buzzing with bees — your bees — and worry about the stings when the time comes.

The Spring Beeyard

Installing package bees and other challenges for the beginning apiarist

by Michael Shook

Curiously you can still meet the occasional old-timer who will tell you that the centre of a bee hive's social structure is the "king bee." If any of his better informed cohorts are about, however, they will say, of the king bee, that "there is no such animal."

Rather, each hive or colony has its queen, whose own royal title, in fact, may be somewhat overblown and misleading. Although she is the most important bee in the colony, the queen's function has nothing to do with ruling or directing the activities of other bees. She is simply a frenetic egg factory — laying up to 2,000 eggs per day to maintain and increase the strength of a colony.

First-time apiarists who start with new hives and packaged bees from the bee-rich southern United States will be interested to know that the queen they receive — seemingly cramped in a tiny wire and wood cage not much bigger than a penny match box — is pre-mated and ready to begin laying.

She has, in fact, led quite a life for her young age. Ready to mate only five days after hatching, the young queen flies off to mate with drones waiting high in the air. This elevated romance, which may last for several days, results in the virgin queen receiving eight to 12 different drones (males) and storing some seven million sperm — enough to fertilize her eggs for well over a year.

The queen's role in the spring is especially critical, for if she loses her fecundity, flies off with a swarm or is killed, the hive's honey production for the year may be lost or lowered considerably. For this reason, it is important that the beginning beekeeper — unlike our forebears who simply plundered honey trees — understand what is going on in his hives.

CRUCIAL BROOD

Except in unusual circumstances, each colony of bees contains only one queen. The worker bees, which number 80,000 or more at the height of the season, recognize their own queen by her highly individual odour and will immediately murder any strange queen who might find her way into the colony.

Unlike the workers and queen, the male drones seem to drift about from colony to colony in the apiary, and are tolerated, fed and cared for wherever they alight. It is quite an idyllic existence, but as the season draws to a close and cool temperatures put an end to the year's matings, the drones must pay for their sloth: They are chased from the warm hive and quickly perish in the cool autumn nights.

The workers, on the other hand, usually die in the midst of their work — most often while gathering nectar in the fields. They are sexually underdeveloped females and the most industrious members of the hive. Any female egg laid by the queen can develop into a queen or a worker, depending on how the developing larva is fed and housed.

When the bees desire a new queen, they select one or more female larvae, build large, specially shaped cells around them and lavishly supply the developing grub with "royal jelly," a substance produced in the hypopharyngeal or brood-food glands in the workers' heads. Milky in appearance, royal jelly has long been sought by men who believed it possessed the power to restore youth — or at least the power to bring a high price from wealthy women. In any case, the worker bee larvae get royal jelly for just two days, while the queen receives it throughout her larval and adult life.

The queen is responsible for the spring build-up of the hive, in which the number of workers soars, giving the hive the strength to gather a surplus of honey. A single, whitish egg is laid in each hexagonally shaped brood comb cell, and it hatches in three days.

Once fully developed, the worker bee spends two weeks in the hive engaged in nursing, janitorial and policing functions. At the end of this time, each bee takes a "play flight," circling about the hive, memorizing every detail of its

The hive tool is a simple machine of many uses, including scraping propolis from the frame edges.

environment and geography as she spirals higher and higher. She then joins the field force in foraging for pollen, nectar and water, and in these pursuits literally works herself to death, sometimes in as few as six weeks during the heavy nectar flows.

Each time the worker obtains a full load in the field, she uses the memory map imprinted during her play flight to make a "beeline" (yes) to her hive. It is an impressive sight to see those thousands of bees, heavily loaded with nectar or pollen, describing that straight-line vector from flower to home.

QUEEN WATCH

The queen is the largest bee in the hive, but it takes a good eye and much experience to find her quickly. Fortunately, it is not necessary to see the queen to know that she is present and doing her job well. We can easily evaluate her performance by checking the brood pattern she is producing. Generally, the three centre frames of the upper brood chamber will give us a good indication: Brood should form a compact circle on the face of each frame. It should not be scat-

tered about randomly, nor should there be a large number of missed cells within the circle, which indicates an old or failing queen. The presence of eggs and young larvae tells us that the queen is alive and functioning.

The queen, shipped in a package from the south, arrives in a small container with several workers, and, because she has been with the packaged bees for at least 48 hours, they usually accept her as the mother of their new colony, even though she was raised separately.

Occasionally, however, a package colony will take into its (collective) head to do away with its royal travelling companion. Also, it can happen that an injured or unmated queen may be shipped from the south. It is thus necessary to check a new colony two weeks after installation to insure that the queen has been released from her cage and is laying eggs. At this time all stages of brood should be found, and nearly every cell within the small brood area should contain an egg, or sealed pupal brood.

BEEYARDS

We have got slightly ahead of ourselves, and this seems the appropriate spot to discuss the installation of package bees.

The bees will arrive in screened boxes, containing from two to five pounds of bees (at about 5,000 bees per pound). A two-pound package of bees, including the queen, may cost about $23; a single queen — about $7. Here in southern Canada, bees should be scheduled to arrive from about mid-April to early May (ask an experienced apiarist in your area about the best time to install package bees in your locale).

Your hives should be painted and ready to go by the time the bees arrive, for the bees cannot be kept long in the shipping containers. A thick syrup solution should be prepared immediately upon their arrival (five pounds table sugar in a clean gallon jug, filled with hot water and shaken). You should also have on hand a feeder pail for each package expected. These feeder pails can be any tight-lidded pail or wide-mouth jar with about six nail holes punched in the lid. Feed the bees immediately by brushing the screens of the packages with sugar syrup, and repeat this twice each day if the bees are not installed immediately.

"Haste makes waste" applies to most beekeeping manipulations, but not to the installation of package bees. A rainy, cloudy or even snowy day is fine for installing them, but herein lies a twist. Don't leave the bees exposed to cold or precipitation any longer than necessary; have everything you will need right on hand, pop the bees into the hive and close them up.

Everything here means a full feeder pail, a bottom board, a hive entrance reducer, one empty hive body with nine frames and a cover for each package. Also, a sharp, pointed knife, a small nail and your bee veil. It is best to have the hives set up with entrance reducers nailed in place before the packages arrive. An entrance reducer, by the way, is simply a small piece of wood the length of the hive entrance with a one-half-by-one-inch notch in the bottom edge. It cuts down the amount of cool air entering the hive.

In place of the hive's normal inner cover, have on hand a piece of cardboard measuring 15 inches by 18 inches with a hole in the centre to fit the top of your feeder pail.

Even on a warm, sunny day, you should move quickly once you reach the hive with the package. Most of the bees that fly during installation will tend to congregate in one hive, thereby reducing the strength of the others.

Place one empty hive body on top of the brood chamber full of frames, then pick up a package of bees and, with the small, commercial feeder can showing on top, give the box a gentle thump to shake the bees down. Now pull out the feeder can and shake any bees clinging to it onto the empty frames. The queen cage is hanging inside by a wire; remove it and, again, shake any clinging bees into the hive. Replace the feed-er can to block the exit of bees in the still-filled package.

One end of the queen cage is filled with candy, while the queen and her attendants occupy the other cavity. Check that the queen is alive, remove the cork from the candy end, push a nail through the candy plug and then pinch the cage, candy end up, between two of the middle frames of the brood chamber. (The candy end is kept upwards, so that a dead worker in the cage doesn't plug the queen's escape.)

If one of your queens is dead, combine her bees with another package, as they are helpless without her. The queen is recognized by her large size and her shiny back — the workers have hairy backs. Any reputable bee breeder should replace a package with a dead queen.

Again remove the feeder can from the package and dump the bees through the feeder hole onto the frames by whatever thumping, shaking or tearing of screens seems appropriate. (If you have just fed the bees with sugar syrup, they will be more docile at this point.)

Now place the cardboard on top of the frames with the hole positioned above the queen cage. Invert your feeder pail over the hole and replace the cover of the hive body. If some bees have refused to leave the shipping package, set the package inside the empty hive body as well. A narrow space between the cardboard and the edge of the hive body should be arranged so that these bees can crawl down to join their companions.

No sooner than a week after the installation, examine the combs for brood. Use very little smoke and cause as little disturbance as possible. If the queen has not managed to escape the tiny shipping cage (worker bees usually eat through the candy plug to free her), set her loose by hand.

If neither the queen nor brood can be found, she has probably been killed. The bees, in this case, should be united with another colony. The best way to accomplish this is to shake all of the queenless bees in front of the colony next door and remove all vestiges of the old hive. Any frames containing honey, brood or pollen can be placed in an active colony once the bees are shaken or brushed away.

If brood is present in the new package colony, but the pattern is spotty (one-third to one-half of the cells missed), if all the pupa have rounded, protruding caps; or if numerous eggs have been laid in a single cell, the colony has some form of queen trouble.

The queen may be injured or inadequately mated or may be missing altogether, so that a worker bee has attempted to take over the egg laying duties (Worker bees can develop laying abilities, but, unable to mate, produce unfertilized eggs that grow into drones.)

In any case, the queen should be found and destroyed and the bees united with another colony. Experience shows that some form of queen trouble may be found in as many as 25 per cent of all packages installed (thus the advantage of having at least two colonies at all times.)

165

Of my 16 beeyards, only six are in ideal locations — on the south or east side of a bush where they are protected from the prevailing winds and exposed to the warming rays of the morning sun (the sooner the bees become warm, the sooner they set off to the fields to work).

In addition to the obvious necessity for bee pasture within two miles of the hive, availability of water is another priority. Bees require a fair quantity of water and, ideally, the apiary should be located within a quarter mile of a lake, river or small stream. If water is scarce, bees will frequent any livestock troughs or swimming pools in the area, and these facilities lead to a lot of drowned bees and a clearly expressed annoyance by swimmers and drinkers.

My least favourable location is in a commercial orchard, because, no matter how careful the operator is with his pesticides, the bee population always suffers during the critical spring build-up (of hive strength).

Another location to avoid is near a town dump, where used honey containers are often deposited. The bees will rob any drop of honey from these jars or cans and carry it home, possibly carrying along bacteria and the start of a bee disease.

ESTABLISHED COLONIES

For colonies that have wintered-over, or that you have bought and transferred to your property, an inspection is in order in April, or even late March if the weather is warm.

On this first visit, we may not even remove the hive covers — anything more than a brief peek into the colonies at this point will only chill the brood. With a little experience, however, one can learn to gauge the amount of honey left in the colony simply by hefting the back of the hive.

The empty equipment of a two-storey colony will weigh about 65 pounds, so if our calibrated hefting hand says 80 pounds, the colony has adequate stores for another two weeks. If, on the other hand, the colony feels quite light, it is time to apply a feeder can of syrup in the same way as described for the package bees.

On this first visit, it is also wise to remove entrance reducers, clear the colony entrance by inserting a stick or wire and scraping out any dead bees or debris that may block the passage. Entrance reducers should be replaced after this operation and remain in place until warm weather settles in.

When afternoon temperatures begin to exceed 15 degrees (C) (60 degrees F), more thorough colony inspections are carried out, but until really warm weather arrives, we must take care not to leave frames containing brood exposed to the outside air any longer than necessary. The developing brood can easily become chilled and die, and early in the season, each and every bee is important to the colony's build-up.

In addition to ensuring that there is a viable queen, I like to ensure that each colony has at least two full frames of honey at all times.

Space does not permit a full discussion of all the various diseases that attack bees and their brood. Local bee inspectors are a good source of information, all the major reference books provide exhaustive treatment of the topic, and provincial and state universities of agriculture should be very helpful in providing information booklets.

Suffice it to say that all apiaries must be registered with the provincial or state inspector, and that an ounce of prevention is far better than the cure (i.e. burning the diseased colony).

The spring prevention consists of a heaping tablespoon of an equal mixture of powdered sugar and Terramycin sprinkled over the top bars of each colony early in the spring.

The bees mix the Terramycin (a trade name for oxytetracycline, an antibiotic that is effective against the spores of American foulbrood) with the honey and pollen that is fed to developing larvae, and this prevents the disease from attacking the brood. The yearly dose of Terramycin should be given at least a week prior to applying the first honey super — this assures that the drug is not stored in honey that will be harvested.

SWARM CONTROL

Through the early spring, the beekeeper's major concern is to keep his bees from starvation and to coax them to increase, gradually, their brood rearing. But suddenly, sometime in May, the days warm up and the fields come alive with the yellow heads of dandelions and the buzzing of bees.

At this point the colonies seem to explode almost overnight as thousands of new bees emerge from the brood cells and the hive fills up with the sweet, fresh honey. Swarming season is upon us.

This most exciting of nature's phenomena is simply the bees' method of reproduction. Though queen bees are prolific egg-layers, the bees that develop from their eggs only serve to further populate the same colony (worker bees do not normally travel from hive to hive). In nature, the only way for bees to generate a new colony is to cast a swarm.

This they will do on a bright, warm day, especially after several days of rain have kept the field bees in the hive, plotting and scheming. When the swarming impulse strikes, the bees make preparations by building a large number of queen cells, so that the bees who remain in the hive will soon have a new monarch.

On the first fine day after these swarm cells have been completed, about half the population of the colony will emerge, along with the old queen, and swarm about — filling the air with an incredible buzzing. This first flight won't take them far, and they often cluster on a branch or post near the apiary while scout bees fly out in all directions searching for a new home.

Once this site is found, perhaps within an hour or up to four or more days after the swarm has emerged, the bees pick up and move there. With them goes the hope for a bumper crop of honey, for half of the workers have left the parent colony. At least two weeks will pass before the new queen can be laying eggs to replenish the population.

It is up to the beekeeper, then, to prevent swarming. There are a number of techniques to achieve this end, but I will describe my own plan of swarm control. It is reasonably effective and a fairly popular method among commercial beekeepers, because it can be applied to a large number of colonies in short order.

No one really knows what causes one colony to swarm while others of the same strength and in the same apiary go quietly about their business. The age of the queen is one factor, and colonies which are re-queened each second year are less likely to swarm than those left to raise a new queen on their own.

In this process, queen cells are started when the old queen begins to fail. The new queen begins to replace, or supercede, the old one. These queen cells are known as supercedure cells, and are built in small numbers on the face of the comb.

Supercedure cells (see photo) are not a cause for concern, and the colony should be left alone to raise its new queen. Swarm cells, on the other hand, are quite numerous and generally located at the bottom of the frames in the upper brood chamber. If swarm cells are discovered, a reversal of brood chambers is in order.

It is the bees' natural inclination to work *upward* from an established brood area. If the lower chambers become packed with brood, pollen and honey, the queen has nowhere to lay eggs, and the workers have nowhere to store honey, the swarming instinct will soon follow.

It is a simple matter to alleviate the crowding by simply reversing the positions of the two brood chambers. Thus the bees will find new space in the now empty combs of the former bottom chamber and will proceed to fill it with brood and honey.

It is not a safe rule to wait until the dandelion flow to reverse brood chambers, however. The best approach is to check the colonies every 10 days or two weeks during May and June and to reverse the chambers when the brood has extended slightly into the lower box. Too-early reversing results in narrow brood patterns and weak colonies, as the bees will not extend their activities to the outside frames, but work upward from a limited base.

Some beekeepers perform a second reversal two weeks after the first, but I do this only with individual hives that show an inclination to swarm. It is an easy matter to check all colonies for swarm cells by tipping up the upper hive body. A puff of smoke between the two brood chambers will clear the bees away so that any swarm cells on the bottom of the upper frames are visible. If swarm cells are present, a second

reversal and the addition of a honey super may get the bees back to work.

However, some colonies will persist in the swarming activities, and more drastic action will be called for. Such a colony can be artificially swarmed as follows:

First, find the queen. If she is nowhere to be found, and there appear to be fewer bees than usual in the hive, put the cover back on the hive and go fishing. They have already swarmed.

If you find the queen, place the frame she is on in a hive body of empty combs on the exact location where the colony originally stood. This will be your swarm and it will benefit by the addition of a couple of frames of brood from the parent colony and second box of empty combs.

Now, carry all the other frames of brood and honey five or 10 feet away and set these up as a second hive, bees and all. The field bees will all return to the original hive stand and carry on under the impression that they've swarmed. *Voilà!*

You should deal tenderly with the queen cells while carrying them to their new location; they are extremely delicate and even a soft bump can maim the occupants. In three weeks a new queen should have been mated and will be laying in this second unit. If the operation is performed early in the season, both colonies should produce a surplus of honey. (Package bee colonies, if properly managed, should have no tendency to swarm the first year.)

And, speaking of surplus honey, we've arrived at the simplest operation in beekeeping — supering. After the majority of dandelion blooms has passed, and the bees have built up to occupy both brood chambers, a super of light combs, foundation or comb sections may be placed above the second brood chamber to hold the surplus honey that will soon be flowing.

Underhill's Leaf Hive, 1865-vintage predecessor to the modern bee hive.

Autumn Beekeeping

*Fall colony management assures healthy bees
and heavy production next spring*

By Michael Shook

One of the surest of the old country signs of winter's approach is the annual scurry of field mice to find a warm, dry place to escape the cold and snow. To beekeepers, this fall migration is a signal that several important fall management activities should be getting under way in the beeyards.

The mice themselves must be dealt with, for they are especially partial to the snugness, heat and ready food supply of a beehive in winter. With the increasing cold, the bees become less and less active, and their unguarded entrances are an open invitation to mousey marauders that can ravage combs, honey and bees.

To foil the mice, apiarists simply add a wooden or metal device called an entrance reducer that narrows the hive opening to no more than one-half of an inch high.

Too, fall is the time to assure that each colony goes into winter with adequate stores of food to survive the nectarless cold season. The vast majority of colonies in Canada and the northern United States are wintered in two brood chambers, and these hives generally require some supplemental feed in the fall, as much of the comb area is devoted to brood rearing rather than honey storage.

For successful wintering, a colony should go into the winter with at least 65 pounds of feed. Although the bees generally will not consume this amount during the winter, they will depend on stored honey to get through the critical early spring period when there is no natural honey flow and early brood rearing steps up food consumption.

The equipment which makes up a two-storey colony weighs roughly 65 pounds when empty, so a winter-ready colony with adequate stores of feed should weigh in at about 130 pounds. Some beekeepers carry a tripod scale into the apiaries and actually weigh each colony, but most of us rely simply on hefting the back of each colony and making an educated guess. When in doubt, it is best to overfeed — the bees won't waste any excess, but use it in their spring population build-up.

The supplemental feed itself is sugar syrup, a fact that may come as a surprise to honey fanatics. The feed process is carried out in the same manner as described in the article on installing package bees but to bring the weight of the colony up to 130 pounds, several feeders may be required in some colonies.

Although I don't object to feeding bees sugar syrup on organic grounds, I do object to the job itself. It is sticky, (the standard mixture is two parts sugar to one part water, by weight) unrewarding work, carried out at a time of year when bee populations are high and tempers short, due to the absence of natural honey flows. One should attempt to complete the feeding operations before cold weather sets in, thus allowing the bees time to arrange their stores for best winter access. At 57 degrees Fahrenheit, the bees start to form their winter cluster, and by 43 degrees they will all be in the cluster and not actively feeding.

Last year we did very little feeding, but chose instead to leave a third super, full of honey, on many of the two-storey colonies. For a commercial beekeeper this practice represents a sizeable investment, as several tons of marketable honey are left out in the apiaries for the bees' winter feed.

Nevertheless, we found the honey-fed, three-storey colonies to be markedly stronger than their two-box neighbours; with the great surplus of feed they were able to build up much faster in the spring. The bonus came from the fact that almost all of these colonies were able to be split at least once, yielding an average of one new productive colony of bees for each extra box of honey left in the beeyards over the winter.

THWARTING FOULBROOD

Another essential part of fall bee management is assuring that no hives are infected with American foulbrood, which means an almost sure death to the colony. As the disease will be further spread by bees robbing honey stores from the dead colony, a preventative programme is essential.

In the control of brood diseases, an ounce of prevention actually converts to a quarter of a teaspoon of sodium sulfathiazole in a gallon of sugar syrup for each colony. This drug is available from any beekeeping supply house, and is effective against foulbrood and other disease organisms. Many beekeepers routinely feed a gallon of medicated syrup to each colony in the fall, whether or not the supplementary feed is needed.

It may seem out of line to recommend feeding honey bees with white sugar and drugs, and some purists may believe the honey produced by such bees to be adulterated. It is not — any drugs or sugar fed in the fall will be consumed during the winter.

In fact, the bees which consumed the artificial feed and up to three succeeding generations of bees will live out their natural lives and die before any honey is again made for human consumption. There is simply no way that these substances can find their way into the honey supers, even in trace amounts. For those of us interested in emphasizing natural methods, the fact to be considered is that the alternative to sugar syrup and sulfa drugs in many cases is dead bee colonies.

WRAPPING OPTIONS

There are several schools of thought regarding the packing of beehives for winter, with success claimed by proponents of each method. The degrees of packing range from none at all to full wrapping to apiarists who truck their colonies to wintering cellars or climate-controlled buildings (an approach that is usually justified only in areas of extremely harsh winter conditions).

The bees themselves are able to generate sufficient heat in the quietly humming cluster to keep the queen at a toasty 94 degrees. Some beekeepers work to conserve this heat by slipping a tarpaper sleeve down over each hive, with a space of several inches left between the hive walls and the sleeve. This jacket is then filled with dead leaves, wood shavings or other insulating material, with a generous layer of insulation placed on top of the hive before the sleeve is folded closed and secured at the top.

However, recent research indicates that this method of wrapping creates a deep-freeze effect, keeping the colony cool on the few warm winter days when flight is possible. The sun's rays cannot penetrate the insulation, and the bees doze on — unaware of the chance to make a cleansing trip outdoors. Studies show that, without the occasional winter flight, feces accumulate in the bees' digestive tract, dysentery can develop and the colony can be severely weakened or dead by spring.

The most used winter packing, therefore, is a thin sheet of builder's felt tied or stapled tightly around each hive. The black surface absorbs the solar radiation, heating the colony and permitting the bees to fly and also to move feed into the cluster area.

SNOW CAVES

Tarpaper and black plastic are not suitable materials for wrapping hives in this manner, for they can seal off the flow of air and suffocate the bees. The function of the porous felt is not insulation or wind protection, as some think, but mainly to provide a dark heat-absorbing surface. Painting the hives black can fulfill the same function, but, as in tennis, beekeeping traditionalists shudder at the thought of anything but clean white. This brings us to the lazy man's method of wintering bees.

I've experimented with the no-wrapping-at-all approach by leaving random colonies unpacked, and depending on the type of winter, have found that there may be a noticeable difference between packed and unpacked colonies — with the unpacked hives often showing the advantage.

In an exposed apiary, the packing may pay off, but in beeyards which are protected or which normally are covered by snow, a layer of black paper may simply serve to block the passage of bees and air into the hive.

I make it a practice to provide an upper exit/entrance for the bees' use when the regular bottom opening becomes blocked with snow and dead bees. Probably the best type of upper entrance is a notch cut in the rim of the inner cover. This will allow moist air and carbon dioxide — by-products of honey consumption and respiration — to escape.

I also keep the second brood chamber pushed back about three-quarters of an inch to provide a middle entrance, and many of my brood chambers have holes bored in the front to provide additional ventilation. The bees are their own best keepers, and if their owner gives them too much ventilation, they quickly seal some of the openings with propolis.

I have dug through two feet of snow to reach the covers of three-storey hives, only to find the bees flying about in a snug little snow cave which the colonies' own heat had melted around the hive walls.

These colonies had young queens, plenty of stored food, good ventilation and no disease. There was no winter loss in that snow-filled yard, and the bees were so strong early in March that I wanted a veil before removing the covers. I hope to winter all my hives under snow this year.

Miriah Is A One Cow Dairy

[Her predecessor was another matter.]

Article and Photography By Hank Reinink

Our first dairy cow was what is known indelicately in the cattle trade as "a three titter" a wild-eyed, undernourished, seven-year-old that we bought at a farm auction where no one else had the nerve (or inexperience) to bid on a cow with only three working quarters.

We named her Minnie, and, although she gave an indifferent amount of milk, she taught us more than a few things about the one-cow dairy business.

For instance, there was the time she came into heat and lost all sense of proportion and propriety and broke out of her pasture to follow us all around the farm, bawling her soul out for something we were in no way equipped to give her. In desperation we ran into the house to phone the artificial insemination unit (a poor substitute from a cow's point of view), and we had to slam the door in Minnie's face or she would have been in the kitchen.

Having been brought up in a herd of Holsteins, Minnie was not accustomed to living as the only cow in a barn full of pigs. She would stand still long enough to be milked, but otherwise it was impossible to get a hand on her. When summer came and we let her out to pasture, she would consent to come to the gate to be milked, but would spend the rest of the day concentrating her efforts on getting through the fences to join the neighbour's herd of cows.

Finally, Minnie's milk production tapered off to a trickle, and she had to make way for Miriah, a purebred Guernsey we purchased from a friend who had raised the animal from calfhood. This was a bovine of a completely different nature — quiet, friendly — a pet — and we jumped at the chance to buy her and have lived happily ever since.

We have not bought any milk, cream, butter or ice cream for well over a year now, and have no intention of doing so until Miriah has to be dried up in preparation for her next calf.

A family cow can be an economical, four-legged food factory if you have the land to sup-

port her — and the willingness to change your life habits to suit her milking schedule.

Susi, my wife, often says that I am very good at starting projects but that she ends up doing the work. I forthrightly admit that that is how our one-cow dairy has worked out. Susi has gradually taken over the regular milking tasks, but this is probably all for the better. For some reason — and you can ask any farmer — cows seem to be more contented with female milkers.

Dairy farmers will also tell you that there is no holiday from milking chores. In their profession it is important to milk twice a day with, as nearly as possible, a 12-hour interval between. The usual schedule is six in the morning and again at six in the afternoon.

In a one-cow situation, however, we have found that it is not as crucial to be on time. We find that three or four hours variation doesn't make that much of a difference, as long as the cow is well fed and in good health. (The exception is the period right after she has calved and the milk flow is most urgent.)

The biggest threat in large dairy herds is mastitis, an infection of the udder, which, if untreated, can permanently harden the udder tissue (this is how Minnie became a "three titter"). Irregular milking can put a strain on the udder, but as long as the interval is no longer than 16 hours there should be few problems (with no other cows around to spread the infection, the family cow is less susceptible to mastitis).

So, in the winter months, when we are often out in the evening, Susi takes on a schedule where she milks at noon and midnight. Dairy farmers may snicker at this, but it works out fine for us and for Miriah. In the summer Susi milks at roughly nine o'clock, but if we have to go out in the evening she may look after Miriah early, perhaps six o'clock, or as late as midnight. So far no problems.

Milking by hand takes a certain amount of practice and getting used to. You have to develop a two-part movement, whereby you squeeze shut the top of the teat with the thumb against the hand before putting pressure on the lower part of the teat to force the milk down and out. The first few times you will have considerable pain in hands, biceps and chest muscles before you are quite finished milking, but persistence will give results any physical fitness programme would be proud of.

In commercial herds the udder is washed with a warm disinfectant solution before the milking starts. In addition to ensuring that no dirt gets into the milk the washing massages the udder, which in turn will cause the cow to "let her milk down". Without the massage she may be reluctant to give any milk at all.

We do wash the udder, although we don't use a disinfectant, and Susi lets the first squirt of milk from each teat miss the wide-mouthed pail she uses, to get rid of any bacteria or dirt that might have entered the opening of the teat.

As soon as she is finished milking, Susi strains the milk through six layers of clean white cloth into one of the three pails she uses on a rotation system in the refrigerator. Just before the next milking, roughly 12 hours later, she skims the cream off that milk with a large flat spoon, and it goes back in the cooler.

Twelve hours later she will skim the cream off again, leaving practically no cream in the milk. It is this twice-skimmed milk that we drink. Any of that milk left unused 12 hours later is fed to the calf. In other words, we never drink any milk that is older than 36 hours.

And we drink a lot of it, summer and winter. We seldom or never have any soft drinks around the house, preferring milk either straight or with a heaping spoonful of Nestle's Quik to make a cold chocolate drink. The four kids love it, and so do we.

A Guernsey cow will give you the best of two worlds. She does not produce as much milk as a Holstein, but you get a lot more cream; she does not produce as much cream as a Jersey, but you get a lot more milk. Miriah provides us with all the milk we can handle, with some to spare for the calf, and she gives us so much cream that we can afford to do all our cooking and baking with butter in addition to what we spread on sandwiches. We get two and one-half gallons (Imperial) of cream from her per week, which produces five to six pounds of butter.

Susi makes butter once a week. The cream, which has been accumulating in a large pan in the fridge, should be slightly sour to make the best butter. In the summer the souring process has already started inside the refrigerator, but in the winter it has to be left out on the counter overnight to get the desired degree of sourness.

The cream should be somewhat warmer than the refrigerator temperature, but somewhat colder than room temperature. So Susi starts churning half an hour or so after she has taken the cream out. She fills the container of the blender about three-quarters full and sets the dial for whipping. The cream will first become whipped cream, but con-

Cleanliness is imperative in any dairy operation, but especially so with a family cow and no pasturization. Straining the milk immediately after each milking, Susi Reinink cools it for 12 hours before skimming off the cream. **RIGHT,** *the once-a-week butter making session which yields up to six pounds of Guernsey butter.* **FAR RIGHT,** *churning with a blender, less romantic than the hand-powered model, but more realistic for this family.*

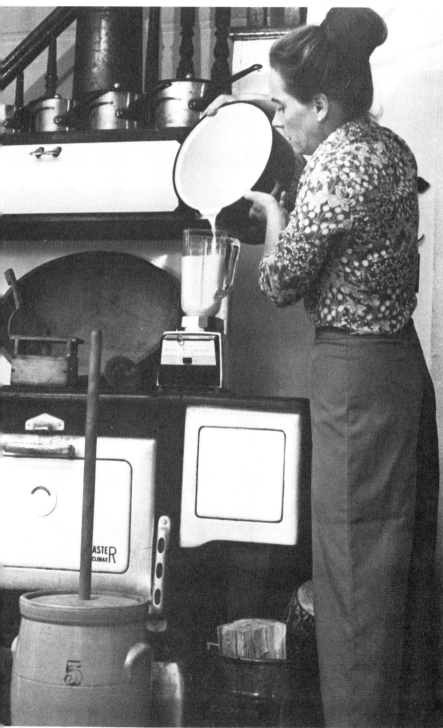

tinued whipping will separate the mass into butter and buttermilk. Susi dumps the whole mess through a fine strainer, leaving the butter in the strainer and letting the buttermilk, also for the calf, run out.

"To me there is no hard-and-fast rule about making butter," Susi says. "Every time I make butter it turns out differently, but I have never lost a batch and we have good butter. When the cream is too sour it is harder to separate the butter from the buttermilk, because the buttermilk is thicker. And when it is too warm the butter is softer and you can get yourself in a mess. So then I just put it back in the fridge for an hour or so before I get back at it. In general I would say it is easier to make butter in the winter than in the summer."

After the butter has been strained from the buttermilk there is still some buttermilk contained in the butter. Washing two or three times in clear cold water and kneading the butter, either with a flat wooden spoon or simply by hand, will finish the separation process. It is during the kneading that Susi adds some salt to bring it to taste. She then divides the butter into containers and puts it in the freezer until needed.

Ice cream making is a family affair. We use a hand-cranked type of ice cream maker that is commercially available from well-stocked hardware stores. You need rock salt (from the local feed store) and crushed ice mixed together in the outer jacket of the machine, while you put the fresh cream, sugar and some flavouring in the inner tub. There are electrical models available to drive the paddles, but in a family of children there will be a battle over who can do the cranking.

A booklet with recipes comes with the ice cream maker, though most of the recipes call for condensed milk rather than cream, just substitute the fresh cream for the condensed milk.

As for the cow herself, there is no need for fancy stabling for the one-animal dairy herd. Minnie, who came to us before we were properly prepared for her, lived in a normal six-by-twelve-foot pig pen. We allowed her to wander freely in that space which meant that she would sometimes lie in her own manure, neither sanitary nor pleasant for the milker.

So when Miriah came we prepared a stable where we could tie her up, preventing that sort of condition. She has a five-foot manger for hay and other feed, and a rubber pail for water. Better yet would be to give her an automatic water-bowl, because the cow will drink a great deal of water (up to 15 gallons) during the day. Since Miriah's pen is in the front of my workshop, we pass her several times a day and fill up the pail with a garden hose. The floor in the cow's pen slopes away behind her, which is important to keep her dry.

Milk producing cows do need some grain in order to get enough protein and minerals, and prepared rations are available from the local feed store. She also needs a good quality hay during the winter months, roughly 100 bales (of 50 pounds each) to see her through the season. Alfalfa hay which contains more protein than grass hay can be bought from local farmers, with the price depending on the area, the hay crop that year and your skills as a bargainer. You may have to figure on $1 to $1.50 per bale.

Sufficient salt is often not present in the dairy ration, and a salt block left in the manger will allow her to lick just what she wants and needs.

We also use some alfalfa pellets, which we buy at the feed store, but it is important not to rely on them too heavily. If the cow gets too many of these good-tasting alfalfa pellets she may refuse to eat enough hay, which in turn may cause one of her stomachs to shift out of position. (Known as a "displaced abomasum" in veterinary terms.) Cows need lots of roughage.

In dairy herds cows in heat will demonstrate their condition by mounting other cattle or by being mounted by them. But in the one-cow situation heat detection can be a problem. Some may go completely erotically bonkers as Minnie did, while others may show only the slightest restlessness when the 21-day cycle reaches the point where she is receptive to the bull.

Either way the cow will have some bloody discharge from the vulva two days after her heat period. It pays to make a note of that on the calendar so you can be prepared to take her to the neighbourhood bull or to call the artificial insemination unit next time around.

Calving can be a great thrill, but it can also turn into a tragedy. If the calf is too big for the cow she and the calf can be lost if there is no experienced help available. For that reason dairy farmers often breed their younger cows to bulls of a breed which will give a small calf. Angus and Hereford bulls "throw" small calves, Charolais calves are notoriously big.

When calving time comes, nine to ten months after breeding, you should be in attendance and if you feel you can't handle the situation don't be long calling the veterinarian or an experienced neighbour.

The first few days a "fresh" cow will produce a much thicker, orange-coloured milk than later on. This "colostrum" is richer in protein and full of antibodies giving protection against a wide range of infections. Colostrum is not used for human consumption in North America, but it is important to the calf's health that he gets enough of it. You might freeze whatever the calf won't drink and keep it in case the calf needs a boost later in life.

We've only had the veterinarian out once for our cow, and that was for prevention rather than cure. He came shortly after Miriah had arrived, only a few days after the calf had been born. He inserted a magnet in her stomach, which will prevent the troubles you can get into if she swallows nails or other sharp objects and he gave her a shot of rabies vaccine. Rabies is quite common in our neck of the woods, so we vaccinate any and all farm animals and pets that might come into contact with wildlife. While here he also castrated (pinched) the bull calf.

Some Advice? *First, don't get a family cow unless you are sure she won't crimp your style. There should be somebody able to milk her twice during the day, even if it is not at very specific hours.*
Secondly: *Don't just buy a cow at a sale. Know the animal's disposition. Make sure she is quiet and friendly. If she is going to fight you all the way, you might better not have her at all.*
And Thirdly: *Read* The Family Cow *by Dirk van Loon, a $5.95 paperback published by Garden Way Publishing, Charlotte, Vermont. Also available from Harrowsmith Books.*

Goat Primer

Hardy and easily managed, the goat is perfectly suited for most North American farmsteads

By Hank Reinink

The goat is an animal of the recession, according to some economic observers who have figures to show that the goat population in North America rises perceptibly in hard times.

Others say that a recent 25 per cent jump in purebred goat registrations in a single year is more than an indicator of the country's economic doldrums. Resourceful, easy to handle (mature does weigh less than 135 lbs.) and inexpensive to buy and feed, the goat is a near-perfect animal for the small farmstead.

With neither elaborate housing nor expensive equipment, a country family can expect to be kept in milk for 10 months of the year by only two does, and during the spring milk "flush" that follows their return to pasture, the dairy goats will produce enough to allow for some simple — almost foolproof — cheesemaking.

"I think there is a terrific future in goats," says Michael Draper, a 31-year-old farmer who keeps some 50 Saanen dairy goats on his farm near Roblin, Ontario.

"They can survive and produce in areas where other domestic animals cannot perform. . . in jungles, deserts and rocky mountain terrain. It may seem strange to us in North America, but it is a fact that 60 per cent of the milk consumed in the world is goat's milk. And the staple meat for most of the cultures in the world is goat meat."

Goats, like giraffes, prefer to eat what they find at eye-level. They will eat grass, but if there is anything green on a tree or bush nearby, they will consume that first. In their browsing they will even eat such apparently bitter evergreens as cedar and juniper.

"Goats will actually do better on browse than on grass," says Draper. "Trees have considerably deeper roots than grass, and as a result there is a higher level of minerals in leaves on trees than there is in grass. I think it may be higher in total digestible nutrients as well."

Goats seem to have lost fewer of their natural habits than other domesticated animals, which may explain why goat keepers report few health problems among their charges. Goats may get lice and most have some intestinal worms, but both types of parasites are easily treated and little cause for concern among healthy goats.

One common skin infection is known as goatpox, which is concentrated around the hairless areas near the udder. It will clear up within four weeks if untreated, and it usually does not affect the same goat twice in her life.

Domestication has not erased one important element of a goat herd in the wild, and the animals still have a distinct breeding season, whereas cows and swine, for example, will breed any time during the year. The doe will come into heat during the early fall, often in September in her first year of life. Healthy, well-grown kids can be bred at the age of seven or eight months, but many goat herdsmen prefer to wait until they are 18 months old (when they are called goatlings) to prevent any chance of stunting.

In some cases it is difficult to detect the heat period, especially if there is but one goat on the premises. The doe will bleat more often than usual, and her vulva will usually be swollen slightly. Other signs of estrus include uneasiness, frequent urination and tail shaking.

Most small-time goatherds don't own a buck themselves, but take the doe-in-season to visit someone else's billy for a modest fee. The heat will only last two or three days, and the second day is usually chosen for the service.

If the owner is unable to get his doe bred in time, or if the mating has failed, the goat will come back in heat every 19 to 21 days after her first heat period. She will repeat this cycle until early January or until she becomes pregnant.

The gestation period is 151 days, plus or minus three, and so most does will accomplish their "kidding" or "freshening" in the late winter or early spring. Twins are the rule, rather than the exception, and triplets are quite common. In some breeds, notably the Nubians, even quadruplets and quintuplets are regularly seen.

As with all mammals, milk production is started by the hormone changes taking place at the time of birth. To sustain high production a goat must kid every year. The doe will increase her daily milk output for the first two months of lactation, which coincides with the best feeding time of the year as well. From a high of four quarts a day, the good average doe will gradually taper off to about two quarts a day by fall. Some genetically inferior animals will only milk

Home Rations

When goats are not in milk production, their nutritional needs are usually met by good pasture and browse. During the winter, good quality hay should be fed free-choice (always available) and a goat will eat from five to 10 pounds per day.

Lactating does do best with one or two pounds of a feed grain mixture per day. Commercial preparations are available, but many breeders prefer to do their own custom grinding (goats prefer a coarse feed). Two do-it-yourself goat rations are:

1. 13.5% Digestible Protein
   (for use with good hay)

Crushed Oats	75 lbs.
Cracked Corn	75 lbs.
Wheat Bran	25 lbs.
Soybean Oil Meal	25 lbs.
Salt	2 lbs.

2. 16.0% Digestible Protein
   (fed with fair-to-poor quality hay)

Crushed Oats	65 lbs.
Cracked Corn	65 lbs.
Wheat Bran	20 lbs.
Soybean Oil Meal	50 lbs.
Salt	2 lbs.

(Linseed oil meal can be used to replace the soybean meal, and some goat herdsmen believe it gives their animals glossier coats.)

for three months, and these should be culled out.

Even though she is still milking, the doe will come into heat again in September, and she is usually kept in milk production during the first part of her pregnancy. Two months before she is to freshen again, it is wise to stop her milk flow in order to assure healthy development of the unborn kids. Simply terminating the twice-daily milkings will stop the lactation.

The oldest doe to freshen on Draper's farm this year was 12 years old. "But that is pushing it," he says. "You can safely expect a goat to produce kids until she is 10 years old, but that is usually the end of her productive life."

"For winter feeding we use mostly hay," Draper continues. "It takes about 20 good-sized bales of hay to get a mature goat through the winter. Some people feed alfalfa pellets as well, which are much higher in protein and which are nearly totally digestible, but it is not a good idea to overfeed the pellets. It is the digestive process in the goat's two stomachs that produces the heat to keep her warm during the colder months, so you always have to make sure she gets some high-fibre feed, like hay."

Some advice for people who have no experience with goats, but who would like to give goat keeping a try?

To help the milk production along, Draper also feeds a grain supplement, which he has the local feed mill make up for him.

"Actually all she needs for satisfactory milk production is rolled oats," Draper says. "But we like to add some molasses, so that it will be less dusty and at the same time palatable. When we feel they need more protein, we have up to 200 pounds of soybean meal added in a ton of supplement. We feed one pound of supplement for every three pounds of milk the goat gives."

For people who keep only one or two milking goats it is hard to make butter, because the cream from which butter is churned doesn't rise to the surface as it will in fresh cow's milk. If larger numbers of goats are kept it may be worthwhile to buy a cream separator, which extracts the cream by centrifugal force. But cheesemaking is no problem.

"Cottage cheese making from goat's milk is pretty near foolproof," Draper says. "Our method gives a fresh type cheese, like a farmer's cheese. We don't make a very exciting cheese, it is rather bland, but you can vary the process to get a tangy cheese."

"We take a gallon jug of milk, fresh and still warm from the goat and add an eighth or quarter of a rennet tablet (rennet is made from the membrane of the rumen of a calf and is the agent which causes cheese to coagulate from milk).

"Then we just let it sit around behind the stove or some other warm place for four to eight hours. The rennet reacts with the milk and forms a curd . . . We then dump it through a strainer, the whey goes through and the cottage cheese is left."

To give the cheese more flavour, the milk can be allowed to sour before the rennet is added. According to Draper, goat milk that has been produced under reasonably sanitary conditions (wash your hands before milking and always use clean utensils) will not develop harmful properties when left to sour.

"I would say the best bet for anyone just starting out with goats would be to buy two just-weaned doe kids," recommends Freda Marlowe, who, with her husband Jim keeps about 75 goats of various breeds. "And definitely, don't just buy one goat. They get lonesome."

Sources

Two sources of rennet and cheesemaking materials are:

Horan-Lally Company Ltd.
26 Kelfield
Rexdale, Ontario M9W 5A2

Charles Hansen Laboratory
9015 West Maple Street
Milwaukee, Wisconsin 53214

One recommended introductory book on goat husbandry is *Raising Milk Goats the Modern Way*, by Jerome Belanger. A quality paperback, 150 pages from Garden Way Publishing, Charlotte, Vermont 05445. Also available from Harrowsmith Books.

This Is A Ewe. The Ram's Name Is Dougal McKenzie.

A look at small-scale sheep husbandry on Cape Breton

By Lynn Zimmerman

April is a mud month in Cape Breton, Nova Scotia: a disconsolate time of seemingly ever-prolonged waiting for spring. The tulips don't dare bloom till mid-May. And so the delivery of wonderfully wobbly-legged lambs during that time has become for us a joyful event on which to focus.

My personal experiences with sheep began casually in the spring of 1973 when a neighbour gave us two, one-week-old lambs which had been abandoned by their mothers. (This happens occasionally in multiple births, or when the mother ewe has a badly infected udder.) We named them Schwartz and Blanche and, unaware of sheep management, diseases or nutrition, we tenderly raised our pet lambs on baby bottles of warm, Jersey milk and an abundance of hay.

Now our small flock numbers six — our gentle ram, Dougal McKenzie, and five ewes (Blanche,

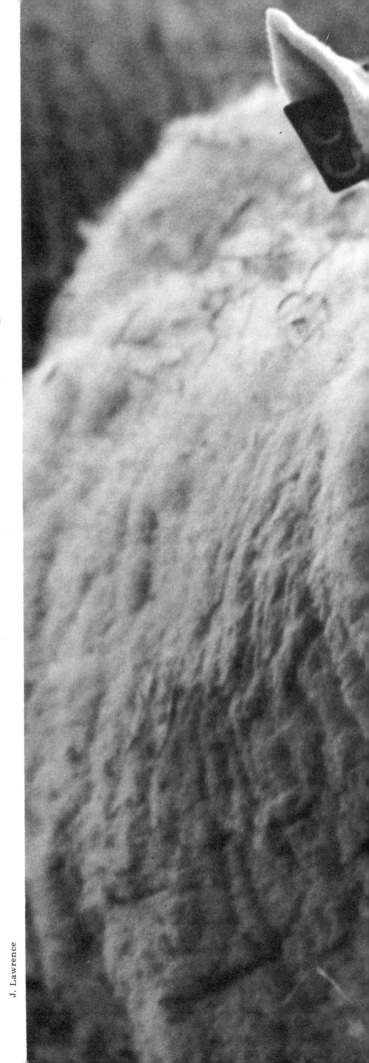

J. Lawrence

Brunhilde, Beulah, Spindle and Abigale) from which we hope for at least five lambs in April. After nearly four years of keeping sheep, we have learned much, and some of it the hard way.

Schwartz's death, the first winter we had the lambs, was a sad and lasting learning experience. One morning he was in the corridor outside his stall. Because he had apparently clambered over the four-foot-high gate and was breathing heavily, refusing to get onto his feet, I surmised (wrongly, as it turned out) that he had broken ribs; I expressed that opinion when I phoned a neighbour who is skilled with animal emergencies. Two days later Schwartz was dead. Too late, we had realized that he had got into a bag of hen scratch (a mixture of oats, wheat, corn and barley) which was stored in the corridor, had eaten, as sheep will do, until he could eat no more, and had soon bloated. Had I not jumped to conclusions but calmly and carefully stood back and gathered facts, perhaps we could have saved him. It was a lesson I won't forget, and needless to say, the grain is now stored elsewhere.

Blanche was bred during her second winter, and in the spring of '75 she presented us with a large ram lamb, but not without assistance. We realized with excitement that she was in labour when we went to the barn in the morning, but as she was on clean hay in a cozy stall, we were content to wait. Four hours later, when only one hoof had appeared, we called our experienced neighbour, Danny Mike Chaisson. "Have a bucket of warm water, some soap, and a disinfectant ready. I'll be right there," he responded immediately. And before I reached the barn he had arrived.

He soon ascertained that one shoulder of the unborn lamb was caught in the birth canal, and he gently manipulated the lamb back into the uterus. Then he patiently felt until he located the two forelegs and gradually led them out of the birth canal. Shortly, a large, limp, wet lamb had entered the world. I was sure it must be dead and was simply glad Blanche would survive. Then Danny Mike began vigorously rubbing the lamb, which suddenly blinked its eyes — alive!

The odour of disinfectant, understandably, does not attract a new mother sheep to her first lamb, and persnickety Blanche, deciding she wanted nothing to do with this strange-smelling creature, forcefully butted him away. We sat her down on her rump and, after squirting out some milk to open the plugs which form at the tips of the teats (the lamb usually takes care of this himself), we allowed the young ram to suck all he could of the nutritious and antibody-laden colostrum or first milk. We then tied her head in a corner of the stall and let him nurse every few hours, and within several days Blanche decided, after all, that this was indeed her lamb and we were to keep away.

An orphaned newborn lamb can be given a temporary boost with colostrum (conveniently frozen until lambing season) or even with dextrose or honey, and warm water. A length of

Abandoned by their mother, as sometimes happens in cases of multiple births, these lambs have readily adapted themselves to the role of farmstead pets.

thin plastic tubing can be inserted down his esophagus, on the left side of the throat, directly into the stomach and the food administered by means of a large syringe.

One way of getting a foster mother ewe to accept an orphaned lamb is to place the orphan in the lambing fluids of a ewe in the process of giving birth.

In cases where you have tried unsuccessfully to find a mother for an orphaned lamb, you will have to feed it yourself from a bottle equipped with a special lamb nipple and containing sheep's milk substitute. *Lamb-Mo* is one which comes in a powder form and is mixed with water slightly warmer than human body temperature.

If you have just assisted with a lamb's birth, you should immediately wipe away any mucus from around its nose and then place it at the ewe's head. She should identify it as her own and finish the clean-up operations herself. If the lamb's navel cord is more than two inches long, knot it carefully and snip it off with sterile scissors below the knot. Dip it in a seven per cent iodine solution to prevent infection.

An inexperienced person should attempt to assist a ewe with an emergency delivery *only* when no qualified help can be obtained within a reasonable time. Fortunately, most ewes cooperate; like most domestic animals, the majority of ewes give birth without trouble.

Still, a time will come when you will be called upon to lend a hand, and the important phrase to remember, we've been told by an experienced veterinarian, is "Push, not pull." The ewe is first placed on her back and the ends of a rope are tied around her rear legs at the hocks. Someone then lifts the ewe's legs, allowing gravity to pull the lamb back into the uterus.

After disinfecting your hands and lubricating them with soap, work one hand carefully into the uterus and feel the lamb to determine its position. Find the head-end — teeth are obvious, the mouth may suck your finger, the head may recoil when you touch the eyes. Find the tail-end. Feel the legs — the joints of the hind leg bend in opposite directions. Also determine whether there are twins or even triplets.

Finally, begin patiently to work the forelegs and the head toward the ewe's feet, never toward her tail. Alternate the lamb's legs so that first one shoulder and then the other is delivered through the narrow birth canal. At length your lamb will be born.

While still very young (some people suggest two or three days of age) lambs' tails should be removed, a process referred to as docking. Most sheep breeds are born with long tails which become traps for dung and breeding grounds for flies, barriers to mating and lambing, and nuisances for shearing.

Using an Elastrator is the safest way to dock lambs. An Elastrator is simply a special pair of pliers which applies an elastic band tightly around the third joint of the tail. Tails fall off within two weeks, after which the stump should be disinfected in a seven per cent iodine solution.

In May our sheep are sheared, drenched for worms, checked for foot rot, and treated for lice and ticks. Sheep are most easily shorn after warm weather arrives in late spring, due to a "rise" of the wool adjacent to the skin; in cold weather the wool does not naturally lift when shorn. Our neighbour, John Joseph LeBlanc, is an excellent sheep shearer and, in fact, he spends most of his spring and summer shearing thousands of sheep from Mabou to Margaree and from Lake Ainsle to Inverness. He usually ar-

rives without warning on a beautiful May morning, and we quickly gather our small flock by shaking a can of grain and calling, "Here lamlamlamlamlam."

John Joseph gently forces a sheep to sit on the floor, kneels behind her and, using electric shears hooked up to an archaic contraption, begins cutting and separating the wool from the neck down the belly to the legs. He keeps the entire fleece intact as he skillfully encircles the sheep's body, carefully trimming one area at a time. His best record was 12 sheep in an hour. "And those were sheared well," he tells us.

An average fleece weighs approximately seven pounds, and today sheep farmers receive about 75 cents per pound for unwashed, unpicked wool of all grades. I take our fleeces to the local carding mill when I've cleaned them, and during the winter I spin the carded wool myself.

After a sheep has been sheared, we immediately dust it with rotenone powder to control lice and especially ticks. Without the wool to hide in, the vermin are vulnerable to the rotenone, which is not poisonous to warm-blooded animals. (It is, however, toxic to fish.)

Sheep are notoriously vulnerable to internal parasites, and unless you want your homestead flock to be in a run-down, disease-prone condition, you must administer regular medication to ward off internal parasites. New drugs such as *Tramisol* (levamisole) do not harm even weakened sheep and destroy a wide variety of internal parasites.

PILLS & DRENCHES

Although administering drugs is a chore you will eventually master without difficulty, it is wise to watch an experienced neighbour before attempting to give sheep medication yourself.

Boluses (large pills) are the safest bet for beginners. (Just be certain that you have bought sheep boluses, not the larger ones for cattle.) Place the bolus as far back in the sheep's mouth as possible, either with your fingers (watch out for the teeth) or with angled, capsule forceps. After the bolus is inserted, hold the mouth closed until the sheep swallows.

Drenching is another method of giving sheep parasite medication. Here a liquid is forced down the sheep's throat through a syringe connected to an aluminum tube. However, if the sheep's muzzle is raised above a horizontal position while drenching, it is possible that the liquid will enter the lungs. The sheep will die as a result, and for this reason drenching is not a procedure recommended for beginners.

Medication, of course, kills only those parasites inside the sheep's bodies. Worm and fluke eggs and larvae will still be present in their pasture, so it is essential that the animals be turned into "clean pasture" after treatment, to avoid immediate reinfection. Clean pasture is that which has not supported sheep for at least the winter and preferably not for one year. Cattle are not susceptible to sheep parasites, and it is therefore wise to alternate pastures from cattle to sheep annually. We repeat medication two to three weeks before breeding and again in December after snow is covering the ground and we have begun feeding them hay for the winter.

A sheep can easily be coerced into a sitting position by turning its head up and around to one side while forcing its pelvis down and around to the other side. In this position its feet can be examined, cleaned and trimmed. Sheep's feet are prone to infection, particularly foot rot, and they should be examined and cleaned twice a year. Dig out mud, dung and stones with a knife and clip away the excess growth of the hooves (a normal occurrence when sheep are not on rocky ground). Stand the sheep for one minute in a foot bath of 10 per cent formalin to kill microorganisms but, as this hardens the hooves, it must always follow cleaning and trimming the feet.

Summer is a relatively carefree time for us with respect to our sheep. They wander within several acres of fenced pasture where both fresh water and shelter from hot sun and rain are available. We do, however, occasionally find a lamb on the wrong side of the fence. Once a lamb has executed his escape, he will soon be joined by his mother. In time, the rest of the flock will find freedom through the opening which grows increasingly wide as each sheep passes through.

Repair fences immediately, and remove telltale bits of wool left by escapees (both those that scoot under your fence and those that leap over) which blatantly announce "exit" for other members of the flock.

In July we make the hay; we figure 15 to 20 bales per sheep for the winter, which for us extends from late November until mid-May. Our local agricultural representative has described for us the nutritional needs of a sheep, and the information is seen in the accompanying box.

In September we sell the ram lambs and separate the remaining lambs and the ram from the ewes, in order to begin improving their condition before breeding in November. We have been told by a veterinarian who is very experienced with sheep that the number of eggs released from a ewe's ovaries during her heat periods depends directly on her physical condition. At this time we again treat all of the sheep with rotenone for ticks and lice, as soon their winter coats will grow too long for the powder to be effective.

In October, before flushing (turning the ewes and the ram separately onto the aftergrass of the hayfields) we inject all the sheep with 2 cc of vitamins A, D and E (in combination) in the muscle of the hind leg. (Incidentally, penicillin is also given in the muscle.) A deficiency of vitamin E is the direct cause of white muscle disease, in which there is a rapid and lethal wasting of muscle tissue. During this month I also make a

Feeding Requirements For Ewes

An average ewe requires, daily, about 0.36 pounds of protein and 2.0 terms of "net energy." (An analysis of your hay by an agricultural college or university should state the percentage of protein and the estimated net energy per 100 pounds of hay.)

For example, if hay is 8 per cent protein and the ewe consumes 4 pounds per day, she takes in 0.32 pounds of protein (8/100 x 4). If hay is only 6 per cent protein and she eats 4 pounds per day, she receives only 0.24 pounds of protein daily (6/100 x 4). In both cases, the ewe's protein intake is below the daily requirement of 0.36 pounds.

And, for example, if hay is analysed to contain 47.5 terms of net energy per 100 pounds, the ewe that consumes 4 pounds of hay per day receives 1.9 terms of net energy (47.5/100 x 4). Or, if hay has only 36 terms of net energy per 100 pounds and the sheep eats 4 pounds per day, she takes in only 1.44 terms net energy (36/100 x 4). In both these cases the ewe's net energy intake is below the daily requirement of 2 terms of net energy.

The deficiency of the hay in these examples can be easily supplemented with grains.

One pound of corn can provide approximately 0.09 pounds of protein and 0.8 terms of net energy; one pound of oats supplies about 0.11 pounds protein and 0.5 terms of net energy.

Feed	Quantity	Protein	Net Energy
Hay*	4 pounds	0.32 pounds	
Hay**	4 pounds	0.24 pounds	
Hay†	4 pounds		1.9 terms
Hay††	4 pounds		1.44 terms
Corn	1 pound	0.09 pounds	0.8 terms
Oats	1 pound	0.11 pounds	0.5 terms

* 8 per cent protein ** 6 per cent protein
† 47.5 terms net energy per 100 pounds
†† 36 terms net energy per 100 pounds

point of feeding the sheep plenty of kale, Brussels sprout and cabbage leaves, as well as beet and turnip tops from the garden.

A ewe's first heat of the fall is usually in October and thereafter occurs every 15 to 18 days until she is bred. Supposedly, she releases fewer eggs in her first heat than during the second or third heat, and therefore her chances of conceiving twins or triplets are greater later on in the fall. Consequently, hoping for twins, we wait until mid-November to turn our ram among the ewes. And, as the gestation period of sheep is 147 days, the lambs arrive in April.

By colouring the chest of the ram with a greasy mixture of artist's powdered pigment and motor oil, the breeder can tell which ewes have been bred (the colour rubs off on the ewe's rump during breeding).

Most large-scale sheep farmers change the colour each week, to help predict with accuracy when the ewe will lamb. Too, if a ewe turns up with no colour, it indicates she hasn't come into heat and has a breeding problem. If she has a number of colours on her rump, it indicates that she is coming into heat but not conceiving and should be checked by a veterinarian. Then again, if all the ewes are sporting more than one colour, it shows that the difficulty lies with your ram.

In December the sheep are brought into the barnyard, where there is a lean-to shelter on the barn. Then we begin feeding hay for the winter. Sheep kept inside during the winter should be allotted at least 10 square feet of floor space per sheep. We feed about four pounds of hay per sheep per day in midwinter, that is January and February. Depending on our annual hay analysis we supplement with grains.

From the beginning of March through lambing, the sheep are fed grain, such as corn, starting with about a quarter pound per ewe per day and gradually increasing, as lambing approaches, to one pound per ewe. The fetal lamb or lambs grow most rapidly during the last six weeks of the gestation period. As the fetus presses against the ewe's four stomachs and prevents her from consuming normal quantities of hay, grain is essential to provide the necessary nutrition.

If the ewe gives birth to a single lamb, you should continue giving her the same grain rations as before she gave birth. If she has twins, double the ration; and triple it if she has three lambs. Sheep rarely give birth to more than three lambs, but if one of yours has four, you will either have to bottle feed one or see that it is adopted by another ewe. Increased grain rations should gradually be phased out after the lamb is eight weeks old and is in the process of being weaned.

Domestic sheep are very prone to lethal diseases caused by clostridial bacteria; these diseases include tetanus, pulpy kidney, "struck," lamb dysentery and braxy. As a prevention, we vaccinate all our sheep with *Covexin*, a combination clostridial sheep vaccine. An initial dosage of 5 cc of *Covexin* is injected under the skin of the "underarm" just behind the foreleg, and thereafter, 2 cc as an annual booster dose is administered about four weeks prior to lambing. Given then it will, via the colostrum, partially protect the newborn lambs for eight to 12 weeks. We also inject vitamins A, D and E at this time.

So, as we prepare the lambing pens in the barn and await the arrival of another group of lambs, winter wanes and spring is near again. Since the unanticipated arrival of our first lambs four years ago, we have become committed to keeping sheep and have accumulated a great deal of information from our many valuable experiences — thanks to Blanche, Schwartz and company.

Woolly Economics

Many inexperienced animal husbanders, eager to wean themselves from supermarkets, dive headlong into livestock raising before they are thoroughly prepared. Too often the result is unhealthy animals and production costs that far outstrip those of the most exclusive butcher shops.

But a small-scale sheep operation, if properly managed, can be an easy and economically viable way to cut food bills and gain knowledge that will prove valuable when the homesteader feels ready to advance to larger ruminants.

Robert Brown, an Ontario farmer who has happily maintained a flock of 50 ewes since drought and low prices forced him out of the cattle business, feels that a five-ewe homestead flock would not only pay its own way financially, but would provide the small-scale shepherd with a steady supply of home-grown wool and a freezer full of chops, loins and roasting legs.

The key to financial viability of a small flock is a ready (and inexpensive) supply of good, high-legume hay. Alfalfa is by far the best. Five bred ewes will consume a total of one bale of good hay per day during the first part of the winter. Six to eight weeks before lambing, their hay ration should be upped to one and one-half bales and they should receive a daily grain supplement.

If your hay is a high-legume variety, corn makes an adequate supplement. Pregnant ewes should get one to one-and-a-half pounds of grain per day beginning six to eight weeks before lambing (the period of greatest fetal growth). If the ewe has one lamb, keep her at the same grain ration for eight weeks following birth. If she has multiple births, the grain ration should be increased proportionately (doubled for two lambs, tripled for three). Gradually decrease the amount of grain fed to the ewe after the lamb is eight weeks old. By this time it should be getting most of its food by grazing. Whole corn, which currently sells for about $3.50 for 25 kg (55 lb.) is better for sheep than ground corn. At an estimated 150 pounds of corn per ewe each year, the cost still comes out under $10.00.

Mineral supplements should also be provided. A licking block of calcium, phosphorous and trace element minerals can be purchased at feed stores for around $3.25 and five sheep will go through one or two blocks per year. But be certain that the mineral supplement you buy is specifically for sheep. Minerals beneficial to other livestock can harm or even kill sheep.

Medication is a necessary but slight expense, and should amount to no more than a few dollars per year for a small flock.

Brown recommends Suffolk, the most popular sheep in North America, for beginners. Although they are not recognized as producers of abundant wool, they generally experience easy births and Suffolk lambs grow rapidly.

In any case, it is wise to obtain a sheep that was born as a twin and comes from a line of known twin producers. Multiple births in April (one and one-half lambs per ewe is considered a good average) will mean more profit and a fuller freezer in the fall.

A clear advantage that sheep have over cattle is that they reach market size in four to six months. (Cattle require 12 to 18 months.) The only sheep that you have to support over the winter are your bred ewes.

A lamb born in April will be market size (90 to 110 pounds) in September. At current prices the lamb will be worth between $85 and $115, or it will give the shepherd 35 pounds of dressed freezer meat.

Although it is more financially advantageous to maintain your own flock of ewes than to purchase young lambs in the spring and raise them for fall slaughter, you need not obtain a ram if your flock is small. Sheepkeepers will lend rams for $5 to $10 per service. But a word of caution here: make sure that the ram you are bringing into your flock is free from disease and parasite infestation and that he has a reputation for siring big, healthy and multiple lambs.

Many small-scale herders purchase one ram to jointly service two or three flocks.

Mature sheep are sheared annually in May. Professional shearers charge approximately $1.25 per head and depending on the sheep breed, you can expect eight to 15 pounds of fleece per sheep that can be sold for 75 cents to $1.25 per pound. Income from Suffolk's fleece (relatively short and lightweight) falls at the low end of this scale.

A neighbour who keeps 10 sheep to sustain a serious spinning habit reports that she spends an average of a half-hour to an hour a day tending her flock — and this includes the frequently time-consuming chore of putting her sheep into the barn every night.

She says that (aside from a streak of stubbornness and a penchant for conquering the most carefully maintained fences) sheep are pleasant beasts to work with.

Further Reading:

Sheep and Wool Science
By M. E. Ensminger
Interstate Printers
Danville, Illinois

(A 1,000-page-plus volume that is recognized as the bible of sheepkeeping.)

Sheepe Dogges

*Neither snipers, a fallen master
nor the wiliest of ewes shall keep the Border Collie
from its instinctive rounds*

By Nancy Martin with photography by Rosann Hutchinson

"This dogge either at the hearing of his master's voyce, or at the wagging and whisteling in his fist . . . bringeth the wandring weathers and straying sheepe, in the selfe same place where his masters will and wishe . . . whereby the shepherd reapeth this benefite, namely, that with little labour and no toyle or moving of his feete he may rule and guide his flocke . . . . "
— Dr. Johannes Caius,
English physician, 1570

There is, on a quiet roadside in the English moors, a monument to a dog named Tip — not an especially heroic name — a humble name bestowed on a country dog by a humble shepherd.

What endears Tip to passing motorists who happen to stop and read the marker is the story of the Border Collie's devotion to the shepherd. Eighty-five years old, in the last winter of his life, the man had set out for a walk in the moors and did not return.

The search party which finally discovered the spot where the old man had died also found Tip, now emaciated and deathly weak, standing vigil over her master's body. The shepherd had been dead for 15 weeks.

Such tales of canine loyalty are, of course, not rare, but this one provides good insight into the character of the Border Collie breed. With origins traceable to the border country between England and Scotland more than 300 years ago, the Border (or Working) Collie today remains one of the most intelligent and least spoiled of dogs.

Alex McKinven of Quebec's Eastern Townships is one of North America's noted Border Collie breeders and handlers and he has a more contemporary story about the breed.

At a sheep dog trial in Killeter, County Tyrone, Ireland in 1973, the Irish Republican Army opened fire on the crowd. When the first shots were heard, the handler who was competing at the time dropped to the ground. One woman was shot and one man wounded. When the smoke had cleared and order was regained, the shaken handler rose to his feet to find that his dog had held the sheep quiet throughout the shooting. Buoyed by the collie's will to continue, the man gathered his composure and completed the course even though the stewards gave him the option to withdraw.

The dog was Maid, dam of Moss, one of the current team of Border Collies at Alex McKinven's Cessnock Farm near North Hatley, Quebec.

THE "EYE"

Set on a hillside overlooking Lake Massawippi, the farm is almost post-card neat, with well-kept grounds, an immaculate barn, 40 ewes and 110

head of sleek, registered Jersey cattle. Prize ribbons from the fairs and animal shows hang in the McKinven dining room and kitchen and it is all the realization of a dream for McKinven, whose father was a shepherd who came to Canada to work on an experimental farm.

McKinven's own reputation as a husbandman has made him an adviser to Sam Pollack, former owner of the Montreal Canadiens and a cattle breeder. Pollack also has one of McKinven's Border Collies.

It is not a bad life, and the farm is a quiet, lovely place to be, but McKinven still calls Scotland home and while we talk about the dogs his accent is thick with thistles.

"A dog that has too strong an 'eye' is just as bad as one without enough," he says, in reference to the Border Collie's inherent ability to cast a hypnotic sort of control over animals it is herding.

In watching the dogs at work with the stock, this trait is very apparent both on literal and figurative levels — the well-trained Border Collie has his eyes on the animals constantly as he works. The sheep seem mesmerized by this continued eye-contact.

The "eye" can be carried to an extreme, says McKinven, with a "double whammy" stare that puts the animals into a state of rigidity from which they are reluctant to move. If, on the other hand, a dog has too little "eye", he will move freely with the stock and may abandon his herding duties and take a drink of water (a move sure to raise the eyebrows of a shepherd or ring judge).

McKinven's dogs are not amiable, tongue-lolling Norman Rockwell mutts, happily padding down the street and content to sniff garbage cans and test the breeze. They have been bred for work and do their tasks with an efficiency and consummate dispatch that astounds people who see them in action for the first time.

In fact, the no-nonsense performance of McKinven's Border Collies led one skeptical farmer to suggest that the Scotsman was using trained *sheep* to make his dogs look good. Watching McKinven directing several dogs at once brings to mind a man commanding a precision squadron of remote-controlled model airplanes.

TWEED, AWAY TO ME

Callum, McKinven's son, stretches the full length of his 14-year-old frame on the grass to watch his father working the dogs. Perhaps just as the Alex McKinven of 12 peered through a fence in Fifeshire, Scotland, to watch the great Jock Murray training his Border Collies. Murray went on to win international championships, and McKinven apparently learned his lessons well, for he is able to work four dogs at a time. This is no small feat, and only one Canadian — Robert Walker of Creemore, Ontario — works as many dogs in public (he shows a five-dog team).

McKinven maintains that an individual can work a sheep dog using only four commands: *Away To Me, Come By, Walk On,* and *Lie Down.*

Enthusiasm for work and instant response are hallmarks of McKinven-trained dogs.

The trainer places his dogs at the corners of an imaginary quadrant in the field before him and then dispatches them, one by one, to go and do their stuff. "Moss, *away to me,*" sends one dog off like a shot to make what is known as an "outrun" to the right. The Border Collie circles wide in order to come up behind the sheep without startling them.

Once Moss is behind the sheep, McKinven commands the dog to drop, and he is instantly, belly and chin, flat on the ground, eyeballing the flock he is about to move. Next, McKinven commands the dog to *walk on*, causing Moss to approach the sheep slowly with head lowered and eyes keen. He stays behind the flock, moving them wherever McKinven indicates.

Come by is the command telling the dog to cast his outrun to the left in order to circle the sheep from behind. The novice handler must remember that *away to me* and *come by* refer to the *dog's* right and left.

The boss canine at Cessnock Farm is Tweed, an eight-year-old male who works alone and is the night watchdog at the farm. For the McKinven family, security is a Border Collie — no arrival at night, by man or beast, goes undetected.

Tweed is prized because he does even more than his job description calls for: more than once, sheep that had managed to get through the fence at night have been found huddled in the dog kennel the next morning where Tweed, undirected, regrouped them.

With three International Supreme Champions (Border Collie fanciers do not mince words when they hand out the awards) in his recent pedigree, Tweed actually does better at home than in the sheep dog trials. (He will be remembered, however, for his actions at the Williamstown Fair, the province of Ontario's oldest exhibition. When the sheep used in the demonstration escaped into the confusion of the crowd-filled midway, Tweed worked through the people to round the sheep up and bring them back.)

DOUBTFUL FUTURE

The Border Collie has just recently been recognized by both the British and North American Kennel Clubs, a development that worries men like Alex McKinven. For centuries, these dogs have been esteemed for their intelligence and working ability, with the finer points of appearance considered secondary. Handlers now fear that all the working qualities may be bred out of the dogs by those mainly concerned with physical conformation to the breed standard.

Another recent change that upsets the most dedicated handlers is the dropping of the requirement that for a pup to be registered in the Stud Book of the North American Sheep Dog Society both its parents must hold certificates of "Proven Working Ability." McKinven says that the high standard in the past worked to cull dogs that had poor dispositions or were faulty workers. A cute dog without brains is of no use to someone who wants a Border Collie to herd sheep, cattle, poultry or swine (the most difficult to herd of all domestic animals).

McKinven sells untrained pups for about $100, while a working dog will cost from $250 to $500, depending on the degree of training. A good trial dog imported from Britain may cost $1,000 or more. Cessnock Farm dogs are sold mainly in Quebec, Ontario and the New England states.

What should a beginner look for in a Border Collie? According to McKinven, the novice should seek out a reputable breeder or handler, and then "wait 'til the pup is at least eight months old. At that age he will begin to show what kind of ability he has.

"Ask the handler to let the dogs out on the field," he advises. "The dog that looks to the stock is the one to watch. If the handler encourages the dogs in any way — by a shout or whistle — there are several things the pup can do. But the one that circles the stock and keeps way off them is the one that will be the easiest to train."

The Border Collie is not a large dog, weighing between 35 and 50 pounds, but he should have a deep chest to provide room for a well-developed heart and lungs. A broad head and good muzzle, say the breeders, give space for brains and ease of breathing. (In the course of a working day a Border Collie may run up to 40 miles.)

McKinven himself can spot a dog with the working instinct at six months of age. The giveaway, he says, is the automatic assumption of the breed's characteristic crouch when confronting stock. When a pup with good working potential, even though still untrained, sees a flock of sheep or cattle in a field, he will pick up his ears, flop belly-down and creep toward the animals.

If the new owner wishes his dog to achieve a Proven Working Ability certificate, he should seek out professional help. McKinven starts his own dogs on sheep, as cattle tend to kick, perhaps injuring the dog or permanently affecting his working attitudes. If a dog begins his training with cows, says McKinven, he will tend to transfer the roughness necessary for dealing with cattle to the sheep he later encounters.

The enthusiasm for work and the instant response to commands is a hallmark of McKinven's training, and one client's Border Collie not only herds cattle, sheep, goats, ducks and geese on the family farm but also keeps reign on the house cats. The man's wife was caused to remark, "If Mac (the dog) could cook, I'd be out of a job."

189

For Love Of Sugaring

You don't have to be a stubborn old Englishman to make the best maple syrup in the country, but it helps

By Russell Pocock

Photography by Peter Hutchinson

If you've never been to the village of Hatley, Quebec, it's a tiny little place of perhaps fifty houses set nicely in the hills of the Eastern Townships — that's east of Montreal and south along the Vermont border. Geographically and architecturally closer to New England than the rest of Francophone Quebec, the Townships were predominantly English-speaking until 60 years ago and its original settlers came mainly from the New England states.

My grandmother happens to live in Hatley, and one day last spring I was visiting her for lunch. The talk turned to sugaring and she served up some maple syrup the likes of which I'd never seen before — crystal clear, lighter than corn oil and more delicious than any I'd ever had.

She said it came from the Curtis place — Edna Curtis was one of her schoolmates 60 years earlier.

Edna Curtis and her husband Clifford, it turns out, are the perennial Quebec Maple Sugar Champions, so on this same day I decided to drop in on the Curtises to see what magic they employ to make the best syrup in the country.

With draft horses and time to feed the birds in their sugar bush, Edna and Clifford Curtis still manage to turn out maple syrup that baffles the grading system.

The Curtis place sits on a high ridge of land that overlooks the back road from Hatley to Kingscroft and in the spring this road is so muddy with underground springs that for days it becomes impassable. On this day we were up to the axles of my old Valiant in mud, but we managed.

It's said that being on this high ridge of land goes part way toward explaining Curtis' quality of syrup. They say the low land will give you a darker, dirtier sap than trees on a high ledge.

Syrup quality is determined by factors such as colour, density and taste. The lighter the colour, the freer the syrup is of sediment and bacteria. The grades range from a high of AA to A, B, C and D. Everyone wants to buy AA, of course, and the lower grades can only be sold in bulk to companies who adulterate it and turn out stuff like "Log Cabin" and "Old Tyme."

Government inspectors visit the sugar houses and check their grading. They distribute coloured water samples in little vials to help people figure out what grade should go on their syrup cans. First run syrup typically rates the best, declining as the season comes to a close.

"AA" AND BEYOND

By coincidence our visit that morning found some government inspectors at the Curtis place curious to know how and why Clifford and Edna raise hell with the official grading system each year. On the window sill of the Curtis' sugar shack sits the little rack of Clifford's different grades of syrup. Even his worst syrup is lighter in colour than the government's AA best.

It must irritate the inspectors somewhat that this old bugger of an Englishman and his wife, using outdated methods such as metal buckets and draft horses, come out with syrup that makes their AA look like molasses, year after year after year. The government is encouraging people to throw away their aluminum buckets and gathering tanks to install plastic spouts and pipelines. Untouched by hands, the sap flows or is suction-pumped directly to the sugar shack where it is boiled down in modern oil-fired, thermostatically-controlled evaporators.

I have a neighbour with a setup along these anti-traditional lines — 5,000 trees connected by pipeline to two oil-fuel rigs. Last year he didn't even sugar. Perhaps he doesn't find it fun any longer; maybe his overhead is so great that it doesn't pay anymore. Either way, zero is a poor return on such a large investment.

Clifford Curtis, on the other hand, says he "loves to sugar." Very simply, you have to love doing it to achieve the quality he manages. "When I get down to AA, I start to get sick. No, I'm not kidding. I'm getting sick of sugaring when I get down to there."

"But," he continues, "there's only six per cent of the producers that get up to AA in the province, so the rest think the standard is too high. How do they do it? What's their secret?"

Answering himself, Clifford says, "Most people tap too many, more than they can handle and it sits around in the bucket too damn long in the hot sun. Well, it's the same with milk. How long's it gonna be before the sap is full of bacteria?

"We tap about 1,500 trees. Do it ourselves. We could tap 3,000 with no trouble at all. The neighbours' boy, he comes and helps us once in a while on weekends. If we had a lot of hired help here, there'd be a lot of trouble. If you have them come to board at your place, they'll make as much work as they do — feeding them, getting good clean bedding for them, washing their bedclothes.

"It keeps Edna working in the house all the time. She enjoys sugaring but not if she's down there washing bedclothes, cleaning bedrooms and all that jazz.

"On a tree a foot through we put one bucket; two feet through, two buckets; and three feet through, three buckets. If you put two on a two-foot tree you don't get twice as much sap though. They're healthier if you don't overtap them.

"An old tree has the sweetest sap. If it's got a bigger top, the sun hits more leaves. All the sugar comes in through the leaves — 100 per cent of it. Did you notice that great big old tree on the corner up here halfway in? Well, that's what blew down — trees like that."

"DAMN PLASTIC"

"We have a good bit of pictures down out of the house that was taken over 80 years ago, and one man in a picture was sugaring over on the hill with a stone arch. Their house was made of stone, flat stones. The stones are over there now, fallen over. He had pans sitting on a flat stove and pieces of railroad rails for grates."

I asked Clifford if he's ever tried plastic buckets or spouts. Some people swear by them and the general trend seems to be towards them. He became almost indignant.

"I don't like plastic. I've won enough of the damn stuff. The champion of the province was getting 300 buckets, covers and spouts every year. Now it's $250 worth of plastic pipe and spouts.

"I had 450 plastic buckets and I give them the same chance as I have with the aluminum buckets. I have two tanks. I put all from the plastic buckets in one tank, all from the aluminum in the other tank, day after day. Give them the same chance exactly. The quality was a whole grade down on the plastic every time, every time you gathered. And they didn't have the same flavour. I don't know why. They're the coldest damn things. If the sap freezes in them, it takes half an hour to thaw it out.

194

"I know a lot of people who are shifting over to plastic and their quality has all gone down, every one. Men I know personally, known them for 20 years, their quality has gone down.

"With plastic pipe you don't have to gather the sap. The sap comes to you. The government is pushing plastic for all they're worth. Anyone that makes a complete changeover to plastic pipe gets 65 cents a tap — that's about one half (the cost). They're looking to get out of the work like anybody else. Yeah. . . they're getting sick of walking in the snow — this brings the sap right into the camp — a lot easier.

"One of my neighbours went to St. — the other day and a man out there was tapping 6,000 with plastic pipe and he had a swimming pool for a storage tank. A plastic swimming pool. And he said that sap was something. He said you could taste the plastic right in the sap before he boiled it. And he tasted the syrup and he said he wouldn't buy a gallon if he was starving for syrup."

The fellow I had helped sugar last year had 500 trees on pipeline, and he gathered the rest with a tractor pulling a gathering tank on a low wagon. It seemed we were stuck in the snow a third of the time. Helen and Scott Nearing developed a method of having small reservoirs located at different points around the bush, all emptying into the camp by gravity. This makes sense, as long as your camp sits on the lowest part of the land.

"I prefer to use horses than a tractor," Clifford continues. "Say you're out there gathering with a man. The damn tractor's way behind you and you got to run back and get it. Just speak to the horses and they come right along, no waiting for anything. Damn tractor, you got to have an extra man just sitting on his rear end on that tractor."

The boiling down operation usually consists of a large covered storage tank for the sap, located outside the shack and running into a big evaporator of nine or 12 feet in length by four or five feet. The large evaporator pan sits on a big arch that contains the fire and allows for the loading of four-to-five-foot logs at one end.

When the sap has boiled down to a consistency approaching syrup in the large pan, it's shifted to a smaller pan on a smaller arch. As the sugar becomes more concentrated, the risk of burning is greater. The final stages are tricky, and you have more control with a smaller arch.

"The hardest part of sugaring is getting it to the right thickness," says Curtis. "I don't care if a man is 90 years old — sugared all his life — he still has trouble to get the right thickness.

"You have to find out what the boiling point of water is at the time you're doing it, because the barometric pressure keeps changing it up and down. It can vary two degrees, at this altitude especially. And this altitude business starts at 500 feet above sea level.

"I have seen water boil here at 209½ degrees. And I've seen it change two degrees from morning to midnight on the same day.

"You have to be right on your toes every second — check, double-check, cross-check, all the time." (The syrup begins to boil at a temperature seven degrees above the boiling point of water.)

One thing in particular about Clifford's pans are their cleanliness. Most rigs aren't kept as clean as his.

"You have to have everything very, very clean to make that quality. If you half-wash everything, why you're going to go down to grade A. It's dirt and bacteria that ruin the colour."

"I usually wash everything thorough when I start, and usually we'll wash the evaporator off after every round. I'll take a wet rag and wash that dry stuff off from the inside before I start again. That looks like dirt, but it's only dry scum. Not half as bad as it looks.

"The pan (for boiling down) is around 18 years old and the arch is 25, 26 years old. . . I went to sleep one night here, seven, eight years ago and burnt both pans into pieces. I'd been going day and night for a couple of days. I sat down on a chair and went to sleep. It was a cold night and the pipe froze — sap didn't come in. Well that was trouble. I thought it was going to come in and I sat down. I woke up coughing and there was blue flames comin' out of the front pan."

For many people, marketing syrup is a problem. A couple of years ago there was a surplus of syrup and producers couldn't get rid of it. I asked Clifford and Edna about marketing their syrup and whether they think the future for the industry is good.

"I don't have a special market for my syrup, just regular customers. I get the going price. Some people advertise in the Montreal paper, have sugar parties and they get $20 a gallon. We could advertise in the paper, too. We could get more money if we asked for it. We try to deal with the consumer. . . we've had these customers for years and they're good customers. They'll pay you when you get it there and they're not fussy about having it today or tomorrow. The consumer is getting it in the neck for everything, for milk and cars and you name it — they're getting it in the neck. If we get another good run this year, we'll do all right.

"You know what's spoiled our market? For one thing, those fake maple syrups made with darker syrups from the last run. People buy it and don't like it," says Edna. "They buy that and say, 'I don't like maple syrup, don't buy it.' And you can't blame them for not paying $12 a gallon for something they don't like.

"People don't know good syrup is light. If anybody gets a taste of it, they're just crazy about it."

Clifford continues on this topic. "Syrup like 'Log Cabin' shouldn't be allowed on the market, but that would do away with a lot of sugar makers. They don't know the grades. I call my stuff lousy. I want to get lighter and lighter. The quality is a lot higher this year than last year. But the government is getting right after them now, especially on density and thickness. Some of them have been selling too thin for years and years and getting away with it. Like those two tanks out there I got about 22, 24 gallons of standard syrup. Well, I can make it too thin and get 30 gallons out of it — at $12 a gallon, there's a big difference."

I help Clifford with a few lifting jobs around the shack, while Edna goes out to feed the birds. She calls to them with bread crumbs in her outstretched hands. Today they're shy because there are strangers around. Nevertheless, a small nuthatch lands on her arm and takes the crumbs, one by one, to be stashed in the bark of a nearby tree. She talks of all the different types of birds inhabiting their woods, birds that she and Cliff know virtually on first name basis. She mentions that there are horned owls, too, that come around. Edna calls louder, and hooting begins a short distance away, echoing through the woods from tree to tree.

You won't get rich making sugar. If you count your time you're lucky if you make minimum wage. It's awful hard work and there's no point doing it unless you enjoy it a bit, too.

Lots of farmers will tell you that it's the worst damn job there is and yet every spring when those first warm days of March start melting the snow you'll see them heading for the bush.

Clifford mentions that one major problem plaguing the maple sugar industry is the fact that so many farmers can't make ends meet and have to sell out. Even his neighbour Rollie Bowen, with whom there has been a friendly rivalry over the years and who has won the title of sugar king almost as many times as Clifford, recently had to sell his property down the ridge.

"Syrup like 'Log Cabin' shouldn't be allowed on the market, but that would do away with a lot of sugar makers."

"The first time I won the cup, in '56, there were over 25,000 sugar places in the province. Now we're down to about 13,000. Of course, some of the big farmers have swallowed up the smaller ones, but a darned lot of them have moved off.

"You know, a lot of farmers are selling out to these city people — Montrealers, eh?" Edna adds, "They buy a whole farm as cheap as they can buy a house, and then perhaps there's a sugar place on it, but they don't use it for farming or anything else. Every day in the summer, people stop by. We're pestered to death — big offers — you know."

Clifford laughs. "But we enjoy it here. We work at it year 'round. In the summer we come up to the woods, and if a tree is down we cut it up with a chain saw and stack it. We come to see our friends (the birds).

"It's nice, you know, to have your privacy. We got to live somewhere and we like it here. It's our home. We wouldn't be happy anywhere else.

"Why consider selling? I think a lot of people who've sold for big money, they're sorry, they're sick."

The Home Sapster's Art

Happiness is a cloud of maple steam on a crisp winter night

By Barbara Keane and Christine Holbrook

But surely rainwater has filled the pails overnight!'' Such is the typical befuddled reaction of newcomers to our sugarbush. It is inconceivable to them that this crystal clear, water-like liquid is maple sap and contains the wherewithal to produce rich, amber syrup without resorting to alchemy.

A taste reveals the faint presence of natural sugars, and it is quickly understood how 30 to 40 gallons of raw sap go into the making of a single gallon of finished syrup.

The Indians, who relied on the annual sap run for a much-needed spring vitamin tonic, called the maple *michton* and had an explanation for the thinness of its sap. One of their ancient leaders, the legend goes, feared that the ease of taking fully thick, sweet syrup directly from the trees — for such was the simplicity of harvest in that time — would lead to indolence among his people. Climbing to the tops of the maples, he is said to have poured water down the top of each, diluting the sap to its present consistency. His visionary action made necessary the long boiling and evaporating process and ended, once and for all, that particular native free lunch.

Making maple syrup thus requires a certain amount of patience and character but, compared to other rural endeavours, is one of the least expensive and uncomplicated tasks for neophytes. Even modestly small batches of syrup will re-

quire spending some crisp nights around a fire bathed in maple steam, but this is for us one of the most satisfying tasks of the year (and a sure indication that spring is on the way).

Trees may be tapped any time from January to April, with the heaviest flow of sap usually starting sometime in early to middle March. Warm days are necessary to bring on the run of sap from the roots to upper branches and bud tips of the maples. The best flow occurs when sharp frosty nights are followed by days with temperatures above 45 degrees (F).

To make maple syrup, of course, you need at least one maple tree. This statement seems simple-mindedly obvious, but for those tapping trees for the first time in their lives it is often the initial stumbling block.

LOCAL PREJUDICE

Armed with a dozen new metal spiles, gallon jugs, a brace and bit and an enthusiasm still remembered, we rushed out to our own first attempt at sugaring and madly began tapping the stoutest maples in sight. This done, we were just standing back to admire the scene when a neighbour came by on his tractor, drawing a small wagon with a load of split cordwood for the small commercial syrup operation he runs.

That look of old-timer's glee crept into his eyes as he spat some tobacco juice and composed the question he would be able to repeat for years to come. "Tappin' some trees eh? Never heard of anybody tappin' a silver maple before. . . "

Now it happens that we should have been ready to defend ourselves — any of the large maples will produce a syrup of acceptable quality, local prejudices to the contrary.

The best sap *does* come from the sugar maple (*Acer saccharum*), but the silver, black, Norway, and other maples can also be tapped.

Leafless, the maples can be a challenge to identify in late winter, but if the elements have left an accumulation of leaves at the base, it should be possible to assure yourself that what you are tapping is not a hickory. If still in doubt, swallow your pride and ask a neighbour who knows his trees.

Although the old-timers will tell you to place your taps on the sunny side of the tree, studies have shown that the total yield per taphole does not depend to any great extent on its relationship to a point of the compass. Avoid tapping dead wood or very close to old tapholes.

The holes are drilled with a half-inch or 7/16-inch bit, angled slightly upward and three inches deep into the white sapwood. Shallow tapping will result in lower yields, and deeper holes neither boost production nor benefit the tree. In smaller maples, the taphole may reach the red coloured heartwood, which does not produce syrup — stop drilling when the curl of wood produced by drilling becomes reddish.

Metal spiles are most commonly used (about 25 cents at hardware and general stores — buying early is advised as they have been in short supply in recent years), although wooden spiles can be fashioned from elderberry, sumac or bamboo. Tap the spile into the hole, and hang the bucket (of at least one gallon capacity) either from a nail or from the metal hook on the commercial spile.

Just how many trees to tap depends on your boiling-down facilities, how much time you have for gathering and processing, and what amount of syrup you plan to consume or give away. Boiling the bulk of your sap outdoors will take approximately a full standard cord of hardwood to serve 75 tapholes. Many small-scale syrup producers build a makeshift fireplace to hold a large kettle or pan. The sap is boiled outdoors until it begins to reach the syrup stage and is then brought inside to finish in the kitchen.

For very small amounts, the entire process can be done indoors, but be sure that the kitchen is adequately ventilated (the steam produced has been known to damage wallpaper in tiny kitchens). For a small family, tapping two or three trees is a good start — expect about one quart or more of finished syrup per tap in a good year.

The sap should be gathered morning and night and should be rendered daily, as it goes sour in 48 hours. Some sapsters manage to get by boiling sap only on weekends by putting each day's yield into the freezer to prevent loss of quality.

A solidly frozen kettle of sap will thaw on the fire in one hour and be reduced to one-fifteenth its volume in another two hours. We then pour the sap into a smaller pot and finish it on the kitchen stove, which takes from 15 to 30 minutes.

Not only is this the most exciting step — the sap really is beginning to look, smell and taste like syrup — but one requiring constant vigilance. The sap will be boiling along at the boiling point of water (212 degrees Fahrenheit at sea level) when suddenly the boiling point will start climbing rapidly. It is bona fide maple syrup when it boils at seven degrees above the boiling point of water. (If you do not live at sea level, check the boiling point of water on your stove).

If you have no thermometer, watch for the sap to start foaming up in the same manner as jelly. Cook for about one minute more or until it has reached the consistency you want. If the sap is boiled too long, it will turn to maple sugar upon cooling — still not a bad price to pay for a mistake. A pot left unattended at this stage is in grave danger of becoming a black, burned, impossible-to-clean mess. (If your intention is to make maple sugar, wait until the boiling point is 20 degrees (F) above that of water and then pour the thick syrup immediately into molds.) At this stage some syrup makers add milk, cream or a beaten egg to coagulate and trap any sediment in the batch. This material is skimmed and/or filtered off before bottling.

Old-timers in the neighbourhood say the next step is to strain the syrup through an old felt hat. Depending on the state of your hat, you might use something else. Multiple layers of cheesecloth or other fabric serve to remove most noticeable bits of ash, bark and other foreign matter. Heavy felt syrup-strainers can be purchased, but these are rather expensive and messy for very small batches. Improvise with thick felt or stiff interfacing material from a fabric store.

Syrup can spoil after bottling, so be sure your jars are spotlessly clean (sterilizing as if for canning is a good idea). Store the syrup in jars in a cool place and refrigerate jars after opening. The latest idea in syrup storage is freezing, but if you've only tapped a few trees, the chances of your needing any storage space are nil (the total yield from one canning kettle of sap is only a cup or two of syrup, which tends to disappear *very* quickly.)

Remember, the best syrup is light golden in colour — not a heavy brown as most of us believe. The early run produces the finest syrup and the colour becomes darker as the season progresses. To achieve the best possible syrup, tap early in the season and boil it down within hours of gathering.

Remember to remove your taps and plug the holes (bang in a tight-fitting piece of wooden stick) when the season draws to a close.

The Maples Of Autumn

"It is no coincidence that our national emblem is not a rising sun, a star, a hammer, a sickle, or a dragon, but a beaver and a maple leaf. Nor is it coincidence that there are more paintings of wilderness lakes, spruce bogs, and pine trees on more Canadian living room walls than in any other nation on earth. We may scoff, we may deny, but the wilderness mystique is still a strong element of the Canadian ethos."

F. Bodsworth

All photography by Donald McCallum, except this page, by John Gregory

Sugar Maples are perhaps the showiest of the 13 species of maple native to North America. As seen in these photographs, all but one of them Sugar Maples, the fall hues within this species (*Acer saccharum* Marsh) can range from brilliant orange to yellow to bright red.

Actually, the intense colour pigments are present in the leaves throughout the spring and summer, but their effect is masked by the presence of green chlorophyll. With the cessation of fresh sap circulation in autumn, however, the chlorophyll (least stable of the pigments) breaks down, leaving the xanthophyll (yellow) or carotene (red/orange) to colour the leaves.

Just what colour a tree's foliage takes on in the fall is determined by the species. Among the maples, yellow predominates in the Manitoba Maple, the Black Maple, the Bigleaf Maple, the Silver Maple and the Striped Maple.

The Douglas Maple, sometimes known as the Rocky Mountain Maple and confined to the Pacific coast regions, turns a dullish red in autumn. Both the Vine Maple and the Mountain Maple, generally shrub-like in nature, can take on either red or yellow colouration.

For those planning to try their hand at making syrup next spring, now is the time to identify prospective Sugar Maples or other sweet-sap producing species.

The Red Maple, above, is noted for the startling intensity of its colouration in the fall. Its seeds, shown on the opening page of this series, are also bright red. *Overleaf:* Close-up of a fallen Sugar Maple leaf, followed by an identification guide to the maples.

Syrup Lover's Guide To The Maples

Tap an elm tree and be remembered for life in your neighbourhood

By Barbara Keane

The successful homesteader, no less than the businessman or the Grand Prix driver, must be fully involved with his immediate task, but all the while sharing his concentration with what lies ahead. The small landholder's deadlines come not from a stopwatch or pressure from rival companies, but from the inexorable change of seasons. When advance preparation is either forgotten or crowded out by more immediate concerns, seasonal opportunities are bound to be missed.

So it happens that people often don't start thinking about maple syrup season until sometime after Christmas. Then the sap suddenly begins to flow, and the leafless winter trees give very little clue as to their identity. In the ensuing confusion, more than one neophyte syrup maker is still remembered for having tapped an elm tree or an old hickory.

Although many trees, including the birch, hickory and sycamore, produce a sugary sap, the maple yields by far the best quality sap. Furthermore, the leaves of the maple are probably the most familiar of any North American tree. The only tree leaf which might be confused with that of a maple is the sycamore, and the unique bark of the sycamore would make such an error unlikely.

Distinguishing among the 13 species of maples in North America (there are 10 in Canada) is, however, slightly more challenging. The best sap, of course, comes from the sugar maple *(Acer saccharum M.)*, also known as hard maple, rock maple, black maple, curly maple and bird's-eye maple. The leaf usually has three main lobes, or segments (see illustration), which are dark green above and pale green below in summer, and red, golden-orange or yellow in autumn. The end lobe appears almost square, and widely separated from the two side lobes by rounded notches.

The black maple *(Acer nigrum)* is classed by some botanists as a variety of subspecies of sugar maple, rather than as a separate species. It is distinguished by its somewhat drooping leaves, usually covered on their undersides and stems by a soft hair. The bark is almost black in colour. It, too, is known as the hard or rock maple, but the confusion shouldn't bother you at sap time: It produces copious amounts of sap high in sugar content. If you aren't sure whether a tree is a sugar or a black maple, tap it anyway. You won't be able to tell the difference in the syrup.

If you are a town or city dweller, it doesn't mean that sugar making is out of the question. If you live in an older neighbourhood, chances are good that silver maples *(Acer saccharinum)* are on or near your property, and they also give a high-quality sap. The silver maple is distinguished by the deep, narrow indentations between the lobes, and the notable silvery-white colour on the undersides of the leaves.

These three species of maple will give you the best syrup, but if you are feeling adventurous (and willing to suffer the remarks of skeptical neighbours), any species of larger maple will give you the opportunity to breathe in the aroma of a bubbling sap kettle over a roaring fire on a frosty spring night. The Indians were said to have used the sycamore's sap, but the resulting syrup is considered much too dark and strongly flavoured by many people. Undoubtedly, it should be kept from your delicate crêpes, but it might add a certain kick to a pot of baked beans.

203

Stalking The Sugar Maple

RED MAPLE *(Acer rubrum)*
Brilliant scarlet colour in fall. Distinguished from the sugar maple by the angled notches between lobes (not rounded) and by the sharp, toothed leaf edge. Three to five inches across, leaf stalks often a bright red.

DOUGLAS MAPLE *(Acer glabrum)*
Usually a tall shrub, but individual trees may reach heights of 35 feet. Coarse, double-toothed edges on leaves, which are three to four inches across.

SUGAR MAPLE *(Acer saccharum)*
Leaves three to five inches across, end lobe squarish and wide, round-bottomed notches between the three main lobes.

BLACK MAPLE *(Acer nigrum)*
Leaves three to five inches across, dense, velvety hairs on the underside, extending along the leaf stalk. Shallow notches between the lobes. Few teeth.

BIGLEAF MAPLE *(Acer macrophyllum)*
A Pacific coast tree, noted by its huge, six to 12-inch leaf. The largest undivided leaf of any species of Canadian tree. The leaf stalk exudes a milky sap when snapped.

STRIPED MAPLE *(Acer pensylvanicum)*
Large, four to seven inches across, with a distinctive, triangular end lobe, and shallow notches between lobes. Very fine teeth along the leaf edges.

MOUNTAIN MAPLE *(Acer spicatum)*
A coarse, small, shrub-like maple of the east. Leaves up to three and one-half inches across, with triangular end lobe, toothed edges. Often grows in clumps.

VINE MAPLE *(Acer circinatum)*
Seven to nine lobes with distinct, V-shaped notches between the lobes. A shrubby, west coast maple, with a short, crooked trunk.

MANITOBA MAPLE *(Acer negundo)*
Unmistakable leaves, divided into from three to seven leaflets. Six to 15 inches long. Native to the prairies, the Manitoba Maple has been artificially spread far outside its natural range.

SILVER MAPLE *(Acer saccharinum)*
Leaves distinguished by having five distinct lobes or segments, with deep notches. The centre lobe is narrow-waisted at the middle of the leaf. Three to five inches across, with a whitish underside.

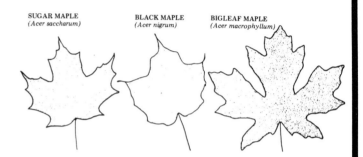

SUGAR MAPLE *(Acer saccharum)* — BLACK MAPLE *(Acer nigrum)* — BIGLEAF MAPLE *(Acer macrophyllum)*

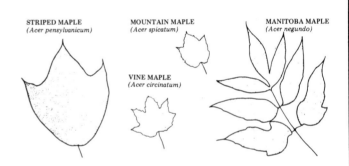

STRIPED MAPLE *(Acer pensylvanicum)* — MOUNTAIN MAPLE *(Acer spicatum)* — MANITOBA MAPLE *(Acer negundo)* — VINE MAPLE *(Acer circinatum)*

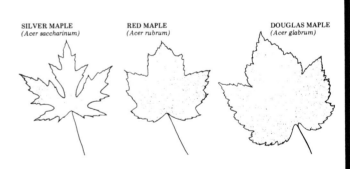

SILVER MAPLE *(Acer saccharinum)* — RED MAPLE *(Acer rubrum)* — DOUGLAS MAPLE *(Acer glabrum)*

If you have an interest in learning more about the maples and other trees, one of the best identification guides available is *Native Trees of Canada*, by R.C. Hosie, available from the Queen's Printer in Ottawa or through most booksellers. Residents of the northern U.S. will also find this profusely illustrated paperback a useful forest guide.

Unripe Is Beautiful

Pickled, fried, composted or just thrown at the moon, green tomatoes are an autumn institution

By Nancy Martin

Seven million bushels of tomatoes are processed annually in Leamington, Ontario, the Tomato Capital of Canada. Yet, each September when I confront our garden and its frost-threatened crops, I get the panicky feeling that I am at the centre of the green tomato capital of the universe.

There are tomatoes everywhere. Many are destined to become red, other low-acid varieties will mature as yellow or white fruits. In the meantime, they are all green. They are ripening on window sills. They are arranged decoratively in bowls and baskets on every available flat surface. They are wrapped individually in newspaper and hidden in brown paper bags to ripen slowly in a dark room. There are whole plants, bearing unripe tomatoes, hanging by their roots in the tool shed.

As they ripen, the reds and yellows are canned (for soups and sauces). I skin them by first dipping them in boiling water, then cooling in a cold bath. For freezing, I cut them in quarters and package the wedges in waxed milk cartons which are then labelled and stapled shut at the top.

Toward the end of the fall ripening period, the jars and milk cartons full of tomatoes make impressive rows under the sink or in the freezer. The warmth of the kitchen is now beginning to be appreciated and the clouds of canning steam forgotten for another year.

I turn to the laggards. They squat in the sun or couch in darkened rooms, defiant — as green and hard as they were the day we rushed to pick them before the first frost. There is the temptation to throw them at the moon, but frugality prevails.

While green tomatoes stocked away carefully will usually ripen right up until Christmas, there is no reason to wait. In fact, considering the versatility of unripened tomatoes, there is every reason to try them in the following recipes:

GREEN TOMATO PIE

This is an unusual treat, somewhat similar to a tart green apple pie, that can be served warm with cheddar cheese, whipped cream or a scoop of ice cream.

Core and slice 4 cups green tomatoes. Combine: ½ cup white sugar, ½ cup brown sugar, ¼ cup corn syrup, ½ tsp. salt, 2 Tbsp. butter, 2 Tbsp. lemon juice, 1 tsp. grated lemon rind. 4 Tbsp. cornstarch, 1½ tsp. cinnamon, ½ tsp. ground nutmeg and ½ tsp. allspice. Bring to a boil.

Line a 9-inch pie pan with pastry and fill with the sliced tomatoes. Pour the cooked mixture over them and add the top crust. Crimp edges and cut slits in the top. Bake at 375 degrees (F) on the bottom oven rack until the top crust is golden and the tomatoes tender (about 45 minutes).

GREEN TOMATO CURRY

To make the base:

Saute a large minced onion in 2 Tbsp. butter. Add 4 large green tomatoes, chopped. Cover and simmer until tomatoes soften. Sprinkle approximately 4 Tbsp. flour and 2 tsp. (or more) curry powder, salt, pepper, a pinch of sugar and cardamon (optional) over the partially cooked tomatoes.

Stir thoroughly, add 2 cups water (or broth) and simmer until thick. Taste and adjust spices.

Add any of the following and heat thoroughly: cooked, cubed lamb, beef, pork or chicken livers; sliced, hard-boiled eggs; or shrimp.

Serve with steaming rice and condiments, or on toast. Serves 6.

FRIED GREEN TOMATOES

This specialty has a following all its own, and one true measure of a greasy spoon is whether or not you can get a side order of fried green tomatoes with breakfast.

Cut green tomatoes in thick slices, then dredge in flour or cornmeal that has been seasoned with salt, pepper and a trace of white or brown sugar.

Pan-fry in 2 Tbsp. butter until crisp and golden on both sides. For special occasions, try adding sour cream to the pan juices, heat, season and pour over the tomatoes. This recipe also works well with red tomatoes.

GREEN TOMATO MINCEMEAT

This mincemeat is not only superb in pies and tarts, but special when heated, set aflame with brandy, and poured over ice cream.

3 lbs. green tomatoes, chopped
3 lbs. tart apples, peeled, cored and chopped
2½ lbs. brown sugar
1 lb. seedless raisins
1 cup suet, chopped finely
½ cup vinegar
1 Tbsp. of each: cinnamon, cloves, nutmeg, allspice and salt

Combine all ingredients and bring slowly to the boiling point. Reduce heat and simmer until quite thick.

Option: Omit the suet, and stir in ¾ cup butter after the mincemeat has simmered.

GREEN TOMATO KETCHUP

8 lbs. green tomatoes
3½ lbs. sugar
2 lbs. onions
1 green pepper
1½ qts. cider vinegar
½ cup pickling salt
In a tightly woven spice bag:
    1 tsp. cinnamon
    1 tsp. powdered cloves
    ½ tsp. cayenne

Wash and stem tomatoes and peppers. Chop onions, tomatoes, and green pepper coarsely and soak overnight in pickling salt. Drain and rinse. Measure tomato mixture and place in a kettle with an equal amount of water plus 1 pt. of vinegar. Cook for 1 hour, making sure that the tomatoes are cooked through. Remove from heat and drain.

Boil the sugar, 1 qt. of vinegar, and the spices for 1 hour. Let cool and mix with the cooked, drained tomatoes. Pack in clean hot jars and seal. Makes 4 to 6 qts.

Vin Sauvage

Rare vintages that are as close as that patch of elderberries

By Louise Price

"Scratch a hippie," the old sixties' saying went, "and you'll find a Porsche."

Scratch a back-to-the-lander, we suspect, and you may find a bottle of Chateauneuf-du-Pape.

More accurately, you may find someone with an appreciation for good wine, but someone in a quandary when faced with either today's commercial pop wines or the stifling prices of imported vintages.

In fact, in toting up our progress toward self-sufficiency five years ago, we found that beer and wine were two expenses that seemed to be hanging on. We both enjoyed a drink, but going to the liquor store seemed increasingly contradictory to our chosen life style.

The knowledge that modern wines are rapidly regressing from the stomped-by-peasant-feet, natural product of the past was a further incentive to seek an alternative. Although they are never listed, chemical additives now lace almost all domestic wines — ingredients that artificially age, enrich, colour, ferment and polish the brew.

Donald McCallum

With all this in mind and a high sense of adventure, we set off for a winemaking supply store and came away with a miniature winery — airlocks, siphon hose, yeast, corks, corker, instruction books and two large glass carboys. Most beginning winemakers can get by with these few tools, and even the carboys aren't really necessary. However, they do hold 10 gallons each, and, in our small house, two of these seemed preferable to dozens of one-gallon jugs scattered around.

That first year we blindly tried many recipes. Luck was with us on most attempts, and the only two wines that proved undrinkable were strawberry and raisin. Even these mellowed after several years of aging.

Cool and refreshing on a hot summer's eve, or hot and spicy after a day's snowshoeing, homemade wine is a great favourite at our house. It has also proved to be an exciting and appreciated gift and is much sought after in a barter system.

For those living in northern climates, the warm months of the year are best for starting batches of wine. Though ecological, a house kept at 60 degrees in winter is too cool for active fermentation, and the spring and summer months provide the best supplies of fresh ingredients.

Natural wines can be made for pennies a bottle, providing you gather most of the ingredients yourself and that you find a free source of bottles. Spring flowers, for example, cost nothing and they make a very special wine that can capture the essence, aroma and memories of sun-filled days — perfect for uncorking in the midst of a snowstorm.

THE BASICS

Whether the wine you choose to try first is dandelion or wild grape, the basic procedure remains the same. There is no such thing as a fool-proof recipe, but most winemaking failures result from three things:

A) Failure to be scrupulously clean. Wild yeasts and bacteria give vinegar and other grimace-producing results.

B) Using too much sugar and/or bottling too soon. Exploding bottles befall people who do not follow the simple rules. Many home vintners simply use too much sugar, and the results are wines that are rough-tasting or cloying.

C) Inattention to the fermentation process. A batch of wine left too long in the open air has little chance of success.

A good recipe book will help you over the rough spots and give some of the refinements for more advanced brewing, but the basic method always follows this form:

1. Place the main ingredient (fruit, flowers, herbs or vegetables) and the sugar in an open container. The initial fermentation can be quite wild, so a container with a narrow neck should be avoided here. Large plastic garbage pails work fine for this.

2. Pour boiling water over the ingredients and stir. This process helps release the essence of the flavouring ingredient into the must (must is the still unfermented mixture of fruit, sugar and water).

3. Allow to cool to a lukewarm 75 degrees (F). As in breadmaking, high temperatures will kill the yeast, while cold will either slow it down or stop its action altogether.

4. Add the yeast. Do NOT use breadmaking yeast, as it does not produce enough alcohol, in many cases, to preserve the wine. True winemaking yeasts are not expensive, they give a cleaner tasting wine and they are less likely to produce a batch of vinegar.

5. While the primary fermentation takes place, cover the container with a cloth and wait the number of days indicated in the recipe. Be sure the cloth is tied tightly around the rim, for fruit flies and other insects may be attracted. Place the whole container in a place with a warm, even temperature (65 to 75 degrees).

6. When the most violent fermentation has subsided, siphon off the liquid into carboys or gallon jugs (secondary fermentors), being sure not to include the layer of yeast cells at the bottom of the primary fermentor.

Fit an airlock over each container. This can be as simple or as ornate as you wish, but its function is to allow carbon dioxide to escape from the fermenting mass while excluding air and bacterial organisms.

7. After filling the carboys, wash out your primary fermentor and add the dead yeast sediment to your compost heap.

8. Leave the wine in the secondary fermentation jugs in a darkened place until all fermentation has stopped (i.e. all bubbling will have ceased).

Now rack the wine (take the liquid and leave the sediment) and bottle it. The bottles should be clean and sterilized (by boiling or with sterilizing solution from the wine supply store). Cork or cap and store the bottles on their sides in a cool place. Try to resist sampling them for one year. All of the following recipes are for one gallon of wine, but remember that larger batches have a better rate of success. Simply increase the quantity of each ingredient proportionately — except for the yeast. A single packet of yeast will usually multiply to ferment five gallons or more (read the label).

BLUEBERRY WINE

This same recipe may be used for huckleberries, raspberries, gooseberries or currants. Blueberry wine is the only "country" wine we've found that is anything at all like grape.

The amount of sugar you use can vary, depending on the sweetness of the berries. With this and other berry wines, we add the sugar in four stages.

I add a quarter of the sugar initially, and when the first vigorous fermentation subsides, add another quarter. Almost immediately the wine will begin "working" again. Three weeks later the secondary fermentation has slowed, so another quarter is added. A month later the rest is mixed in. This process has given us our top wines.

2 - 3 lbs. blueberries
2 - 3 lbs. sugar
½ oz. citric acid (or substitute sliced lemon and orange)
yeast (plain wine yeast is fine, but try Burgundy if you can get it)

Add yeast nutrient (see package for quantity) to this and ferment for five days in an open pail. Use the following method:

1. Bring water to a boil and add sugar.
2. Allow to cool to lukewarm.
3. Add the crushed berries.
4. Add the yeast, nutrient and citric acid.

This type of wine leaves quite a bit of sediment after the first fermentation (berry seeds and skins) and extra care should be used when siphoning off. (It is always better to leave behind a few ounces of wine than to have the whole batch clouded with sediment.)

During fermentation, the berry skins may rise and coat the surface of the must. This "cap" should be broken several times a day and the skins mixed back into the liquid to help prevent spoilage and the formation of acetic acid.

PLUM WINE

I've used this recipe with both damson and prune plums, the first giving a pale yellow wine and the second a rich pink. Both are excellent if allowed to age the full year, although prune wine is tasty after six months.

4 lbs. plums (pitted and diced)
2½ lbs. sugar
yeast (Burgundy or Port yeast is best)

Use the same procedure as for blueberry, allowing five days for the primary fermentation.

DANDELION WINE

Dandelion wine can be terrible, but it can also be terrific. If you can remember tasting some of the horrible kind in your earlier years, don't let that memory put you off. A common oversight that produces bitter wine is the cleaning of the flowers. Snip off all green stems, leaves and blossom tips, leaving just the ripe flower.

The picking itself can be tedious — make it a social occasion or picnic. As with all flower wines and wild foods, don't pick along heavily used roadsides — dust, exhaust and even herbicides will ruin the wine. I find the best way to gather dandelions is to pick the heads in the field or yard and then trim the green parts with scissors at home.

2 qts. dandelion flowers (packed firmly)

2½ lbs. sugar
2 chopped oranges
1 packet (or part thereof) wine yeast
(read the label)

Ferment the mixture in the open pail for one or two days. Siphon off and follow the method as given above.

GOLDENROD WINE

This is a truly golden wine. I was needlessly skeptical about this one for several years, but it has turned out to be the type of wine that, if you make a gallon, you'll later wish for 10.

1 gal. tightly packed goldenrod flowers
(trim away the stems)
2½ lbs. sugar
½ oz. citric acid (or lemon and orange slices)
yeast (you might try a Sauterne yeast, but a general wine yeast will do)

Some home winemakers like to add some yeast nutrient and Campden tablet to this wine. Ferment in the open pail for five days.

ROSE PETAL WINE

This is an especially delicate wine that has been made for centuries and was especially popular in feudal Great Britain and France.

½ gal. rose petals (tightly packed)
3 lbs. sugar
½ oz. citric acid (or substitute fresh sliced

orange or lemon)
yeast (Chablis, Champagne or a general purpose yeast)

The primary fermentation takes about six days, after which the wine should be put in the carboy and airlocked.

NETTLE WINE

Nettles are known mainly for their ability to irritate the skin, but they can be used to make a fine, delicate wild wine. Use gloves and long sleeves when gathering to avoid being stung. We have just opened our first batch of nettle wine and it is delicious — certainly an experiment worth repeating. (Many people have already discovered the worth of young nettles as a spring potherb.)

3 qts. bruised young nettle tops (rub the tops vigorously between your gloved hands)
3½ lbs. sugar (use 2½ if you prefer a really dry wine)
½ oz. citric acid (or add a few oranges and lemons, sliced with peels intact)
yeast (a Hock yeast is suggested, if you can get it)
1 Tbsp. ginger powder per gallon (or slice up some whole ginger root)

Place in primary fermentor for five days before racking to the airlocked carboy or jugs.

While harvesting your garden, you'll find many vegetables that can be made into wines. I have tried only two types at this writing, both of which are still aging in the wine cellar. Turnip,

from two years ago, is still known as "Old Root Cellar" here, but we hope that one day it will mellow enough to drink.

All vegetables are deficient in the balancing ingredients essential to good wine, so tannin, acid and sugars will have to be added. Experiment with such vegetables as beets, broad beans, cabbage, carrot, celery, kohlrabi or turnip, using the general guidelines and ingredients of the recipes above.

The flower wines also offer a fertile area for experimentation — chamomile, made with one quart of the blooms, is one of our favourites.

WILD WINES

What to do with all that rhubarb

Recipes by Magret Paudyn

RHUBARB WINE

This is a pale amber wine with an unusually vinous quality. It should be aged for at least a year, and longer gives better results.

5 lbs. rhubarb
1 gal. water
3 - 4 lbs. sugar
½ lb. diced raisins
1 lemon, cut up
1 orange, cut up
½ oz. yeast
1 slice of toast

1) Cut the rhubarb stems into 1-to-2-inch pieces and place in crock or pail. 2) Pour sugar into crock, mix with rhubarb and let stand overnight, covered. 3) Boil water and pour over the rhubarb/sugar. 4) Let stand for 6 to 8 days, stirring once each day. 5) Strain through heavy cloth until the pulp is very dry.

6) Heat liquid on stove and bring close to the boiling point to kill any foreign yeasts. 7) Add raisins and the lemon and orange to the hot liquid. 8) Dissolve yeast in ½ cup warm water. 9) Place yeast on top of toast and set afloat on the must. Cover, do not disturb.

10) When traces of foam appear on top of the must, remove the toast (this takes about 4 days). Complete fermentation takes 14 days. 11) Strain into jars and stop with fermentation locks. 12) Rack at least 3 times before bottling — after 2 weeks, 6 weeks and 3 months.

RED OR BLACK CURRANT WINE

Red currants give a flavourful, still-rose-type wine, while the black currant produces a rich, heavy wine much like port. Age at least one year.

8 to 10 qts. fresh currants
2 gal. boiled water

7 to 8 lbs. sugar
½ oz. yeast
1 slice of toast

1) Place currants in crock and mash well. 2) Boil water, pour over fruit. Let stand 24 hours. 3) Strain mixture through cloth until the pulp is very dry. 4) Warm the liquid on stove and add sugar. Cover and place in a warm spot.

5) Dissolve yeast in ½ cup warm water. Spread over slice of toast and set afloat on the mixture. Cover and set in a warm place. 6) When slice of toast has completely "fallen away," after about 4 days, remove it. Re-cover the mixture and leave it alone for 2 to 3 weeks. 7) When fermentation has stopped, strain wine to glass bottles with fermentation locks. 8) Rack after 2 weeks, 6 weeks and 3 months. Bottle when clear.

WILD GRAPE WINE

Wild grapes may not always have the best qualities for eating, but they make a very dark wine — almost blue-black — that is distinctive and unlike any other I've tasted.

2 gal. wild grapes
2 gal. water
5 lbs. sugar (4 lbs. for drier, 6 for sweeter)
½ oz. yeast
1 slice of toast

1) Fill a 2-gallon container with ripe wild grapes with stems. This doesn't take as long as it might sound. For sweeter berries, wait until after the first frost. 2) Wash to remove all bugs, but do not let the grapes remain in water too long. 3) Place in crock and mash with wooden mallet or hands. (Be advised that the Indians used the juice from wild grapes to dye leather.) 4) Boil water and add to grapes. 5) Add sugar, stirring until dissolved. Let stand 4 hours or overnight (covered). 6) Strain through heavy cloth or sieve. Remove as much juice as possible from the pulp. 7) Dissolve yeast in ½ cup warm water and spread over the toast. Set this afloat on the must. 8) Ferment 14 days. 9) Siphon into secondary fermentation vessel with airlocks. 10) Rack in 4 weeks and again after 8 weeks. Rack once more if necessary before bottling.

ROSE HIP WINE

Rose hips are the fruit of the rose and are one of the most concentrated sources of vitamin C found around the home or farm. They are seen when the petals fall, and appear as a bright scarlet when ripe. The pods should be gathered before they start to darken in colour.

Rose hip wine is clear, light and rose-like, of exceptionally good taste.

2½ qts. rose hips
6 lbs. sugar
2 gal. water
½ oz. yeast

If using dried rose hips, soak 2 quarts of them in clear water until they soften. (This may take 2 days.)

1) Mash rose hips between your fingers or grind them. 2) Place in cooking pot with 1 quart water (more if needed), and bring to a boil. 3) Add all sugar while still boiling and stir to dissolve. 4) Cover and simmer for 1 hour, removing the lid at intervals. 5) Transfer to crock, add water and stir well. 6) Add yeast as per instructions on package. 7) Ferment for 2 weeks or until all activity ceases in the must. 8) Strain through sieve. 9) Let stand for 1 week, do not stir. 10) Siphon into glass container with small neck opening and airlock. 11) After 3 or 4 weeks, rack very carefully into bottles, trying not to disturb the sludge on bottom of the jar. 12) Rack again at intervals if necessary.

MAGIC METHEGLYN & THE ELDER MOTHER

Winemaking without sugar

By Gillian Thomas

Metheglyn is the subtlest of wines, but the first time I made it, a hammer, an axe and a butcher's cleaver had to be brought into play. Honey is the base for metheglyn, and anyone who has dealt with good, solidified honey, a year or more old, will know the problems I faced.

What most people know as a runny, sticky liquid transforms itself into an incredibly unspreadable, unpourable lump. A friend had bought a 35-pound bucket of honey in its seductive summer state, and now, many months later, had broken several kitchen knives in attempting to hack out small quantities for cooking.

It seemed a perfect time to make a huge batch of metheglyn, so we agreed that I would have unlimited free honey to experiment with in hopes that we would be able to share any resulting drinkable wine.

Metheglyn is an old Welsh word for medicine. Basically it is a *mead* (simply honey wine) combined with various herbs. Some writers suggest that it may well be the oldest form of winemaking known. At any rate, it seems to have been very widely used in northern European countries where the climate wasn't sufficiently warm to grow good grapes.

Herbs that can turn ordinary mead into the magic of metheglyn, can be selected from a huge range of wild and cultivated plants. Almost any herb that you enjoy in a herb tea will probably make a pleasantly flavoured wine.

Here is an interesting experiment which can lead to many delightful "instant" herb wines, and which will give an indication of whether you'll want to use a particular herb to make metheglyn.

Buy a bottle of the cheapest, *dry* white wine your local liquor store offers. Pour a cupful into a clean, pint-sized Mason jar and add a table-spoonful (fresh or dried) of the herb you wish to test.

Cover the jar and keep in a cool place. After 24 hours strain off the herbs and taste the wine, which will now have absorbed about the same amount of the herb's flavour as a "finished" wine made from the same plant.

The bonus here is that you may well discover a herb, or combination of herbs, which redeems an otherwise undistinguished wine. One of the less drinkable wines sold in Nova Scotia liquor stores becomes a very pleasant apéritif when steeped with a little woodruff for a day.

Despite metheglyn's venerable history, honey is a trickier substance to use in winemaking than ordinary sugar, so it helps to understand something about the basic process. "Wine" is produced by the action of yeast on sugar. Under suitable conditions, fermentation takes place and the sugar is gradually converted into alcohol. Whether the result is drinkable depends on the ingredients used and the conditions during fermentation.

THE PROCESS

Stores specializing in winemaking supplies have all the equipment and ingredients you'll need for making metheglyn (or any other type of wine, for that matter). I've marked (*) the ingredients which are most easily obtained through a specialist store. However, you can substitute for most of these items in the ways I've suggested, without serious loss of quality.

EQUIPMENT

— large saucepan (at least 3 quarts capacity)
— a 2-gallon plastic bucket (the primary fermentor)
— 2 one-gallon capacity glass jugs (secondary fermentors)
— about 2 yards of plastic tubing with a siphoning attachment*
— a piece of gauze or cheesecloth to use as a filter
— plastic funnel
— plastic or glass fermentation lock*

Have in reserve, but do NOT add to the wine, a package of sodium metabisulphite in powder form and a package of sodium metabisulphite tablets (Campden tablets*).

INGREDIENTS

— 1 teaspoon yeast nutrient (ammonium phosphate)*
— one 15 mg Vitamin B1 tablet (crushed)
— ¼ tsp. Epsom salts
— any wine yeast*
— 2 pounds honey
— 1 pint white grape concentrate*
  OR 8 oz. sultanas
— small pinch grape tannin* OR
  ½ cup strong cold tea
— herbs (see suggestions below)
— water to make mixture up to one gallon

PROCEDURE

The first essential is that your equipment be as clean as possible. This is to prevent the introduction of wild yeasts which exist everywhere and which will usually produce a very sour wine if allowed to enter a fermenting mixture.

The best way to sterilize equipment is to make up a sterilizing solution by diluting the sodium metabisulphite (about one ounce to a gallon of water) and use it to clean your glass jars, plastic bucket and plastic tubing. This solution gives off carbon dioxide, so it's advisable to use it in a well-ventilated area. You can rinse the equipment with boiled water which has been allowed to cool slightly.

Dissolve the honey in two quarts of water in the large saucepan over low heat. Don't bring it to a boil, or some of the subtler flavours of the honey will be lost. Put the herbs in a plastic bucket and pour the warm honey and water over them. Add all other ingredients, except the yeast. When the liquid has cooled to about 70 degrees F (21C), sprinkle the wine yeast on the surface. Cover the top of the bucket with a plastic bag and keep in a warm place for four or five days. As the mixture ferments, the yeasty, fizzy crust rises to the surface and should be very gently stirred back into the liquid several times a day.

Strain the fermenting mixture into a clean gallon jug and cap it with a fermentation lock, half-filled with sterilizing solution. This keeps out undesirable wild yeasts and wine flies. (Either of these will turn your wine to vinegar.) It has the added advantage that the two levels of liquid within the lock indicate that fermentation is taking place. When the liquid stays at one level you know that fermentation has ceased.

After two or three weeks in the secondary fermentor (it may be much longer if your house is fairly cool) fermentation should have come to an end. The wine will look clear and inviting and will contain about 10 to 15 per cent alcohol. Don't imagine that it is in any way drinkable, however.

Some of the best wines I've ever made have smelled and tasted appalling at this stage. With practice you get better at distinguishing the promising from the doomed quite early on, when they all seem uniformly disgusting to the novice.

Siphon the wine into your second clean gallon jar, being careful to leave the bottom half-inch or so of sediment behind. Inevitably, you also leave behind about a cup of clear wine, but this sacrifice is preferable to a cloudy mess of stirred-up dead yeast cells. Top up the second jar with water to eliminate any air space.

At this point, most winemaking books recommend adding a Campden tablet to kill off any further fermentation before allowing the wine to mature. This certainly saves explosions and is essential if you plan to bottle the wine in screw-top bottles.

Wine which decides to re-ferment in these containers produces instant shrapnel which can fly a considerable distance. If you fancy taking a risk, however, and plan to use proper wine corks you can skip the Campden tablet and may be lucky enough to produce a delightful *petillant* (slightly sparkling) wine.

CRAFT & MYSTERY

That, basically, is all you need to know to make an experimental gallon or two. If you plan to go beyond that I'd strongly recommend that you borrow or buy a basic winemaking book which gives instructions on some of the more esoteric aspects such as the use of hydrometers, yeast starters and the like.

However, even with the "scientific" approach such books usually promote, the difference between two batches of the "same" wine can be quite startling. The old mead-making guilds recognized the rather chancy nature of the thing when they spoke of learning "the craft and mystery of mead making."

If you plan to gather wild herbs for wines, you should, of course be absolutely certain that you have positively identified the plant in question. This is especially important when gathering plants for wine, rather than tea, because there are many herbs which yield their constituents more readily to alcohol than to boiling water. Consequently, you need to be sure that you know what you have picked and that it's safe to ingest.

These are some of the herbs I've used with consistent success in making metheglyn:

RED CLOVER (Trifolium pratense)

Use from two quarts to a gallon of the fresh flowers, or about two cups of dried flowers. Red clover can be found very widely on cultivated land, but you might bear in mind that in some of these locations you will be gathering chemical sprays along with your herbs. Clover is believed by some to benefit the liver as well as to soothe bronchial disorders. The resulting wine is smooth-tasting and sometimes develops a pinkish colour.

HAWTHORN (Crataegus oxycantha)

The flavour is similar to clover, but a little stronger and more aromatic. You will need about two quarts of the fresh flowers or a cup of dried. You can often find hawthorn trees forming hedges on old farms, but beware of the sharp thorns which seem to project in every direction. Despite its thorns, hawthorn has a very high reputation among herbalists as a remedy for high blood pressure and heart disease. It also seems to have mildly sedative properties.

GOLDENROD (Solidago)

You can choose from more than a hundred known species and pick goldenrod by the bucketful on "waste" ground in every province. Two or three handsful of fresh flowers will be enough, however, to make a fine wine. It tastes and smells particularly noxious in the early stages, but when aged for a year or more produces a lovely dry wine rather like the prohibitively expensive *Tio Pepe* sherry.

YARROW or MILFOIL (Achillea millefolium)

After the dandelion, yarrow is the most commonly occurring Canadian weed. Evidently, it established itself in North America after being brought here by early settlers who found it useful in staunching bleeding wounds as well as for colds and fevers.

Yarrow produces the most "herbal" tasting of the wines mentioned here, but makes a very consoling drink when you're choked up with a head cold. Use about two cupfuls of fresh flowers or one cupful of dried.

ELDER (Sambucus nigra and Sambucus canadensis)

Northern European peasants thought that a female nature spirit, the Elder Mother lived within the elder tree, and they carefully asked permission with this rhyme before using any part of the tree:

Owd girl, owd girl, give me
some of thy wood
And I will give thee some of mine
when I become a tree.

In Russia it was believed that the Elder Mother's spirit then accompanied whatever parts of the tree one took and that she drove away evil influences and gave long life. All parts of the elder have been used in folk medicine and in magic, but the elder flower is the champagne of herb and flower wines. If the flavour alone weren't enough to recommend it, those who have used it can vouch its remarkable effect on the early stages of a head cold.

The wine needs only two cups of the whitest, freshest flowers you can find or one cup of dried flowers, Remember, by the way, that neither you nor the birds will have the benefit of elderberries if you pick all the flowers.

You can experiment with innumerable combinations of herbs. Clover and hawthorn go well together. Elder and yarrow combine splendidly. You can also, of course, use these herbs and many others substituting sugar for honey. It's cheaper and fermentation using sugar tends to be surer, but the flavour is usually less subtle.

Despite the added trouble and expense, there's an odd satisfaction in producing a wine whose connection with the insect and plant world is so complex and mysterious.

Besides, I have a feeling that it pleases the Elder Mother.

Suggested Books

Bryan Acton: *Recipes for Prizewinning Wine,* Amateur Winemaker, 1971.
B.C.A. Turner: *Recipes for Home-Made Wine, Mead, and Beer,* Mills and Boon, 1972.
S.F. Anderson with R. Hull: *The Art of Making Wine,* Longman, 1974.

Fear Of Fungus

---

*Stop quivering and read this Harrowsmith guide
to the edible wild mushrooms*

By G. I. Kenney, Jo Frohbieter-Mueller and the
staff of Harrowsmith Magazine

"He was a bold man,
Who first eat an oyster. . . "

--Jonathan Swift,
Polite Conversation, 1738

"Toadstools," sniffed our uptown friend recently. Oblivious to Swift's observation, he slipped back a glistening Malpeque Bay oyster, swimming in its own gelatinous juices on an iced half shell. "Wouldn't be caught dead eating wild mushrooms."

Even more recently a friend slightly better attuned to wild foods and country ways scanned a light table filled with colour transparencies of wild fungi. "God, I hate to think what would happen to you if you ate one of these things," he said.

The slide he had picked showed a pair of morels — mushrooms so prized by Central Europeans in Medieval times that they burned whole forests in order to bring flushes of the delectable fungi the following spring.

With cases of mushroom poisonings still commonplace, however, it is easy to understand the trepidation with which many people approach the idea of picking and eating wild fungi. The sinister image is, of course, overblown, as anyone who has discovered the art of mushroom hunting knows. The results are delicious and the search itself lends new purpose and a sense of challenge to walks, day excursions and camping trips.

Like many of the folk arts, mushroom hunting is a skill that in the past was handed down from one generation to the next. Unfortunately for many of us, a disinterested generation or two may have slipped in. Perhaps the best way to learn about edible fungi is at the elbow of an experienced mycophagist (mushroom eater), but as many have learned, you can easily teach yourself the secrets.

Several years ago a friend and I became interested in wild mushrooms and found a copy of *Mushroom Collecting For Beginners*, a Canadian Department of Agriculture booklet. We searched and searched through fields and woods, but with no luck. It was June — we now know that it was too early in the season for the summer and fall fungi and too late for the spring mushrooms.

Our lack of success at the time made us lose interest, and we soon forgot the whole project.

Then one day my friend's young son, Craig, rushed into the house shouting, "Daddy look what I found!" He proudly held aloft a beautiful, fleshy, whitish mushroom.

"Where on earth did you find that?" we asked in disbelief ingrained by hours of our own futile searching.

"On the front lawn," Craig beamed.

Sure enough, there on the lawn were a half dozen more mushrooms, just pushing their heads above the thick grass. We took them inside, got out our government booklet and after careful examination decided they were meadow mushrooms (*Agaricus campestris*), a close relative of the common cultivated mushroom and eminently edible.

We cooked them in butter and ate them with toast — but not without the typical beginner's apprehensions; we reread the booklet several times after the meal just to reassure ourselves. But they were delicious, no one died and that was the start of a fascinating pastime that has yielded many a fine meal.

There is no magic involved in the identification of mushrooms. It's simply a matter of familiarity, in the same way that raspberries and strawberries are known and picked by almost everyone, while other red and perhaps poisonous wild berries are avoided. We started with a 35 cent government pamphlet and were soon able to recognize a number of common edible mushrooms and to reject the poisonous ones.

Familiarity in this field does not, however, breed contempt. Some species — very commonly found species at that — are so deadly that even one small bite is enough to kill.

For North Americans, the primary concern is the *Amanita* family, which has been estimated to be responsible for 90 per cent of all mushroom-eating deaths. They are common in pub-

Right, *the prized morel, perhaps the best of the spring mushrooms, often found in open woods or burned-over areas.*

Below, *Pleurotus ostreatus*, the Woods Oyster or Oyster Shell, is a common mushroom that is both delicious and easily identified. Found from spring through fall, it is distinguished by its white colour, its white spore print and the lack of a stalk. It grows on fallen trees and stumps, and Europeans have been known to water fresh stumps to encourage its growth. **Right,** the Golden Chanterelle is one of the most prized and popular of the edible wild mushrooms. It has a yellowish spore print, gills that extend down the stalk and, when mature, a dished cap. Do not confuse with the poisonous Jack-O-Lantern, which has a white spore print and which has the ability to luminesce in the dark.

Right, the stately Parasol mushroom, *Lepiota procera*, regarded by many as one of the finest edible mushrooms to be found. It grows in weeds, lawns and wooded areas from summer through fall, and has both white gills and a white spore print. The cap may be up to 10 inches in diameter. It is identified by having a ring which can be moved up and down the stem and by the scales which are an integral part of the cap (not detachable and wart-like as in the *Amanitas*). Check your mushroom guide and know the *Amanitas* before eating this one. **Above,** *Pholiota squarrosa*, a large fleshy fall mushroom that is found on hardwood or conifers. It is edible and has a brown spore print.

216

Left, *Hericium coralloides*, a Coral or Lion's Mane fungus that is not widely known as an edible mushroom, but which is delicious when sautéed in butter. It grows in heavy clumps on logs in late summer and fall, and is distinguished by its white, fleshy body and numerous "teeth" which point downwards. **Below,** a member of the genus *Boletus*, mushrooms which have pores rather than the more common gills. The only poisonous members of this genus show blue when bruised or have red pores (or both). These specimens, photographed in the Muskoka region of Ontario, appear to be *Suillus cavipes*. Positive identification would show them to have a hollow stalk and a dark brown spore print. They are found under tamarack or larch and are eminently edible.

G.I Kenney

NFB Phototheque

Bruce Gunion

Bruce Gunion

Left, the False Morel, *Gyromitra esculenta*, which is eaten without trepidation by some, has been known to cause sickness and death. It is avoided by knowledgeable mushroom hunters. Compare it to the photograph of the true morels on previous page. It can be seen that the false morel has a wrinkled cap that has the appearance of being deformed, while the true morel is pitted with ridges. **Above,** *Lycoperdon perlatum*, one of the many species of edible puffballs, which range from marble size to five-foot monstrosities. Never eat a puffball that has coloured centre flesh, or which shows a stalk or gills when cut in half, top to bottom.

*Just emerging **Amanita muscaria**, with the characteristic tufts of volva clinging to the future cap.*

An adult specimen of the same species, which was once used to make house fly poison and sometimes known as the fly agaric.

*The poisonous **Amanita muscaria** in its lemon-yellow form — colours vary widely in this mushroom.*

Left, *early spring bounty of huge morels and fiddlehead greens. Harvesters often return to the same harvest spot each year, keeping the location a well-guarded secret.*

---

lic parks and in one season caused the death of more than 30 foolhardy mushroom pickers near New York City.

The *Amanitas* are particularly insidious, in that they are said to taste very good and produce no symptoms of poisoning for eight to 12 hours. By this time it is too late for the victim, as the toxins have entered his bloodstream and, converted by liver enzymes, are destroying liver cells.

At this point it is of no use to pump out the stomach and the victim often experiences extreme pain, blurred vision, wretching and lethargy. A period of relief may follow as a second group of slower-acting toxins begins to take effect, but severe pain resumes for up to six days, often followed by death. Even those who do not die are ill for several weeks and may have to live with the handicap of permanent liver damage.

As frightening as these prospects are, they shouldn't turn you away from mushroom hunting. In fact, if you follow one simple rule there is no reason ever to fear a dish of wild mushrooms. *Do not, ever, eat a mushroom which you have not positively identified.*

In some ways, it actually can be more reassuring to learn your mushrooms one by one through reliable books. There are people who claim to know their mushrooms and who have "foolproof" ways to test for poisons. Others employ European knowledge to North American mushrooms — sometimes with disastrous results.

Certain *Amanitas*, for example, have long been eaten and esteemed in Europe. Unfortunately, the species of *Amanitas* found in North America far outnumber those in Europe, and several of the deadly types closely resemble those that are edible.

The notion that there are quick and easy tricks for detecting mushroom poisons is absurd and should be debunked right now. There is no scientific or practical evidence to back the old wives' tale about poisonous mushrooms causing a silver spoon to blacken on contact or when dipped in the cooking pot.

It is all too commonly said that a poisonous mushroom cannot be peeled, but the virulent, *Amanita virosa* — aptly named the Destroying Angel — can easily be skinned.

POISON-PROOF MUSHROOMS

Neophytes will take heart in knowing that there are several mushrooms that are both delicious, relatively common and impossible to confuse with any poisonous variety.

We have chosen to include the morel, the puffball, the shaggy-mane and the woods oyster, as four foolproof mushrooms that every beginning mycophagist should know. These fungi are so distinctive that, having once seen a photograph

or an actual specimen, there should be no problem in recognizing them in the field.

The Morels

Scientific name: *Morchella esculenta.*
Common names: Morel, Sponge Mushroom, Pine Cone Mushroom.
Season: Spring only.
Spore Print: Yellowish.
Size: 2 to 6 inches high, cap ¾ to 1½ inches at thickest part; huge morels — in excess of 8 inches — are sometimes found and are perfectly good eating.
Colour: Tan, whitish or yellowish; or brown cap with tan or whitish stem. All edible (see text).
Features: Cap resembles pine cone in shape, with a latticework of irregular ridges and furrows or pits. Both stem and cap are hollow and slightly brittle.
Habitat: Open woods, creek banks, old fields and orchards, burned-over areas.

Morchella is the Greek word for mushroom, and, to many people, the morel is the only true mushroom. It is not only the easiest to identify — being almost impossible to confuse with any other fungus — but is generally regarded as one of the very finest to eat. All of the many attempts to domesticate this mushroom have failed.

Look for morels when the spring rains begin to soak the land and the sun begins to warm the earth. This, of course, may be in April or, in more northern areas or at high altitudes, well into July. At about the time spring flowers begin to appear, morels can be found in old orchards, woodland hillsides, along streams and in damp, low-lying areas.

They can be found by the bushelful in earth that has been recently burned over by forest fire.

Mushroom hunting requires alertness, as the earth-tones of many species help camouflage them, and this is especially true for the morel. Until you become familiar with their growth habits, they can be difficult to spot as they stand — brown or tan or yellowish — against the brown, mottled leaves of the forest floor.

As morels tend to reappear in the same spots each spring, don't be surprised if experienced hunters won't share their stalking grounds. This is typically a well-guarded secret — I would rather give someone my full day's find of morels than tell where I found them.

My best spot happens to be just downhill from an ancient privy that sits in the woods. Fed by the woodland humus and the nutrients leached from the old backhouse, morels here are so large and fleshy that they can trip you up (under such conditions morels can reach eight inches or more in height). Big, thick-skinned morels are excellent stuffed with a mixture of fried onions and sausage mixed with bread crumbs.

It is wise to cut each morel collected lengthwise to reveal any insect which may have taken lodging in the hollow interior.

Meadow Mushroom

Shaggymanes

There actually are several species of morel found in North America, all choice eating. *Morchella angusticeps*, or the Black Morel, has a white or light tan coloured stem and a brown cap in which the ridges and pits are all the same dark shade. This mushroom is found even in the Yukon and Alaska, especially in burned-over areas.

The Common or Yellow Morel (illustrated at the opening of this article) is a uniform tan or yellowish colour and is known properly as *Morchella esculenta*.

The White Morel *(M. deliciosa)* has a white stem and ridges, with dark or black furrows, while the Cow's Head Morel has a shallow, light brown cap that covers not much more than the tip of the stem.

The only other fungus that might confuse a neophyte mushroom hunter is the False Morel, or *Gyomitra esculenta* (see photograph). It is generally shorter than the common morels, and its head appears deformed or wrinkled and convoluted, in contrast to the latticework surface of the morels. It, too, is hollow and comes up in the early spring, although it can be found throughout the summer and fall.

The False Morel is one of the more perplexing mushrooms for professional mycophagists, for it is widely eaten and enjoyed in Europe. The Germans call it *lorchel*, and J. Walter Groves, author of *Edible and Poisonous Mushrooms of Canada* says, "I have seen it on sale in grocery stores in Finland and have eaten it myself when it was served by friends there, and the flavour is excellent."

Nevertheless, hundreds of deaths have been attributed to the False Morel, both in Europe and North America. It remains unknown whether there are actually different species with similar external appearances, one poisonous and the other edible, or whether some people are simply vulnerable to the poison and others immune.

A case reported in 1911 further confuses matters, for a family reportedly ate part of a collection of False Morels with their evening meal and had no ill effects. When they ate the remaining mushrooms the next day, however, all became seriously ill and one person died.

For a time it was believed that parboiling removed the poison. This has been proved false, and mushroom hunters are advised to avoid the False Morel.

Puffballs

Scientific name: *Calvatia maxima, Calvatia caelata, Lycoperdon perlatum*, and others.
Common names: Puffball, Snowball, Devil's Snuffbox.
Season: Spring through fall.
Spores: Brown, olive-brown, lilac, mustard yellow.
Size: ½ inch to more than 5 feet in diameter.
Colour: Exterior white or tan, darkening with age.
Features: Range from marble size to the diameter of a basketball to newsworthy monstrosities. Interior pure white while prime, becoming coloured and inedible as the puffball matures. Skin may be smooth, irregular or covered with small spines or scales.

Habitat: Woods, fields, parks, hedgerows, on logs and stumps or on the ground.

For sheer size and ease of identification, the puffball ranks as one of the favourite North American mushrooms — rare is the year in which one or more of the major news services does not carry a photograph of a small child dwarfed by a giant puffball.

Actually, recent puffball finds are no match for a behemoth discovered in New York State in 1877. Mycophagist J. Ramsbottom reported that it measured 5 feet, 4 inches in length and 4 feet, 6 inches in width. It was mistaken at first sighting for a sheep. Ramsbottom calculated that even a 16-inch-by-12-inch-by-10-inch puffball, if each of its 7 trillion spores were to germinate and produce a new puffball of the same size, would produce fungus that would outweigh the earth by some 800 times.

Actually, there are many varieties of puffball, and the smaller ones, though less dramatic, are probably more common and better eating.

The foolproof rule of thumb for puffballs is that they are all edible as long as the flesh inside is pure white. The only poisonous puffball known has a purple centre even in its early growth stages. All puffballs are members of the general fungus group known as the Gastromycetes. Gastro means stomach, and these mushrooms are characterized by internal spore production. The flesh that is white and edible is known as the *gleba*, or spore mass, and as the puffball grows old it dries out in preparation for release of the spores.

The second ironclad rule for eating puffballs is to cut them open — never to eat them whole. All specimens should be sliced in half to check for insect or worm damage, and, most important, to assure that small button-like "puffballs" are indeed puffballs and not the immature stage of a poisonous *Amanita*. If the halved specimen shows signs of a stem running from bottom to top or any indication of gills, discard it. True puffballs are a homogeneous white inside.

Puffballs may be found throughout the growing season, but are most prevalent in the fall. Although many people believe that they spring up overnight, the larger specimens actually only develop at a rate of 3 to 4 inches per day.

The most popular method of preparing puffballs is to slice them in half-inch thick slices and fry in a small amount of butter. They may also be broiled or breaded with egg and crumbs and fried. Experienced mushroom hunters also use the puffball as an extender for days when other varieties are hard to find. Strips of puffball fried with meadow mushrooms, for example, will take on the *Agaricus* flavour and provide ample servings for all.

Woods Oyster

Scientific name: *Pleurotus ostreatus*.
Common names: Woods Oyster, Oyster Shell, Oyster Mushroom.

Season: Early spring through late fall.
Spore Print: White or pale violet.
Size: 3 to 8 inches in width, viewed from above.
Colour: Gills white; cap white or ivory to tan or grey-brown.
Features: Almost no stem present, gills run from the margin of the cap to the point of attachment; grows in clumps, often overlapping. Is said to resemble the shape of an oyster shell or kidney when viewed from above.
Habitat: Always on wood, usually fresh stumps or logs.

A widespread and delectable mushroom that can be gathered with assurance, even by beginners, is the Woods Oyster or Oyster Shell. It is almost unmistakable in that it has a white spore print and grows only on wood — usually on freshly cut stumps of maple, elm, willow or poplar.

The distinguishing feature that sets this mushroom apart is the off-centre and nearly nonexis-

Sources

It doesn't take much to collect mushrooms: a sharp knife, a basket and a field guide. A good book with identification keys to the edible and poisonous mushrooms will not only ease your mind, but help introduce you to many more collectable varieties.

Mushroom Collecting For Beginners
By J. Walton Groves
Available free from:
Information Division
Agriculture Canada
Ottawa, Ontario K1A 0C7

Foraging For Edible Mushrooms
By Karen and Richard Haard
A 94-page paperback that is especially strong on western mushrooms.
Cloudburst Press, 1974. $3.95

Common Edible Mushrooms,
By Clyde M. Christensen
A clearly written and illustrated basic guide, available in paperback.
University of Minnesota, 124 pages. $3.45

Mushrooms of North America,
By Orson K. Miller, Jr.
Regarded by many as *the* classic in modern mushroom books, this well-organized, 360-page reference book (not a field guide) has 292 colour photographs. A must for the serious mycophagist.
E. P. Dutton, Softcover $9.95

tent stem. The gills extend from the edge of the cap right down to the point where the fungus springs from the tree or log.

This mushroom can be found anytime from early spring to late fall, so a productive stump can be revisited at wet periods with some assurance of success. In Germany, Woods Oyster gatherers actually water these stumps to promote the growth of this delicacy.

The only drawbacks involved in picking this fungus are that it has a short life and should be picked in its prime. Old specimens become discoloured and tough, and a small species of black beetle will often beat you to the booty. Pick

Right, Destroying Angel (Amanita virosa), *the mushroom species that has caused more North American deaths than any other. A single bite can bring death or serious illness, and the mushroom hunter should learn to recognize its distinctive characteristics (see below).*

Distinguishing Characteristics of the Deadly Amanitas and the Edible Meadow Mushroom

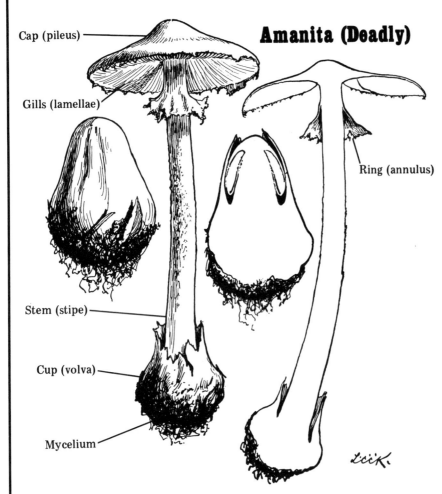

Amanita (Deadly)

Cap (pileus)

Gills (lamellae)

Ring (annulus)

Stem (stipe)

Cup (volva)

Mycelium

Meadow Mushroom (Edible/delicious)

Amanitas

1. Grow in woods, very rarely in old meadows or lawns.
2. Usually quite tall and conspicuous.
3. Caps of *Amanita muscaria* are bright lemon-yellow to orange; in *Amanita virosa,* white. Others, yellowish, brownish, grey, olive brown and blackish.
4. Surface of the cap may bear traces of the volva in the form of detachable warts or patches.
5. Gills do not touch the stems, and are usually white, although some have reddish or lemon-yellow gills.
6. Spore deposit is *always white.*
7. Stems are usually long, and, in most species, bulbous at the base. Remains of the volva are almost always present, either as crumbs, concentric scales or a more or less distinct bag.

Field Mushroom

1. Grows in old meadows and fields, lawns. *Never in the woods.*
2. Short and squatty in stature.
3. Cap is white to brownish in colour.
4. Surface of the cap may be smooth or scaly, with the scales difficult to remove.
5. Gills do not touch stem, are *pink* at first, soon turning a *dark chocolate brown* or *purple-brown.*
6. Spore deposit is *always dark purplish or chocolate brown.*
7. Stem is short, the base never has any trace of crumbs, scales or bag.

Mary W. Ferguson

only the uninfested mushrooms — a mass of Woods Oysters will contain both young and old specimens. Collect only the young ones and leave the bug-infested fungus behind. Even with selective picking, you should have a good harvest as they grow profusely and it is not unusual to collect a bushelful from a single stump.

Oyster mushrooms can be soaked in a dilute salt solution to assure that all bugs are flushed from the gills, and they can be prepared in a variety of ways. Some cooks fry them in butter, others add them to Chinese dishes and still others dip the mushrooms in egg and bread crumbs and cook them in hot fat or butter.

Shaggymane

Scientific name: *Coprinus comatus.*
Common names: Shaggymane.
Season: Spring to fall.
Spore print: Black.
Size: Generally 4 to 8 inches high and 1½ to 2 inches in diameter, giant specimens approaching 20 inches in height have been collected.
Colour: Whitish with tan or brown scales or tufts on the cap; turns black with age; gills white, then pinkish, then black.
Features: Cylindrical cap similar to an umbrella folded around the stem; shaggy tufts on the cap; has a loose ring around the stem in early stages; cap turns into a black, inky liquid as it matures.
Habitat: Hard ground, such as lawns, parks, fields, along streets and boulevards and in town and country fields.

The Shaggymane is the most noteworthy member of the Inky Cap group of mushrooms, whose caps dissolve into a slimy, black mess with age. Shelley noted the process in rather dramatic tones:

> *"Their mass rotted off them*
> *flake by flake,*
> *Till the thick stalk stuck*
> *like a murderer's stake,*
> *Where rags of loose flesh yet*
> *tremble on high,*
> *Infecting the winds that wander by."*

Scientists know this process as *deliquescence,* or autodigestion. An enzyme present in the mushroom quickly breaks the fungus down, so collectors must be quick to find these specimens and get them into the pot. Once cooked, the enzyme is destroyed and the digestion process stopped.

Despite this unappealing characteristic, the Shaggymane is a highly regarded culinary mushroom and its distinctive shape cannot be confused with any poisonous variety.

The shaggy cap and the cylindrical or barrel shape combined with the tendency for the gills or cap edge to deliquesce make for positive identification. The inky edge of the cap can usually be trimmed away before cooking, if the specimens are still reasonably fresh.

Cooking time for this delicate mushroom is short. As with the other mushrooms, it can be fried lightly in butter and served on toast with salt and pepper, or it can be mixed with eggs or added to gravies. If served soon after picking, it can be eaten raw in salads.

Meadow Mushroom

Scientific name: *Agaricus campestris.*
Common names: Field Mushroom, Meadow Mushroom, Pink Bottom.
Season: Fall (occasionally in spring).
Spore print: Chocolate brown or dark purplish.
Size: Cap, 1 to 4 inches or more; stem, 1/8 to ½ inch thick and 1½ to 3 inches high.
Colour: Cap whitish, gills pink at first, becoming purple-brown.
Features: Stout stem that is easily separated from the cap; cap at first convex, becoming almost flat. Gills do not attach to the stem; veil present on stem in early stages.
Habitat: Open fields and lawns; *never in the woods*, prefers well-manured or fertilized lawns and pastures; common both in urban and rural areas.

This is the closest relative to the supermarket or commercial mushroom, although years of cultivation have made the domesticated mushroom a variety which is believed to have no exact counterpart in nature.

By no means a foolproof mushroom, it is still distinctive enough that anyone taking reasonable care can quickly learn to recognize it. The greatest danger for North Americans is that a deadly *Amanita* might be confused with a field mushroom. (See the accompanying box for a comparison of the two.)

The meadow mushroom is best known by its pink gills during early stages, and the chocolate brown or purplish spore print. It is a stocky, meaty mushroom very similar to fresh store-bought mushrooms.

Cool, wet fall weather usually brings these mushrooms out in their greatest numbers, although they can be found in the spring or during long wet periods of the summer.

An inflexible law for the beginner should be: *Never pick button mushrooms.* Until the mature cap has separated from the stem, breaking the veil, there is little that the inexperienced mycophagist can do to distinguish between a choice meadow mushroom and an *Amanita virosa.* Be patient: If the gills are pink upon opening, the mushroom is safe — many collectors prefer well-opened larger specimens anyway, believing they have a better-developed flavour.

City dwellers shouldn't have to travel far afield to find *Agaricus campestris.* Author Jo Frohbieter-Mueller says her favourite collecting spot is a university campus in the middle of town.

"The lovely white caps are easy to see against the neatly clipped lawns, and on many occasions I have filled my sweater, purse and finally my hat with the abundant mushrooms."

Mushroom Cookery

By Jo Frohbieter-Mueller

Sooner or later, you are going to have a really good day in the field and find yourself in the enviable position of having more mushrooms than you can use immediately. Fortunately, they can be preserved by drying, freezing or pickling.

To dry, slice them and lay the slices one layer thick on a rack in the sun, on a hearth, over a heat register, or in an oven with a pilot light. After they are *completely* dry, store them in an airtight jar. Mushrooms can be kept for months like this. Before using, soak the dried pieces in water and they will become plump again.

To freeze mushrooms, wash and pat dry. Then lay them *separately* on a cookie sheet and freeze. When frozen, put them into a plastic bag. Individual mushrooms can be removed when needed. Do not defrost before cooking.

Wild mushrooms can be pickled by placing them in a pan without water, sprinkling a little salt over them and leaving them at room temperature overnight. In the morning, cover them with cider vinegar or wine vinegar, add mixed pickling spices, a tiny red pepper and gently simmer for 5 minutes. Pickled mushrooms keep well in the refrigerator, or they can be sealed in small jars and stored.

The following recipes are the most popular ways to cook wild mushrooms. Actually, wild varieties can be used in just about any recipe calling for mushrooms, although it would be a shame to use them with foods that mask their flavour.

CREAMED MUSHROOMS

4 to 6 Tbsp. butter or margarine
2 cups cream
1 pound mushrooms
flour
¼ tsp. nutmeg
salt

Sauté mushrooms in butter for 5 minutes. Add cream to the mushroom liquid and then add flour to thicken, stirring constantly. Season with nutmeg and salt, and serve over toast, rice, noodles or baked potatoes. Mushrooms make an excellent main dish.

FRIED MORELS, WOODS OYSTERS, OR FIELD MUSHROOMS

2 eggs, beaten
1 cup cracker crumbs
1 pound mushrooms
butter or bacon drippings
salt

Dip the mushrooms into the egg and then roll in crumbs. Heat the butter or drippings in an iron skillet and fry the mushrooms, turning only once. Cook each side about 2 minutes. Salt and serve hot.

SCRAMBLED EGGS WITH MUSHROOMS

This is a fine recipe if you don't find quite enough mushrooms to make a good-sized dish.
1 cup mushrooms, sliced or chopped
4 Tbsp. butter or margarine
4 eggs
½ cup milk or cream

Sauté mushrooms in butter for 5 minutes. Beat eggs, add milk or cream, salt and pepper and pour into a hot skillet with the mushrooms. Scramble. This makes a great dish for campers.

MUSHROOMS IN SHERRY

mushrooms, sliced if very large
butter or margarine (5 Tbsp. per pound of mushrooms)
sherry (½ cup per pound of mushrooms)
salt

Heat butter in skillet and add mushrooms. Cook 5 minutes, pour in the sherry and cook another 2 minutes. Salt and enjoy.

Fit To Be Dried

Resurrecting the ancient art of drying fruits and vegetables

By Wm. W. Rowsome

A generation ago, during the autumn and winter, farmhouse kitchen ceilings above woodburning stoves were festooned with loops of apples — cored, sliced and hung to dry. Braids of onions and garlic hung in cellar stairways and pantries. Bundles of sage and parsley hung upside down from rafters while racks of vegetables occupied every available space in the kitchen, spare room and attic as families preserved the harvest to augment the winter diet of potatoes, turnips, cabbage and salt pork.

The practice of air drying to preserve fruits and vegetables is as old as the history of mankind, but it was a technique largely lost in the commercial food processing revolution. Today, as we discover that life can go on without *Green Giant*, *Swanson* and *Chun King*, dehydration is still an efficient and healthful method of food storage. It uses less energy than either canning or freezing, it concentrates natural flavours, retains most of the vitamins and minerals, and does so without resorting to preservatives.

Drying fruits and vegetables not only provides new snacking and dessert foods, without added sugar, but it can also alleviate storage problems for those with a shortage of freezer space. Dried foods can be eaten out of hand or reconstituted, while bits of dried apple and apricot add new flavour dimensions to a granola breakfast. Preparing "leather britches" (dried green beans), fruit leathers or a pungent basketful of rosemary not only imparts a rich appreciation of our heritage, but is, for us, a particularly satisfying family project.

Today, thanks to our discovery of food drying, homemade pizzas and spaghetti sauces are enlivened with dried tomato and pepper slices, our guests are treated to crispy cucumber chips and our children are weaning themselves from junk food snacks.

Bananas were our first attempt at drying after building a simple dehydrator. We found that their flavour becomes more intense through drying and that a dried quarter-banana stick has a surprisingly chewy consistency.

There is no other easy way to store this exotic fruit, yet the extreme fluctuation in prices encouraged us to try some form of preservation. Periodically, bananas are on sale for 10 to 15 cents per pound and a shopping cart full can be dried at home before any of the bananas become too ripe.

While pushing around a shopping cart heaped with bananas, my daughter counters raised eyebrows and questions with the explanation that they are for our "gorilla," her imaginative nickname for the monster which silently stands in our dining room reducing raw fruits to uncommon treats, warming the house and spreading old-fashioned aromas throughout.

THE ART & THE SCIENCE

Food drying can be done with equipment as rudimentary or sophisticated as you wish. The essentials are that the fruits or vegetables be exposed to low heat in the presence of a good draft to carry off the escaping water vapour.

Drying outdoors in the sun is the simplest form of dehydration, and in the right weather and climate it cannot be improved on from the point of view of energy efficiency.

Mould is the enemy of all dehydrating food, and outdoor drying is impossible in humid conditions. If the daytime temperature reaches 100 degrees (F), without high humidity, food drying is possible.

In northern climates this temperature can be found most easily on rooftops or areas where a solid surface is reflecting solar radiation (e.g. paved driveways).

As in all forms of drying, the fruit or vegetable pieces should be placed so as to allow free circulation of air on all sides. This is most easily accomplished by placing them on a frame covered with cheesecloth or nylon mesh (avoid metal screening which can affect the purity and flavour of the food) to prevent moisture from building up on the bottom surface. Individual pieces should not overlap or touch each other as mould will form at the contact points.

Insects, birds and animals are all enemies of sun-drying foods, so a loose covering of cheese-cloth or nylon mesh, propped up so as not to touch the food, will probably be necessary. Bring the trays or racks indoors at night or at the first hint of damp weather.

Depending on the thickness of the fruit or vegetable slices and the weather, outdoor drying may take two or three days — the finished product should be stiff, leathery and with only the slightest feeling of springiness at the centre. (As fruit seems to be moister when it is hot, allow a few pieces to cool before testing them.)

Drying indoors ends all problems with rain and nature's marauders.

Great differences of opinion exist about the best temperature for drying, but the range is from 90 to about 140 degrees (F). Hotter temperatures destroy enzymes, thereby allowing long-term storage, but cause scorching and destruction of vitamins; lower temperatures preserve more vitamins but allow mould to form. Foods dried at the moderate temperature of 110 degrees will keep for six to nine months.

Kitchen ovens are suitable for drying experiments, but many ovens show no temperature graduation under 200 degrees. You will be wise to employ an oven thermometer to adjust the temperature so that it doesn't exceed 140 degrees, and the door must be left slightly open to allow moisture to escape. The whole process may take from four to 12 hours.

The food may be placed directly on the oven racks (if the slats allow the food to fall through, use nylon mesh or cheesecloth). About five pounds of fruit at a time seems right for the average size oven.

The oven is fine for a few initial traysful but is probably as energy-inefficient a way to preserve food as you would care to try.

GORILLA TACTICS

For anyone considering food drying on more than a test-and-see basis, an indoor dehydrator set-up seems the best alternative. Controlled heat dryers retain food colour, flavour and vitamin content better than sun-dried foods.

Many cabinet dryers are now on the market (from about $100 to $200), but making one requires only the most basic carpentry skills and a few materials.

Our "gorilla" is an adaptation from commonly seen dehydrator plans and from ideas passed on by a friend who built a dryer that employs light bulbs and a light rheostat instead of the more common heater and thermostat.

I changed, adapted and improvised liberally to keep costs to a minimum and to utilize materials which I either had on hand or could get easily, and my teenage son and I constructed it during a weekend. The total cost, including a $20 thermostat, was less than $40.

The basic design allows placement of a heat source under a box which will accept racks of prepared food. The top should allow moist, hot

air to escape, and a thermostat is an essential option unless you are willing to check the temperature constantly.

The sides of our dryer are three-eighths-inch unsanded plywood reinforced as indicated (see diagram) with one-by-three-inch pine.

My design was carefully worked out on paper to get maximum size and convenience of operation. The door is removable rather than hinged, thus allowing for easy insertion and removal of trays. The top is removable and is easily adjustable for draft.

It is always left one-quarter to one-half open to allow air circulation and to permit my youngest son's arm in — he is the official tester of progress. Strangely enough, the top tray is often only lightly loaded by the time drying is complete.

The dryer is two feet wide, two feet deep (front to back) and three feet high, plus nine-inch legs. It is braced as illustrated with the tray supports fashioned out of one-by-one-inch pine and the trays out of one-by-three-inch material.

The source of heat is an old 660-watt cone portable heater, purchased at an auction sale. I wove iron stove wire around and through the guard wires and placed two aluminum pie plates above this. The wire helps to hold the heat and modifies the extremes, thus reducing the cut-ins and cut-outs of the thermostat.

The pie plates spread and distribute the heat more evenly, and perform the same function as the tin sheet found in the bottom of most dryers. The thermostat has a range of 30 degrees to 110 degrees (F), and is connected in series with the heater and generally set at 100 degrees. (The same heater and thermostat serves triple duty: in the dryer, in the root cellar to keep it from freezing in midwinter and in my cold frame to keep plants from dying in early spring.)

Any kind of small heater will probably serve with this type of dehydrator, including a brooder lamp bulb which sells for under $3 at farm supply and hardware stores.

Six or eight 100-watt bulbs will also serve, and an inexpensive light dimmer (less than $5 during periodic sales) may be used in place of a thermostat. By adjusting the dimmer, the temperature in the cabinet can be kept relatively constant. In our house, however, the night temperature in the dining room drops considerably, and the thermostat eliminates the need for constant vigilance.

CARE & FEEDING

Our dryer will handle 24 pounds of bananas at a time, or the equivalent in apples, peaches or tomatoes. The basic secret of drying is the constant movement of warm air around the slices of food, which we cut at a uniform thickness of one-quarter inch. The removal of water reduces the volume of fruit and vegetables by about 80 per cent, and a tough skin is created to prevent the intrusion of decay-causing bacteria.

Our procedure with bananas is fairly typical of the treatment of other foods. The fruits are first washed, then peeled, cut in half lengthwise, then into quarters. The strips are arranged on the racks, unsliced side down (to allow the sticky upper side to dry).

All the family likes to feed the "gorilla" — mind you, not all the bananas make it as far as the dryer — but the prime purpose of our involvement is to substitute natural foods for refined sugar. Thus "snigging" during preparation and processing is not only tolerated, but encouraged. Don't forget to add the banana skins to the compost heap; get everything you can from the bargain cart of fruit.

Our dryer takes about two days to process a full load. The cured pieces of banana and apple are firm and rubbery, while peach slices and tomato chips, when finished, are rather brittle.

Every six to eight hours the trays should be rotated one step downwards to keep all drying at an even rate. (With this type of dryer there is a temperature and humidity difference between the top and bottom shelf areas.)

Several books recommend sulphur fumes and commercial dips for some fruit; it is not necessary from our point of view. We try to process our food as naturally as possible.

Among the fruits commonly dried are: apples, apricots, bananas, plums, grapes, oranges, peaches, pears, pineapple, cherries, tomatoes and berries of all kinds. Rhubarb should not be dried because it contains an acid which becomes toxic when it is concentrated.

Sweet or sour cherries should be pitted before drying, and fruits such as apricots and plums are usually cut in half and the halves turned inside-out to allow thorough drying of the moist interior. Seedless grapes are dried whole to make homemade raisins. Strawberries should be halved, but other berries can go into the dryer whole. (Crack their skins slightly, either by nicking with a knife or by a quick dip into scalding steam or hot water. This cracking allows inside moisture to escape.)

VEGETABLE DRYING

While many people prefer to preserve their vegetables by freezing, drying can give your family some eating pluses that no other method can: inexpensive foods for backpacking and camping; vegetable powders for blending into sauces, dips and cheese spreads; carrot, zucchini, turnip and tomato chips to dip with; pumpkin flour for baking with; and easy lunches of quick stews and instant soups.

Onions, leeks and garlic can be sliced or diced and dried, but most other vegetables should be blanched before going into the dryer.

Blanching helps preserve the colour and stops enzyme action that can ruin the vegetables in storage. It also makes them easier to reconstitute before use.

Blanching simply means suspending them above a pot of boiling water so that the steam can heat the vegetables. A metal colander or mesh basket works well for this — it should fit inside a pot with about two inches of boiling water and a cover should be placed over the entire apparatus during blanching.

Dipping the vegetables in boiling water serves the same purpose and is faster, but allows nutrients to leach into the water.

The following listing of blanching and preparation procedures from Rodale's new version of

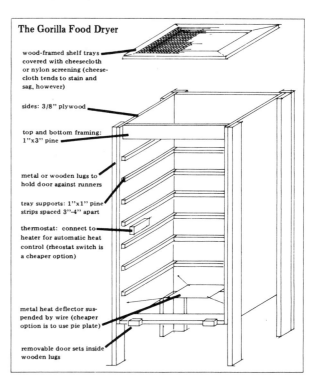

The Gorilla Food Dryer

wood-framed shelf trays covered with cheesecloth or nylon screening (cheesecloth tends to stain and sag, however)

sides: 3/8" plywood

top and bottom framing: 1"x3" pine

metal or wooden lugs to hold door against runners

tray supports: 1"x1" pine strips spaced 3"-4" apart

thermostat: connect to heater for automatic heat control (rheostat switch is a cheaper option)

metal heat deflector suspended by wire (cheaper option is to use pie plate)

removable door sets inside wooden lugs

Stocking Up will serve as a guide to get you started.

Dried fruits and vegetables must be sealed in airtight containers — either glass jars, plastic bags or tin or plastic containers. Some people advocate placing the food in brown paper bags before sealing in either plastic or tin, to prevent contamination or off-flavours. When using plastic bags and glass jars for storage, it must be remembered that the food's colour and nutrients deteriorate with exposure to light. Check inside containers for moisture and store in a cool, dry place.

The "gorilla" provided my wife with an unexpected benefit when she found that its bottom shelf is an ideal spot for cultivating homemade yogurt. Others use their drying cabinets to raise bread dough, grow sprouts, dry sunflower and pumpkin seeds, herbs, bread-crumbs and homemade noodles, sour buttermilk and cream, and start warmth-loving plants such as tomatoes and eggplants.

PREPARING VEGETABLES FOR DRYING

Asparagus: Use only the top 3 inches of the spear. Blanch until tender and firm, about 10 minutes.

Beans, Lima and Snap, Soybeans: Shell and blanch 15 to 20 minutes.

Beets: Remove tops and roots and blanch about 45 minutes or until cooked through. The time will depend upon the size of the beets. Cool, peel, and cut into ¼-inch cubes or slice very thin.

Broccoli: Trim and slice into small (½-inch) strips. Blanch 10 minutes.

Brussels Sprouts: Cut into lengthwise strips about ½-inch thick; blanch 12 minutes and dry until crisp.

Cabbage: Cut into long, thin slices and blanch 5 to 10 minutes.

Carrots: Wash and slice thinly. Blanch 8 to 12 minutes.

Celery: Remove leaves and cut stalks into small pieces. Blanch about 10 minutes.

Corn: Husk and remove the silk, then blanch the whole cob 10 minutes to set the milk. Cut the kernels deep enough to obtain large grains, but be careful not to cut so deeply as to include any cob.

Mushrooms: Peel and cut off stems if they are tough. Leave whole or slice, depending upon their size. Do not blanch, but dry while still raw.

Onions, Garlic, Leeks: Peel and slice into small strips, or peel and grate. Blanch onions and leeks 5 to 10 minutes if you plan to use whole, as in a cream sauce or casserole. If for seasoning, do not blanch.

Peas: Shell peas and blanch (15 minutes if steamed and 6 minutes if boiled).

Peppers, sweet: Clean and slice into thin strips. Blanch 10 minutes.

Peppers, hot: If possible, do not pick until they are mature and fully red. However, if frost threatens, harvest your crop even if some are still green; many should ripen while drying. String the peppers by running a needle and thread through the thickest part of the stem. Hang them outdoors or in a sunny window to dry. They will shrink and darken considerably and will be leathery when they are dry. Although dried hot peppers can be kept in storage containers, they are best left hanging in a dry place.

Potatoes: Wash and slice into ¼-inch rounds. Peeling is optional. Blanch for 5 minutes in

Sources

Dry It — You'll Like It
1973 ($5.00 Canada, $3.95 U.S.)
Living Food Dehydrators
Box 546
Fall City, Washington 98024
Has general instructions, specific recipes for drying fruit, fruit leathers, vegetables, meat and fish, plus detailed plans for a dryer. Available in some health food stores.

Home Drying of B.C. Fruits
Free for a stamped, self-addressed envelope to:
Food Processing Section
Agriculture Canada Research Station
Summerland, B.C.
Details of equipment, predrying treatment, fruit preparation, drying methods, and simplified plans for a cabinet dryer.

Stocking Up, How To Preserve The Foods You Grow, Naturally
Edited by Carol Hupping Stoner,
Rodale Press, Revised Edition (1977) $14.95
Available from Harrowsmith Books

Putting Food By
Hertiberg, Vaughn & Greene
(The Stephen Greene Press, 1975)
A Bantam Book available in the food section of any large bookstore. Has a chapter on drying methods, preparation and details for particular fruits, vegetables and meats. $5.75.

Home Canning — Preserving, Freezing, Drying
A "Sunset" Book available in large bookstores. Has a chapter on drying, preparation, drying guides for many fruits and vegetables.

Dehydrating for Food and Fun
Equi-flow Dehydrators, 1976
($3.95 plus 50 cents for mailing and handling)
514 State Street
Marysville, Washington 98270
General how-to, plus recipes — approximately 150 — with everything from granola bars, fruit ambrosia, pemmican and jerky to applehead dolls and potpourris.

All of these books, and there are undoubtedly many others, are easily read and will encourage you to experiment.

steam and then soak in ½ cup lemon juice and 2 quarts cold water for about 45 minutes to prevent the potatoes from oxidizing during drying.

Pumpkin, Winter Squash: Clean and cut into 1 inch strips and then peel. Blanch about 10 minutes, until slightly soft.

Spinach, Swiss Chard, Kale: Cut very coarsely into strips. Blanch spinach and Swiss chard about 5 minutes and kale about 20 minutes. Spread not more than ½-inch thick on trays.

Summer Squash, Zucchini: Do not peel, but slice into thin strips and blanch about 7 minutes.

Tomatoes: Wash, quarter, and blanch for about 5 minutes. Run through a food mill to remove skins and seeds. Strain out the juice through a jelly bag or several layers of cheesecloth. Use a little hand pressure to extract more water, then spread the remaining pulp on glass, cookie sheets, or pieces of plastic. Turn the drying pulp frequently until it becomes dry flakes.

Turnip: Wash and cut into thin slices. Blanch 8 to 12 minutes.

PREPARING FRUIT FOR DRYING

Apples: Pare, core and cut into thin slices or rings. Don't peel unless the apples have been heavily sprayed.

Apricots: Cut in half, remove pit, and leave in halves or cut into slices or pieces.

Bananas: Peel and slice thinly. (Or just cut into quarters, lengthwise.)

Berries: Halve strawberries and leave other, smaller berries whole. Crack skins by quick blanching or nicking with a knife.

Cherries: Pit and remove stems. (If you don't pit them, the dried cherries will taste like all seed.) Let drain until no juice flows from them.

Grapes: Remove stems and crack skins by blanching quickly or nicking with a knife. Drain until no juice flows.

Peaches: Cut in halves and remove pits. Skin if desired, or just remove fuzz by rubbing briskly with a towel. Then slice.

Pears: Skin, remove core. Cut into slices or rings.

Plums: May be pitted or left whole. Crack skins by quickly blanching or nicking with a knife.

Prunes: May be pitted or left whole. Like plums, they may be blanched quickly or nicked with a knife to crack skins.

Rose Hips: Cut off blossom ends and stems. Crack skins by quickly blanching or by nicking with a knife.

Reprinted from STOCKING UP, by the editors of Organic Gardening and Farming. *Copyright © 1977, Rodale Press, Inc., Emmaus, Pa. Available from well-stocked bookstores or Harrowsmith Books.*

PEACH LEATHER

Fruit leather is a chewy, candy-like natural food that can be made from almost any fruit or fruit combination.

1 gallon pitted peaches
1½ cups unsweetened pineapple juice
honey to taste
chopped nuts (optional)

Cook peaches and pineapple juice over low heat until the fruit is soft. Drain well (and save the juice to freeze or drink fresh). Blend the peach pulp and nuts in a blender until smooth and sweeten with honey to taste. Spread the pulp ¼" thick on cookie sheets covered with freezer paper. Cover with cheesecloth (to keep dust and insects out) and dry at 120 degrees (F).

When the leather is dry enough to peel it up from the trays in sheets, move it to cake racks to facilitate drying the underside. When no longer sticky, wrap in clear plastic and roll it up. Store in brown paper bags or freezer paper. To serve, cut in pieces with scissors.

Jams Session

*From gooseberry to strawberry-rhubarb,
a special summer guide to jamming and jellying*

By Jean Cameron

Pig's head brawn, oatcakes, rag rugs, handknitted stockings and homemade long underwear were only part of the output of my Scottish grandmother's kitchen. The long johns were fashioned out of harsh, itchy wool and winter was a time of scratching, but Grandmother, who reared me, was a strong woman with a profound mistrust of everything made in a factory or bought in a store.

Her garden grew every fruit and vegetable which could be bullied into surviving the bitter winters and chilly summers of northeast Scotland. The picking, preparing and weighing of fruit, followed by bottling, pickling and jamming dominated our lives from June through October. Sugar was stringently rationed during the 1940's and well into the 1950's and one of my aunts was heard to mutter: It's a wonder Mother doesn't grow sugar beets.''

"Don't go putting ideas into her head," shushed another aunt quickly.

I was co-opted into this family labour force at the early age of seven, Grandma's creed being that anybody could do anything if they just put their minds to it. In those days before bottled pectin, some jams — strawberry, for instance — were too difficult for novices, while the jellies (apple, white currant, red currant) called for an expert hand with the heavy "jeely" bag. Gooseberry, black currant and the traditional Scottish rhubarb-ginger, however, were the foolproof, "beginner" jams, and they became my jams.

Our day-to-day cooking was mostly carried out on trivets swung over the open kitchen fire; had more modern equipment been available, Grandmother probably would have scorned it.

Miller Services

Now, it is next to impossible to make jam over a sulky peat fire, and so we brought an ancient, flaring primus stove into use.

One had to prod constantly at the burner with a primitive instrument known as a "stove needle", a metal bar with a spike jutting out of one end at a 90-degree angle. At the same time one had to keep a sharp eye on the seething contents of the copper preserving pan which balanced above. It is a tribute to Grandma's sublime confidence (exhibited in the face of auntly prophecies of disaster) that the whole works never tipped over, the house never burnt down and "the bairn was never scalded half to death."

With all our modern gadgetry, jam making is easy enough today, but where are the obliging never-fail fruits? Black currants and gooseberries seem to have vanished even from farmers' markets; rhubarb is cheap and plentiful in supermarkets, but Grandma would mutter darkly, "You never know where it's been." Or what's been sprayed on it.

Too, jam making is only economically worthwhile when the fruit is home-grown and almost free. Happily, the easy-to-jam fruits are also the most obliging about growing in our northern climate — even in a small, urban garden we can "grow our own jam."

A gooseberry bush, for example, blends decoratively into any ornamental border that is well shaded. Gooseberries thrive in heavy soil, with little sun, and seem impervious to the worst Ottawa winters. Our most productive bush came from McConnell's Nursery (see Sources) four years ago and is planted at the northwest angle of two brick walls where a ray or two of sunshine *may* penetrate on a midsummer's day. We picked almost five pounds of gooseberries from this one bush last year. Two neighbouring bushes catch a little more sunshine and are only slightly less prolific.

The gooseberry is our earliest shrub to flower and leaf out in the spring and it ignores late frosts. The only pest I've ever noticed is a small, green caterpillar which attacks and strips the leaves late in summer after the berries are harvested. The plant apparently isn't seriously troubled by the tenants, but a stripped shrub is unsightly. Several applications of rotenone, over and under the remaining leaves and on any webs, will take care of the visitors.

The black currant, too, is an extremely hardy fruit in most parts of the country. Our five-year-old bushes flourish in the same heavy, well-nourished clay loam as the gooseberries but enjoy rather more sunshine. Although the black currant is a charming shrub throughout most of the year, especially when the ripe black fruits hang down like Christmas decorations, I would hesitate about including black currants in an ornamental group. In a wet fall, after the fruit is safely picked, our black currants invariably are the victims of mould. This, too, seems to do the bushes no harm whatsoever, but their appearance is wretched, resembling a row of decaying corpses. At such times, I'm glad that they're well

out of sight, planted around an old compost heap at the end of the garden. As black currants may serve as an alternate host for pine blister rust disease, their cultivation is forbidden in some parts of this continent.

Rhubarb, that cheerfully exuberant plant, enjoys heat and lots of sunshine. Our nameless variety, the gift of a neighbour, thrives along the south wall of the house, poking little snouts up through a protective mulch of maple leaves with the first spring sunshine. With a downspout from the roof at hand, the rhubarb gets plenty of moisture and is the prime beneficiary of our recently invented winter composting system.

Throughout most of the long, cold, Ottawa winter, the regular compost bins (vinyl-covered mesh fencing, supported by two-by-two's) are buried in snowdrifts.

It was irritating to see valuable debris go out for the garbageman from November until March. We devised a very successful compost arrangement using two old and battered metal garbage cans with extra holes punched in the lower half. They are painted black to absorb any heat and are placed against a white, south-facing wall which reflects sunshine. The distorted lids have to be kept in place with a brick apiece, but we have had no trouble with rodents despite the fact that our yard adjoins open land which has a large wildlife population.

Every two or three days someone treks around to the cans (on snowshoes in the depths of winter) and empties a six-pound honey container full of kitchen refuse (of vegetable origin only — nothing to attract animals). We keep a small shovel handy and whenever a bit of bare earth is visible a little is scattered on top of the accumulation.

By mid-March, the contents of the cans are beginning to degenerate and I add a couple of inches of fresh earth and a sprinkling of organic compost starter, replace the lids and forget about the whole thing until late April. By then the cans contain a sludgy mess which smells exactly like the middens from the kitchen of my childhood and therefore must be *just right*. Choosing a day when the neighbours are out, we hastily spread this evil mixture around the rhubarb which is mercifully close at hand. What's left goes to the gooseberries and black currants. We cover the ripening compost over quickly with a layer of sweet, fresh earth and only the plants are any the wiser.

This process may be neither scientific nor hygienic, but it works splendidly.

The rhubarb is gathered throughout the spring, and the plants are then left alone through the summer, when the stalks are no longer at their best. Impressively large seed stalks sometimes spring up from the rhubarb crowns, and these should be removed quickly, as they sap the strength of the plants. Rhubarb leaves, by the way, contain large amounts of poisonous oxalic acid and should never be eaten.

GRANDMOTHER'S RECIPES

These jams do not have the often cloying sweetness of, for instance, strawberry or raspberry jams. Gooseberry and black currant jams have a marked tang, while rhubarb-ginger is spicy and slightly syrupy.

My grandmother insisted on meticulous weighing of the fruit and sugar, but I have achieved good results by adapting her recipes to cup measurements. Our old copper preserving pan required scrupulous care and cleaning to remove poisonous, greenish-blue deposits of verdigris (copper sulphate). Today we need only make sure that the jamming pot (of stainless steel, aluminum or enamel, free from any chipping) is large enough — not more than half-full when the fruit and sugar are put in, or vigorous boiling action may result in boiling over. Rubbing the inside and rim of the pan with a little butter helps prevent sticking and reduces scum.

Old commercial jam jars may be recycled for use, if washed well, scalded with boiling water and inverted to drain thoroughly. I also scald the ladle, which may be overdoing things.

We once lost a whole batch of gooseberry jam by setting the full jars on cake racks to cool. One jar tipped and brought down the rest in a gooey mess of jam and melted paraffin wax. Now we use several thicknesses of old bath towel to protect the counter top.

To "test for set," remove the pan of boiling jam from the heat. Scoop a small amount of jam onto a cold saucer. On a hot day it is best to put a few old saucers in the refrigerator to cool before starting jam making. Allow the jam on the saucer to cool for a few moments. If the setting point has been reached, the surface of the test

patch will wrinkle when pushed with a fingertip. Otherwise, return the pan to a vigorous boil again and test frequently. Gooseberry, black currant and rhubarb jams are all easy setters.

GOOSEBERRY JAM

7 cups gooseberries, still green and slightly under-ripe. (Over-ripe gooseberries produce a jam which does not set.)
1½ cups water
6 cups white sugar

Top and tail the berries, i.e. use kitchen scissors to cut off the stalks and blossom-end tufts. Wash and drain well. Place in a large pan with the water. Simmer until the berries are softened, up to half an hour. Add the sugar and stir, simmering, until dissolved. Bring to a rapid boil and boil furiously for 10 minutes. Remove pan from heat and test for set. Remove any scum which may have formed, ladle into sterilized, warm, dry jars and seal.

BLACK CURRANT JAM

5 cups black currants, stalks removed. (No need to bother with the blossom-end tufts.)
3 cups water
6¼ cups sugar

Place the currants in a large pan with the water. Simmer slowly until the skins are well softened, at least half an hour. Our tough, unknown variety of black currants needs a good 45 minutes to simmer or the end product contains hard black pellets. Stir to prevent sticking. When the fruit is thoroughly soft, add the sugar, simmer and stir to dissolve. Boil rapidly for 10 minutes, remove pan from heat and test for set. If the setting point has not been reached, return to the boil and test frequently, removing pan from heat while testing. Remove any scum, ladle into sterilized, warm, dry jars and seal.

RHUBARB-GINGER JAM

This jam was so extensively eaten throughout the most impoverished parts of Scotland, that I can only conclude that preserved or crystallized ginger must have been readily available and cheap. These days it is hard to find in the supermarkets and constitutes the most expensive part of the jam. *Dried Cargo Ginger*, in gaily printed packets at Chinese groceries, is satisfactory but may well cost up to $1.50 for six ounces, enough for two batches of jam.

8 cups rhubarb, washed and cut into chunks, approximately 1 inch long
7 cups sugar
Juice and rind of one medium-large lemon, grate or chop the rind finely

Combine all the above in a china or earthenware bowl — *not* metal and preferably not plastic. Let stand 12 hours or overnight. Transfer to a large pan, bring to a boil and simmer uncovered until thickened (1 hour or a little less). Add ¼ cup of finely chopped preserved or crystallized ginger, and simmer for 10 to 15 minutes. Ladle into warm, dry, sterilized jars and seal.

Into The Jam

By Elinor Lawrence

Rather than being drawn, I was pushed into the world of preserves, when my Old World grandfather-in-law thrust a hamperful of freestone prune plums into my protesting arms, saying, "Take. It's good. You gonna like. Sure, sure, sure."

Less than two hours work found me with four dozen still-warm, gleaming jars of homemade jam, not to mention a great sense of pride and the thought, "Can this be all there is to it?"

Despite the reputations enjoyed by those *real* cooks who make their own jams and jellies, the whole process is far simpler than the uninitiated would ever suspect.

1. Wash, rinse and sterilize enough jars to accommodate the recipe you will be using. Although small canning jars made especially for jams and jellies are available, recycled ones will do fine. (The sealing qualities of the lids are not critical here, as in vegetable canning, and old covers may be used.) If the lids have been lost, waxed paper rounds and elastic bands will do.

To sterilize, either invert in two inches of water and boil for 15 minutes, or place on an oven rack at 225 degrees (F) for at least 10 minutes.

2. Melt paraffin wax in a double boiler.

3. Wash and clean the fruit of any soft or spoiled parts (these can ruin a whole batch). Peel, pit, chop or grind the fruit.

4. Make the jam in a large pan, according to the recipe chosen. This only takes about five minutes, stirring continuously with a metal spoon. To reduce foaming, add ½ teaspoon butter.

In order to jell, a jam must contain pectin, sugar and acid in definite proportions to the fruit. Some fruits, such as currants, tart apples and quince, have enough natural pectin and acid. Others, such as strawberries, raspberries, pears and peaches, need added pectin or acid (from lemon juice), or both. *(Pectin is a naturally occurring carbohydrate found in certain fruits — most notably green apples. It has the ability to emulsify and thicken jams and jellies.)*

5. Remove from heat, add the pectin, and stir and skim for an additional five minutes to prevent floating fruit (stirring assures a uniform jam, rather than having all the fruit pieces on the top layer).

6. Pour the jam into a scalded jar, leaving ½ inch at the top for paraffin. In order to keep any bacteria from working into the jam, do not wait until all jars are filled, but pour and seal each jar in its turn.

7. Spoon a thin 1/8-inch layer of paraffin on top of the hot jam. Tilt and turn the jar gently to ensure a good seal at the edge.

When jellies contract and expand, some of the jelly may seep up around the wax and mould. A thin layer of paraffin adjusts better. (Jams and jellies can also be preserved by sealing in regular canning jars with rubber-edged lids and screw bands rather than with wax.)

8. Cool, then cover (to keep out dust) and label.

Many fine cooks have found that the jam and jelly recipes wrapped under the label of *Certo* (a commercial pectin) are thorough and give excellent results. New Canadian editions are bilingual and contain about 40 recipes, while the U.S. Certo booklets have over 80 recipes. These change from year to year, so you may wish to save them. (Their wine jelly recipe is innocent enough beside roast duck, but the alcoholic cooking vapours almost landed this jam maker on her keaster the first time she tried it.)

JAM PROCEDURE

STRAWBERRY JAM

This recipe also serves for blackberries, dewberries and boysenberries.

3¾ cups prepared berries
(start with about 2 qts. or 2½ lbs. ripe berries)
¼ cup lemon juice
7 cups sugar
½ bottle fruit pectin

Mash the fruit completely and measure 3¾ cups into a large saucepan, along with the lemon juice (omit with the tart berries and add an additional ¼ cup of fruit).

Add sugar, mix and bring to a full, rolling boil and hold for 1 minute, stirring constantly.

Remove from heat, stir in the pectin and skim. Stir and skim for 5 minutes. Yield: About 12 one-cup jars.

PEACH OR PEAR JAM

4 cups prepared fruit (about 3 lbs. ripe fruit)
¼ cup lemon juice
7½ cups sugar
1 bottle fruit pectin

Prepare the fruit by peeling and either pitting or coring. Grind or chop finely or use a blender. Place fruit, lemon and sugar in a large saucepan and boil hard for one minute. Then follow the procedure outlined above.

To make spiced peach or pear jam, add ½ to 1 teaspoon of allspice, cinnamon and ground cloves, or any combination of the three.

STRAWBERRY-RHUBARB JAM

1 lb. red-stalked rhubarb
1 qt. strawberries
6½ cups sugar
½ bottle fruit pectin

First chop or slice the rhubarb thinly and stew it in ¼ cup of water until soft (about 1 minute). Mash the fully ripened strawberries and combine with the rhubarb. Measure 3½ cups of fruit and add sugar. Bring to a full boil for 1 minute and proceed as above. (For rhubarb jam — without strawberries — follow the same recipe but use 2 lbs. of ripe rhubarb and 5½ cups sugar.) Yield: About 11 one-cup glasses.

BLACKBERRY JELLY

Jellies are slightly more challenging, and require a jelly bag or one square yard of cheesecloth several layers thick.

3¾ cups juice (from about 2½ qts. ripe berries)
7½ cups sugar
¼ cup lemon juice
1 bottle fruit pectin

Wash, then crush thoroughly (or blend) the ripe berries. Place in the dampened jelly bag or cheesecloth and squeeze out the juice. Measure 3¾ cups of this juice into a large pan and add ¼ cup lemon juice (strained).

Add the sugar, mix, and stirring constantly, heat to a boil. Immediately add the pectin, bring once again to a boil and hold for 1 minute. Remove from heat, skim off foam and pour quickly into jars. Cover at once with 1/8 inch hot paraffin. Yield: About 11 one-cup glasses.

The same recipe can be used for strawberry, red raspberry, boysenberry, dewberry or loganberry jelly. When using tart berries replace the lemon juice with an extra ¼ cup of berry juice.

FREEZER JAMS

Now that you've proven you can hold your own with anyone's Aunt Emma, try one of the newer freezer jams. The fruit in these tastes so fresh and light (because it hasn't been cooked) that you might never look back.

STRAWBERRY FREEZER JAM

1¾ cups prepared fruit
(about 1 qt. strawberries)
4 cups sugar
2 Tbsp. lemon juice
½ bottle fruit pectin

Mash the berries thoroughly. Measure 1¾ cups into a large bowl and mix in sugar. Let stand 10 minutes. In a separate bowl mix the lemon juice and pectin and then add this to the fruit/sugar mixture. Continue stirring for 3 minutes. Some sugar crystals may remain undissolved — ignore them.

Pour quickly into prepared jars or plastic containers, cover at once with tight lids (no wax). Allow to set at room temperature for 24 hours, then label and freeze. Containers that will be dipped into within 3 weeks may be kept in the refrigerator. Yield: About 5 one-cup glasses.

PEACH FREEZER JAM

2¾ cups prepared fruit
(start with 1½ qts. fresh ripe peaches)
6½ cups sugar
1/3 cup lemon juice
1 bottle fruit pectin

Peel, pit and grind or mash the peaches, or use a blender. Measure 2¾ cups into a large mixing bowl, and follow the procedure set out in the above recipe.

The Early Bite

*As the snow melts and
garden fresh salads are still weeks away,
wild potherbs and untamed greens
are prime for the harvest*

By Louise Langsner

British farmers have a hollow-sounding epithet for the tail end of winter. When the hay loft's bulk has shrunken and storage bins grow light, while lush spring pastures are still no more than a hint of green on thawing earth, the "hungry gap" looms before the barnyard animals and fowl.

However, a cheery antidote exists for this grim situation. The shrewd farmer has fall-seeded oats or rye to provide an "early bite." After a winter of soup beans and sauerkraut with the pantry shelves lined with empty jars and the spinach barely planted in the garden row, I am ready for an early bite myself. This is the time to do a bit of harvesting in nature's garden of wild edibles.

Long before the earliest of domesticated garden crops reach picking size, a host of tender wild greens have sprung up with careless disregard of late frosts (or even snow) and are at their delectable best. Dark green, succulent salads can be had for the picking, if you know where to find them.

Of course, the hunt is half the fun. Some of these wild salad ingredients may grow within feet of your doorstep. Others pop up in the garden or in empty lots, while still others will lead you farther afield. Meadows, cropland and woods all hold promise for the early spring forager.

First, be able to recognize the plant you seek, and have an idea of its favoured habitat. Guidebooks and local herbal lore should get you off to a good start. Search out sheltered places — spring and stream banks, hedgerows, south-facing slopes, edges of woodlands and up against buildings — where early leaf growth is encouraged.

Filling the hungry gap and providing sport are not the only benefits of wild spring greens. Longstanding folk traditions claim tonic qualities for many, and for good reason. These plants are full of trace elements and minerals, as well as being rich sources of vitamins A and C (see box). Thus fortified, the potent greenery is reputed to cleanse and tone the human system and make it fit for the work of the coming season.

WILD GREENERY

Violet *(Viola papilionacae).* One of the earliest and most welcome signs of spring is the bright green, heart-shaped leaf of the common blue violet. Euell Gibbons called this plant of the hedgebanks and woodlands "nature's vitamin pill" in view of the high vitamin content of both leaves and blossoms. Their tart flavour is attributed to vitamin C. Gather the young, just-uncurled leaves, as they are the most tender. Both leaves and blossoms are delicious raw in tossed salads or may be gently sautéed or stir-fried in a wok.

Curled Dock *(Rumex crispus).* A true early bite is provided by the various docks, curled dock being one of the mildest and most tasty. Its leaves emerge one after another from the centre of a basal clump, growing up to a foot long and one or two inches wide. Edges are wavy or crinkled. Dock is an exceedingly healthful and vitamin-rich herb. To enjoy it at its best, pick the newly-grown leaves while days are still cold and nights

G.I. Kenney

Right, *Fiddlehead fern,* Matteuccia struthiopteris. *Except where noted, the photographs for this section were taken by Alana Kapell.*

Nettle
Urtica dioica

Wild Lettuce
Lactuca canadensis

Dandelion
Taraxacum officinale

Lamb's Quarters
Chenopodium album

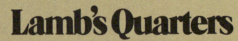

Wild Leeks — Albert Hoffman
Dandelion — Donald McCallum

frosty. Chop finely for salads, steam or sauté; any acidity is tamed by cream sauce.

Nettle *(Urtica dioica)*. Yes. The stinging nettle — my favourite potherb. Nettle leaves are amazingly rich in vitamins, minerals and protein, and they are delicious. Gather the tops of young, first-growth plants before they are a foot tall. (You will probably wish to wear gloves to avoid being stung.) Although nettles are widespread,

the choicest grow in rich woodland coves and along streams. After cooking, which destroys all stinging properties, the nettles may be incorporated into many fine dishes (see cooking notes).

Chickweed *(Stellaria media)*. Another favourite. Chickweed likes rich soil and may be found creeping over the ground in last year's fields and gardens. Its many branched stems may grow over a foot long and are covered with bright green,

Wild Leeks
Allium ursinum

Purslane
Portulaca oleracea

maranth
aranthus retroflexus

Woody Nightshade
Solanum dulcamara   *(Poisonous)*

one-quarter-by-one-half-inch, pointed leaves in opposite pairs. Pinch off the top few inches of each stem for a fine salad green. Chickweed is excellent cooked gently, especially when mixed with more pungent greens.

Mustards *(Brassica).* Five species (of endless variations) are widely distributed over North America, and are easily recognized for their bright yellow, four-petal flower. The leaves of young plants — those best for eating — are slightly rough (though some varieties are smooth), fine to coarsely toothed and indented, and have a familiar mustardy odour and sharp taste. Early leaves provide a spicy salad or potherb, while flower buds may be eaten raw or steamed like broccoli, and the flowers are a cheerful addition to vegetable dishes.

Milkweed
Asclepias syriaca

Red Clover
Trifolium pratense

Curled Dock
Rumex crispus

Plantain
Plantago major

Shepherd's Purse *(Capsella bursa-pastoris)* bears much resemblance to mustards in form and peppery taste. The low-growing rosette is variable in form, the radical leaves being irregularly cleft or entire. Used very young to avoid bitterness, the leaves make good early eating.

Lamb's Quarters *(Chenopodium album)* is another high achiever, growing upwards of six feet in fertile soil. Its wedge-shaped, scalloped leaves are a silvery blue-green and reminiscent of spinach, a relative. When small, the whole plant is tender and tasty; top leaves may be picked continually from older plants.

Amaranth *(Amaranthus retroflexus)* is a voracious weed, thriving on rich soil such as is found near a pigsty. It is also known as pigweed and pigs would certainly find it a succulent treat. Oval, opposite leaves are often tinged red and stems and roots are bright red (another name is red root). Leaves may be prepared alone or added to more tangy herbs.

Wild Lettuce *(Lactuca canadensis)* resembles a lighter green, deeply-indented dandelion leaf; another variety *(L. scariola)* has wavy-edged leaves that could pass for the first two leaves of garden leaf lettuce when young. It must be young to make prime salad fixings. After a week or two, *Lactuca* turns bitter, with little of the sweet, tame-lettuce quality remaining. Broken leaves exude a white juice, hence its name, *Lactuca*.

Winter Cress *(Barbarea)* is found growing in damp ground, along streams and in old fields. The deeply-cut basal leaves are another sharp-tasting member of the family Cruciferae. For salads and sandwiches.

Spring Cress *(Cardamine bulbosa)* is a small fountain of many, finely-cut basal leaves, also growing on stream banks and fields. Delicious!

Dandelion *(Taraxacum officinale)* is well-known for its dark green circle of lion-toothed leaves. In rich soil, the foliage grows lush and sweet. Both young leaves and flower buds may go into salads or cook pot.

Chicory *(Chicorium intybus)* is a plant of fields and roadsides that may also be mistaken for dandelion at an early stage. However, the basal leaves are not cut, but finely-toothed. Young leaves add a pleasantly bitter, dark green zap to mild salads.

Ramps (Wild Leeks or **Wild Onions)** *(Allium ursinum)* have broader, lily-like leaves growing from a strongly-scented bulb. They are found (by those who know where to look) in rich woodland coves near springs. An indelicate delicacy, ramps are highly esteemed by mountain farmers for their garlic-like potency. Cooking mellows the bite. The cylindrical bulb can be used like onions and can be frozen for off-season use.

Branch Lettuce *(Saxifraga nicranthidifolia)* is a plant with four-to-eight-inch, shiny green, succulent leaves with scalloped edges. Branch (brook) lettuce grows at springheads, on rocks in streams, or on river banks. Ramps and branch lettuce cooked in bacon grease are a Southern mountain favourite.

Wild Garlic *(Allium tricoccum)* is a field and roadside plant with tall, very thin, onion-type leaves. Garlic odour is unmistakable. Small bulbs as well as greens may be used.

Vetch *(Vicia sativa)* volunteers over fields and pasture land. The tender fiddlehead tips may be pinched off as they uncurl for a distinctive and tasty vegetable.

GREEN CUISINE

During a year's travel through Europe, Greece and Turkey, we were continually impressed by the extensive knowledge and use made of wild herbs by the rural householders with whom we often stayed. Never had we tasted such a variety of delicious greens pulled from fields and pasture land.

Preparation is generally very simple. When young and tender, most greens may be eaten raw in salads. At most, they need only be wilted in a hot skillet or wok, or steamed ever-so-slightly. Tougher greens will require a little more cooking time. I find that combining the complementary flavours and textures of various herbs makes a more interesting and tasty dish. Once cooked, the greens may be dressed Greek-style with olive oil and lemon juice, or with hot bacon grease and vinegar as in the southern Appalachians.

Other popular dishes combine the cooked greens with boiled potatoes or rice, sometimes with the addition of tomato sauce and grated cheese. This combination is great over pasta, as well.

One memorable Turkish meal consisted of a large dish of seasoned, cooked greens, a pile of chapati-like bread, green onions, and yogurt or sour cream.

Wild greens can work their way into any number of recipes from soy burgers to quiche. They are always a treat. Commenting on their outstanding flavour (and food value), Euell Gibbons said, "I have concluded that the reason wild foods taste better is that they *are* better." It's that simple.

Spring Tonic

Vitamin A. Among the richest sources of this vitamin in the vegetable kingdom are: Dandelion greens (14,000 I.U./100 grams), dock (12,900 I.U./100 grams), violet leaves (8,258 I.U./100 grams), and nettle (6,566 I.U./100 grams). Also rich are lamb's quarters, cress, amaranth, chicory and watercress.

Vitamin C. Many greens are surprisingly good sources of vitamin C, often surpassing oranges on a weight-for-weight-basis. These include violet leaves (210 mg/100 grams) and blossoms (150 mg/100 grams), winter cress leaves (152 mg/100 grams) and curled dock (119 mg/100 grams). Lamb's quarters, dandelion, amaranth, watercress and nettles also provide respectable amounts of this vitamin.

Calcium. Chicory, dandelion, mustard, plantain, watercress and lamb's quarters are plants rich in this important mineral.

Copper. Chickweed, chicory, dandelion, and garlic, contain this mineral — an aid to the digestive organs.

Iron. Chicory, dandelion, dock, nettle and watercress are storehouses of iron.

Magnesium. Dandelion, poppy greens.

Phosphorus. Chickweed, watercress.

Potassium. Dandelion, nettle, plantain.

In Search Of The Wild Leek

By Albert Hoffman

Drew Langsner

Above, *author Louise Langsner gathering an early spring salad.*

When the first of the spring's home-grown spinach, lettuce and scallions are still but a gleam in our gardening neighbour's eyes, I can usually be found walking through the woods with a large bucket in one hand and a small digger in the other.

For those of us who have not yet lost all of our food-gathering instincts to evolution, the early spring is one of the most satisfying times of year.

By midsummer most potherbs and wild salad ingredients will have grown bitter, tough and often difficult to find in the profusion of green foliage.

While many wild foods can be found growing very near the house (or as "weeds" in the vegetable patch), I head for wooded areas to find one favourite foraged food.

Wild leeks, also known as wild onions or ramps, are often found in the shade of trees in heavily forested areas. The characteristic long lily-like leaves are often seen sprouting in fairly large patches near spring runs and swampy areas in the woods, and a single wild patch will often yield enough to fill all containers and pockets. In addition to being a highly prized taste sensation in the very early spring, the wild leeks are also most easily found at this time when their green sprouts stand out in the brown leaves of the forest floor.

Once you've located a well-populated patch, remember it, for it will yield nicely year after year and can be revisited throughout the warm months. One of my favourite leek spots is handy to a brook, where I can wash the harvest before heading home. The annual ritual is complete that evening when the fresh leeks are dipped in salt and served with venison steaks.

Wild onions may be used in the same manner as their cultivated relatives, and we enjoy them boiled and served with vinegar, salt, pepper and butter. Surpluses can be washed, bagged and frozen for use in cooking during the winter months.

Wildfood Guides

Stalking the Healthful Herbs
Stalking the Wild Asparagus
by Euell Gibbons
(David McKay Co.) $3.95 each

Herbal Handbook for Farm and Stable
by Juliette de Levy
(Rodale) $3.95

Feasting Free on Wild Edibles
by Bradford Angier
(Pyramid) $1.75

The Edible Wild
by Berndt Berglund and Claire Bolsby
(Scribner's) $3.45

Foxfire 2
edited by Eliot Wigginton
(Anchor) $4.50

A Weedy Bounty

By Alana Kapell

Last spring when we departed for Halifax, our animals, house and garden were left in the care of friends. Although we would be gone a month, the garden seemed well under control: everything was planted, well-cultivated and many plants were already up and thriving.

Coming home from a month of cool, damp weather in the Maritimes, where gardens were just being planted, we were taken aback by the growth that had gone on at home. The herbs were already in blossom and ready to harvest; the tomatoes were coming into bloom and the spinach was going to seed.

Our friends had nurtured the plants through a very dry month and the vegetables looked green and healthy. Unfortunately, the weeds were equally lush.

Being new to gardening, our caretakers had been confronted with the problem of recognizing the weeds from the planted vegetables.

While most of the uninvited herbage went directly to the compost heap, some also found its way to the table in the form of boiled greens, salad herbs and teas. Especially in the early stages, any garden will yield a variety of wild plants that make fine additions to the intentional vegetable fare. The following are easily identified and common in North American gardens. Good weeding and eating.

Milkweed *(Asclepias syriaca)*. Considered a delicacy by some northern Indian tribes, immature milkweed still makes a fine wild vegetable when gathered before the stems and leaves become filled with milky juice. Young shoots, unopened flower buds and young pods (picked when the size of a walnut or smaller and before any seed silks develop) all make fine potherbs. Some people steam them, but we find the flavour milder when boiled in three waters. The mature flowers, by the way, make a fragrant addition to a summer morning's bath.

Purslane *(Portulaca oleracea)*. This is a common, ground-hugging plant frequently seen in gardens. The succulent leaves and stems can be eaten raw or cooked, and they can be frozen or pickled. It is high in iron and its okra-like mucilaginous qualities make it useful in thickening soups and stews.

Plantain *(Plantago major)*. The bane of lawn care fanatics, plantain is a favourite of children who love to strip the tightly-packed seeds from the plant's distinctive single seed stalks. Tender young leaves can be added to salads or boiled in salted water and served like spinach.

Wood Sorrel *(Oxalis violacea)*. This tiny, delicate plant adds zest to spring salads and is high in vitamin C. It should be eaten only in moderation, however, as it contains oxalic acid and can prove toxic when ingested in large quantities. The leaves can be boiled in water for 15 minutes to make a refreshing tea — especially good when sweetened with honey and chilled.

Red Clover *(Trifolium pratense)*. Foraging animals love clover, and for good reason. Clover flowers and leaves make a vitamin-rich and unusual addition to salads and omelettes, or they can be steamed as potherbs. White clover is milder in taste and can be put to the same uses.

Fiddlehead *(Matteuccia struthiopteris)*. Resembling the tuning end of a violin or the curled top of a bishop's crozier, this delicacy is one of spring's most prized finds. Found in damp, shady places, the sprouting fronds occur in clumps and are encased in a loose, brown protective skin.

The fiddlehead or ostrich fern is distinguished from most other common North American ferns by its meatiness — it is rather substantial in comparison to most of its delicate relatives. Fiddleheads can be eaten raw or may be simmered in salted water and served with butter. (Some cooks prefer to change the cooking water once or twice to give a milder flavour.)

POISONOUS. Woody Nightshade *(Solanum dulcamara)*. As with mushrooms and other wild things, one does not pick and eat indiscriminately. A common poisonous plant that should be recognized is the Woody Nightshade, whose red berries help distinguish it from the related Deadly Nightshade (which has black fruit). Both should be avoided. (Deadly Nightshade is also known as *belladonna* — meaning beautiful woman. This name can be traced to the reputed use of this drug to dilate pupils of the eyes and create a skin pallor — considered beauty traits by women of ages past.)

Here's To Thee, Old Apple Tree!

Whence thou may'st bud, and whence thou may'st blow,
Hats full! Caps full!
Bushel-bushel-bags full!
And my pockets full too!
Huzzah!

By Drew Langsner

Old English toasts such as this were used to wassail fruit trees on Christmas Eve. The ceremony consisted of the farmer, with his family and labourers, going out into the orchard after supper, bearing with them a jug of cider and hot apple cakes. The latter were placed in the boughs of the oldest or best bearing trees, while the cider was flung over the limbs after the farmer had made his toast and passed the jug.

English farm families had good reason to drink to the health of their apple trees. In times past, cider was made in great quantities on every farm. Apples were crushed in a pounder and pressed by a massive, oftentimes wooden, hand-turned screw, and the juice fermented in oak barrels or vats. North American colonists brought their favourite trees and cider-drinking habit to this continent where apples became a staple food and cider the daily drink.

In North America the term "cider" is commonly applied to apple juice as it comes straight from the mill. Now, this is a fine beverage — one of my favourites — and is well appreciated when mulled over the fire on a winter's evening, or as a cool summer drink. However, if the juice is not preserved by canning or freezing, it will naturally ferment, reaching an alcoholic content between two and seven (rarely eight) per cent. The Europeans use the term "cider" for this fermented drink. It may be bottled to achieve an effervescent "champagne" quality, or left in the keg — a light, refreshing not-so-sweet beverage that may be enjoyed in large, cool draughts.

The great ciders of England and France are the product of a long-standing tradition of craftsmanship with orchard trees planted expressly for the superior cidermaking qualities of their fruit. From three to six varieties of these apples are mixed to yield a cider balanced in acid (about .6 per cent) and tannin (.10 to .15 per cent). While these rates are higher than in most commonly available apples, a very good cider (or apple juice) may be made using such varieties as *Baldwin, Winesap, Northern Spy, Cortland, Russet, Rome Beauty, Gravenstein, Yellow Newton, McIntosh, Spitzenburg, Jonathan* and *Roxbury*. Sometimes excellent varieties may be found in old orchards or around homesites; a good cider apple is tart and very juicy.

Cider apples may be divided into three types: sharp (high in acid, lacking tannin), sweet (less acid, more tannin), and bittersweet (high in tannin; a bittersweet apple may exceed .2 per cent tannin, while a *Baldwin* or *Roxbury* has about .06 per cent and *McIntosh* or *Northern Spy* about .08 per cent).

Acidity plays a vital part in determining cider quality. Lack of acid results in poor fermentation and poor character. (You may test for acidity using Narrow Range pH indicator paper — aim at a colour reaction equivalent to between pH 3 and 4. However, we have never found this necessary.) Acidity may be raised by adding the juice of one or two lemons per gallon of juice.

Tannin, found in the skins and stems of fruit, is another essential ingredient in a successful cider. To tannin is attributed the mouth-puckering "bite" of a zestful drink. Tannin also gives cider (or wine) an impression of dryness after drinking. If your apples are short on tannin, add a few oak leaves or one tablespoon of strong black tea per gallon. We add five to 10 per cent crab apples or fence row volunteers to give additional tannin and tartness.

Tales are told of old-time orchardists who harvested dessert apples with gloved hands and transported them home on beds of straw. Such

precautions are not called for in cider-making. Windfall apples are considered best, as they are fully ripe and coated with the yeasts that promote fermentation. Some French cidermakers dump their apples into a vat for a few days before beginning crushing. Others beat their apples with clubs, then leave the bruised fruit for 24 hours to develop colour and flavour.

We are perfectly happy to use naturally bruised fruit, as is the case with the windfalls we gather from the ground. Shaking the trees was the preferred method of harvesting in old England. You can also use an apple picker, consisting of a loop of stiff wire attached to the end of a long pole. No matter how you get your apples, they should be fully ripe for the best flavour and highest yield.

Hopefully the fruit has fallen on clean sod and need not be washed. Natural yeasts on the skins are what will set your cider bubbling. (Washing is probably advisable if you wish to inhibit fermentation for refrigerated storage.) Rotten fruit should be discarded, but worm holes and such simply enhance the flavour.

Though our southern Appalachian neighbours are traditionally quite handy when it comes to homemade stills, many have never heard of a cider mill. Upon learning of our plans to build a cider press, one woman hastened to tell us how her pappy made cider the "old-timey" way.

"He just took a piece of roofing tin and punched it full of nail holes. Then all the young-uns scraped apples over the sharp edges of the holes. The juice dripped down the tin into a tub."

They then put the shredded apples in a flour sack to squeeze out the remaining juice. Perhaps one has to live among these mountaineers to appreciate this scene.

There are many ways to proceed with expressing the juice. Pick a method appropriate to the quantity of fruit to be processed and suitable to your pocketbook or skill as a craftsman. You can spend a fair amount on cidermaking equipment, but it should last for generations, and can also be used for making other juices or wine. A press without a crusher is useless for apple cider, since you must chop or crush the apples into a pulp before the juice can be squeezed out.

The simplest way to make a small quantity requires no special equipment. Chop the apples into very small pieces. Put the pulp (the British call this "cheese") into an enamel or stainless steel container. Add just enough water to cover. Tie a cheesecloth over the top to keep out flies. Let the mixture set a few days. Then put the pulp into a clean cheesecloth bag and squeeze tight.

Another kitchen approach that is faster and more efficient is to quarter the apples and then grind them through a meat grinder. This makes a nice, mushy cheese that's fairly easy to squeeze or press.

Electric juicers do the work in one easy step. *Champion* juicers automatically separate the juice from skin, seeds and pulp. The liquid comes out quite thick and cloudy. This is fine for juice,

but you would want to filter it for cider. Centrifugal juicers, like *Acme* and *Braun*, are more efficient, and make a clear, fine juice. However, you have to stop about once per quart to clean out the pulp.

PRESSING NEEDS

We personally enjoy making cider with an old-style cranked crusher and screw press. A family can make more than twenty gallons in a day, including the canning or freezing. And we don't need to use electricity (which we don't have). Cider presses have been manufactured in many types and sizes, from a peck capacity (too small for anything serious), to several bushels (for old-time, extended farm families). Our hand-press holds crushed pulp from one bushel of whole fruit. Units are available new, used and in kit form.

Making an apple press is not especially hard. You can copy the traditional woodwork and use a screw or a small hydraulic jack. A bushel press requires a one-ton jack, or a solid screw measuring at least one inch in diameter. *Acme* threads

Peter Hutchinson

with wide, square shoulders are strong and fast to use.

Building an efficient crusher is more difficult. The best manufactured crushers are fitted with a heavy flywheel and reduction gearing. Plans for building a simple, but workable crusher and mill are found in *Cloudburst* (edited by Vic Marks, Cloudburst Publishing Company, Mayne Island, British Columbia) and *Build It Better Yourself* (edited by William Hylton, Rodale Press, Emmaus, Pennsylvania, U.S.A.).

Kits and complete, new presses are available from American Village Institute, (Marcus, Washington U.S.A.) and Garden Way, (Charlotte, Vermont, U.S.A.) The AVI American Harvester has two separate baskets for crushing and pressing, and features heavy iron castings, machined gearing, and solid oak framework.

If you decide to buy an "antique" mill, make sure that all the iron parts are working or repairable. Broken gears or cracked castings are difficult to repair, and missing parts may be impossible to duplicate.

The method of pressing is determined by the type of crusher available. Crushers that chip the

fruit are easiest to use. The cheese is then simply dumped into a slotted basket. A follower (see diagram) is placed between the screw and fruit. The screw is turned down until the juice starts flowing. You wait for the juice to stop running, then turn down some more.

Crushers that really mash out pulp give higher yields than chipper types, but with a very fine pulp it is necessary to line the slotted basket with cheesecloth or nylon mesh. That is an extra step in the production and cleaning process for each batch, but you will get more juice and far less waste. The type and condition of apples will determine yield as much as any other factor. Two to three gallons per bushel is considered a fair yield.

You can also make a second (or even third) crushing. Dump the spent cheese into a vat and cover with water. Wait overnight, then press again. The product is poor for apple juice, but with the addition of some sugar will ferment into an acceptable cider.

If you press a small amount of juice you may be able to drink it before fermentation sets in. Fresh apple juice is a great party drink. Indeed, making the juice can become a party in itself. Apple juice will begin fermentation quite quickly. Refrigeration retards the rate. We generally drink copiously the day we press, then can some or begin making cider. The best way to keep plain juice is by freezing.

Canning fresh apple juice in *Ball* or *Mason*-type jars is relatively fast and easy. Pour the strained juice into sterilized jars, leaving one-quarter-inch headroom. Screw down the lids. Process in a hot-water bath (185 degrees F) for 30 minutes (pints or quarts).

We put up between 20 and 25 gallons of juice and cider by this method each fall.

Our family enjoys canned juice straight, iced, diluted with water, or mixed with tea or other fruit juices. Warmed apple juice is an excellent treatment for winter colds, especially sore throats. You can also add cinnamon and cloves, or make an old-fashioned wassail (see recipes) for chilly winter nights.

APPLE SPIRITS

Hard cider, along with perry (the equivalent made from pears) is probably the easiest of all fermented beverages to make. Cider begins to ferment on its own within 24 hours of pressing.

In fact, fermentation begins so quickly that it was with trepidation that we delivered several jugs to some teetotal neighbours. Of course, we could not warn them that the innocent-looking drink was "working." Luckily, most was consumed in good order. But one woman stuck the jar in a back corner of her refrigerator and opened the door a week or so later to find a bubbling brew spewing merrily over her frankfurters and string beans. Happily, she has not let it come between us.

Cider makes itself — the result of the action of yeasts naturally present on the skins of the fruit. You can make a delicious cider by simply setting a loosely capped jug of apple juice in the refrigerator for several weeks. This is a foolproof method because the main enemies of cider — vinegar flies — are absent. However, you will probably want more than a jug or two of cider.

Batches of cider made outside of a refrigerator require some precaution, but there is really very little to go wrong. Hopefully, you will be starting with fully ripe apples gathered from clean grass and well-endowed with bruises, worm holes and yeasty skins.

Put the freshly pressed juice into a vat or wide-mouthed crock. Do not fill to the top since the brew bubbles tempestuously at first and may overflow. If the apples were washed you may want to add a wine or champagne yeast (bought from amateur beer and wine shops). We've done well enough with natural yeasts on the fruit. Cover with a cheesecloth so that fruit flies can't get through and turn it to vinegar. Place the container where you can check it several times a day. A quick primary fermentation will occur at 60 to 70 degrees (F), but French cidermakers prefer a slower fermentation at 40 to 50 degrees. (Our best cider has been fermented in gallon jugs set in a cool spring box.)

Be forewarned; the open vat of cider bubbles loudly. We once spent an uncomfortable afternoon visiting with a straight-arrow, God-fearing abstainer while raucous hiccoughs emanated from behind the stove. We have now moved our cider operation to an unheated storeroom.

As a result of this move we have had the op-

CIDER WASSAIL BOWL

3 apples
1 cup water
1 cup sugar
1 piece whole ginger, broken
1 3-inch piece cinnamon
½ teaspoon nutmeg
3 whole cloves
4 coriander seeds
2 whole cardamon seeds
3 whole allspice berries
½ tsp. mace
1 lemon rind
3 eggs, separated
4/5 qt. sherry
4 cups hard cider

Bake apples in a shallow baking dish in a little water for 45 minutes at 350 degrees (F). Combine spices, water, sugar and lemon rind in a saucepan. Bring to a boil, then reduce heat and simmer 10 minutes. Add sherry and cider and reheat mixture to just below the boiling point.

Meanwhile, beat egg whites in a bowl until stiff. Beat yolks until thick and lemon-coloured, then add to egg whites.

Strain hot mixture, then pour slowly into eggs, stirring constantly.

Pour into a punch bowl and float apples on top. Serve *hot*.

BAKED APPLES WITH RUM AND CIDER SAUCE

6 large apples
2 Tbsp. raisins
24 whole blanched almonds
9 Tbsp. brown sugar
2 Tbsp. butter
1/3 cup water
1 Tbsp. rum
1 cup sweet cider
1 Tbsp. cornstarch
½ tsp. freshly grated nutmeg
Whipped cream (optional)

Wash and core apples, leaving about ½ inch of apple at the bottom to hold the filling. In each apple place 1 tsp. raisins, 4 whole almonds, 1½ Tbsp. brown sugar and 1 tsp. butter. Score the skin at the tops of the apples into sixths. As the apple cooks, it will open like a flower.

Place the apples in an open, ovenproof dish big enough to accommodate them without touching each other. Combine the water and rum and pour in. Bake at 375 degrees (F) until the apples are tender, about 30 to 50 minutes depending upon the apple variety. Baste periodically with the pan juices.

Shake the cider and cornstarch together in a jar. Combine with the pan juices. Cook until thick and clear. Stir in nutmeg.

Pour some sauce over each apple. Serve the rest in a jug. Pass cream or whipped cream if desired.

MULLED APPLE CIDER

1 tsp. whole cloves
1 tsp. whole allspice berries
1 cinnamon stick, cracked
8 cups fresh cider
½ cup firmly packed brown sugar

Tie the spices in a piece of cheesecloth. Simmer together with the cider and brown sugar for 20 minutes. Discard the spice bag. Pour into mugs.

portunity to taste another apple beverage — apple jack. When the temperature dipped down below zero one night, we discovered that a small crock of cider had frozen. We lifted off the ice and found the "jack" below. Potent stuff! And a fine addition to an evening's wassail bowl.

Once the action has slowed down, siphon the cider into a barrel or glass carboy. Bung the neck with a water seal (also called an airlock), half-filled with water and plugged with cotton if it is the old type. This is a cheap glass or plastic device which allows gas to escape but excludes fresh air, dust and insects. (Available at beer and winemakers' shops.) The secondary fermentation lasts two to six months. You can track the rate of fermentation by watching bubbles of carbon dioxide escaping through the water seal.

If you are concerned about bacteria in the juice, you may sterilize by stirring one Campden tablet into each gallon. This kills undesirable bacteria, but does not harm wild yeasts. Other preventative measures are:

1. Strain the juice after pressing to remove seeds, yellowjackets attracted to the party, and other foreign matter.

2. Wash fermenting vats or crocks with scalding water.

3. Keep carboys topped with juice from a spare jug.

Begin drinking the hard cider any time you desire, from a few days old up to two years. The best ciders are usually about six months old. For a draught, or still beverage, siphon off cider as you wish to drink it.

Sparkling cider can be bottled at a hydrometer reading of 1.005. Some makers bottle after two months fermentation, some after six. Use real champagne bottles (cheap wine bottles may burst) which are sterilized and rinsed with a solution of sodium metabisulphite (sold as Campden tablets) and a quarter teaspoon of citric acid dissolved in a pint of water. You may prime with a little sugar before carefully siphoning the cider to the midpoint of the bottle necks. Use wine corks or champagne stoppers, and wire securely in place. Bottles should be aged three to six months before serving.

To make apple cider vinegar ferment cider in an open cask past the alcoholic stage until it turns acidic. Dilute the brew to desired strength with fresh spring water. Bottle with herbs of your choice, such as dill, tarragon, marjoram or basil.

Store apple juice, cider and vinegar in a dark, cool place — a temperature between 40 and 65 degrees (F) is preferred.

Hard cider, in sum, is a mild drink. It's easy to quaff in large draughts, and tends to grow on you. Sweetness varies from bottle to bottle — cider, like wine, may be either sweet or dry. Perhaps it is fair to say that cider tastes like apple ale. Bottled, it might be likened to apple champagne. Actually, cider is in a special class by itself, and a very fine drink it is.

Each year we fill two five-gallon glass carboys and five or six one-gallon jugs with cider and top them all with water seals. Because we like cider at all stages of fermentation, we start in on the gallon jugs after waiting a week or so. By the time we've finished the small jugs, the large ones are coming into their own. We never bother to bottle cider, just leave it under the water seals.

This, to us, seems a mild enough addiction, and we hope eventually to acquire several 10-gallon barrels to provide a year-round stock of cider. Huzzah!

Hard Cheese

Eighteen illustrated steps to distinctive homemade cheese

By Virginia Naeve and the Harrowsmith Staff
Photography by Rosann Hutchinson

In the deserts of Saudi Arabia a primitive method of making *kishk* cheese is still used today: camel milk, allowed to sour and curdle in the scorching sun, is squeezed, salted and then set to dry on the goat-hair roofs of nomadic tents.

In the southeast corner of France, Trappist monks keep the recipe for their *tamie* cheese a closely guarded secret.

Mankind makes more than 400 varieties of cheese, using milk from cows, goats, sheep, camels, reindeer and even the water buffalo, and has given these cheeses some 800 names.

Little wonder then, that the homesteader is cowed by the mystique of cheesemaking and masters nearly all other aspects of food self-sufficiency before attempting to make more than cottage or simple soft cheeses.

Virginia Naeve had not made hard cheese until three years ago, when her cow Minnie began regularly producing four gallons of fresh milk per day — this for just one couple.

As attested to by the accompanying series of photographs of her kitchen cheesemaking procedure, Virginia has mastered the technique. She now teaches hard cheese making in the Eastern Townships of Quebec and offers the following basic step-by-step instructions as an answer to the need for an illustrated cheesemaking guide.

"Right now I'm making a cheese a week with a neighbour," says Mrs. Naeve, "so that she can learn to make her own. I am keeping her cheeses on my shelves because, for a new cheesemaker, the biggest temptation is to cut into a new wheel before it is aged at all. A brand new cheese is pretty drab in flavour, so *try* to wait until it ages at least a month."

YOU NEED:

A. *1 rennet tablet (from Horan-Lally Co. 26 Kelfield St. Rexdale, Ont. Hansen's Rennet Tablets — $4.35 (for 25 tablets.)*

B. *2 Tbsp. of salt (optional — 1 Tbsp. of garlic salt substituted for 1 Tbsp. of plain salt.)*

C. *2 soft milk filters (6½"). A box of 100 filters can*

be purchased at a local feed store.

D. *Large sieve, enamel canner and lid, plus a larger enamel pan to act as bottom of make-do double boiler, 1 Jiffy cloth to use as wrapping, long handled wooden spoon, long knife.*

E. *Dairy thermometer — $6.50 from Horan-Lally Co.*

F. *A homemade cheese press.*

STEP 1

One 3½-pound to 4-pound cheese takes 4 gallons of cooled, whole milk. Jersey milk is richer and makes a yellower cheese than Ayrshire or Holstein. Cheese from a hay-fed cow is different from that produced when she is on fresh pasture. Adding garlic salt, sage or caraway varies the basic cheese recipe.

STEP 2

Pour 4 gallons of cooled milk into the canner on the stove. Put it into the larger enamel pan which is a quarter full of hot water, enough to come up around the sides of the canner. Bring the milk to 86 degrees (F) or cheese on the dairy thermometer.

STEP 3

When milk reaches 86 degrees remove canner from water and stove and set aside in warm place. Only then crush one rennet tablet into ½ cup of cold water. Then pour the rennet into the milk and stir with a long wooden spoon for 2 minutes, or 120 times, to mix rennet thoroughly. Cover container and let sit for 45 minutes. Test for firmness with finger after 45 minutes.

STEP 4

Cut firm, curded milk using a knife long enough to reach the bottom of the canner. Make half-inch cubes with knife.

STEP 5

Retrace cuts in curds, but slant knife diagonally, criss-crossing the curded milk.

STEP 6

Stir slowly with hand, breaking up curds that are too large. Make them all more-or-less uniform in size. Do not squash curds. Stir for about 15 minutes.

STEP 7

Return canner with curds to stove again putting it into the large pan with medium-hot water. From this point on, make sure your heat is at no higher than simmer (on a wood stove pull your pans to the cooler side of stove). Heat curds very slowly, 2 degrees every 5 minutes, taking from 40 minutes to 1 hour to reach 102 degrees. Do not hasten this process or you will have Indian rubber balls, not curds. Stir every 5 minutes so the curds don't form a mass and stick to the bottom of the canner. After reaching 102 degrees turn off the heat and remove pan from stove. Leave in whey until curd is firm. If it breaks apart and shows little tendency to stick together, it is ready.

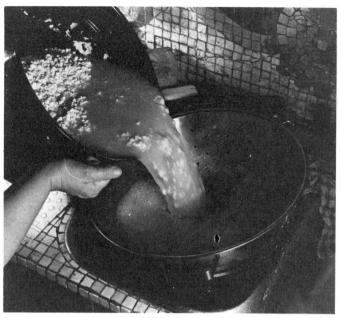

STEP 8

Pour curds out into large sieve, catching whey in pan under sieve. Remove whey in pan and use in cooking or give it to your farm animals. Chickens, geese, dogs, cats, pigs and cows like the whey while it is still warm. Or pour it on your compost piles. Don't waste it down the sink drain. (Scientists at Oregon State University have come up with a new use for whey. They add yeast cultures causing the whey to ferment. The result: wine which compares favourably with apple wine. No energy is required and all the whey is utilized.)

STEP 9

Cool curds, breaking up into medium-sized lumps. When they reach 90 degrees, mix in 2 Tbsp. salt, one at a time, mixing well after each spoonful. (At this point I sometimes substitute 1 Tbsp. of garlic salt or add ¼ cup of crushed caraway seeds.)

STEP 10

After mixing in the salt, wait until the curds reach 85 degrees.

STEP 12
Fill press can with cooled curds.

STEP 11
Put one clean milk filter in bottom of pressing can (made from a gallon tin can with holes punched through the bottom from the inside).

STEP 14
Put pressing can into cheese press, lowering the follower (see photo.) Then put four bricks on the top of the follower for 10 minutes. Remove the follower, pouring off excess whey that has gone into the catching pan at the bottom of the press. Resume pressing, this time with eight bricks, for one hour (eight bricks or 40 lbs.).

STEP 13
Place second filter on top of curds. The filters make it much easier to remove the cheese after the first pressing.

STEP 15
Remove cheese from press can, first tapping can. When cheese comes out, remove the filters from both ends. Wash the cheese in warm water and dry.

STEP 16
Take a Jiffy cloth (dampened) and wrap the cheese in it. Return one filter to the bottom of the press can, and put the wrapped cheese in can. Next, replace cheese in press, insert follower and press with eight bricks for another 18 to 24 hours.

STEP 17
Remove cheese from press and unwrap. Wash cheese in hot water or whey to form a firm rind. Put cheese in a cool dry place — turn each day and wipe if necessary. I put my cheese on a double piece of waxed paper and change paper if it gets moist. (Don't use brown paper bags as I did to conserve waxed paper — they stick to the cheese!)

STEP 18
After a week or so I butter the cheese and keep it on the shelf at least a month, turning it every day. I find buttering the rind keeps it so I need not paraffin my cheeses. If I have too many cheeses I freeze them after they have aged.

Salad Days And Mulching Nights

Free for the taking, seaweed brings new life to gardens and summer kitchens

By Lynn Zimmerman
Photography by George Thomas

While wandering along the quarter-mile of beach below our house after a fierce storm several weeks ago, I met a good neighbour who is both a farmer and fisherman. During the weekend, unmercifully strong, northeast winds had battered and smashed hundreds of lobster traps, and the rolling surf had tangled as many as 30 pots together and deposited the unmanageable mess of lines, buoys and traps on the sandy shore. My neighbour was undertaking the grim task of searching his buoys out from the snarled wreckage.

But when he paused to talk, it seemed he was more interested in the tons of spongy seaweed -- a blackish mixture of rockweeds, kelps, Irish moss, dulse and other sea plants — piled a full two feet high along the edge of the sea.

With a connoisseur's admiration, the fisherman pointed out that the fetid mass was real seaweed, and suggested that if I wanted to grow good cabbages, I should plant them in ground mixed with real seaweed.

We have used seaweed to fertilize our garden for years (and take some deserved pride in our cabbages), but it turns out that our neighbour's advice is rooted in a more firm medium than local lip service. Government bulletins and gardening guides all claim that cabbage is a heavy feeder, and potassium is one ingredient it particularly craves. Seaweed ranks high on the list of natural sources for this hard-to-get plant nutrient.

The old people must have known this when they carted away partially decomposed aquatic plants from the beach by means of horse and wagon.

Today, a few of our neighbouring farmers employ tractors to gather the free fertilizer, but sadly the practice has for the most part been forgotten, and the soil-building seaweed is neglected, while expensive, commercial chemical fertilizers are extolled.

Sea plants have been used as fertilizer for hundreds of years. In fact, in the mid-seventeenth century it was decreed by royal command that only certain seaweeds be collected from the coast of France, and the ways in which they could be applied were specified. Some experts now attribute the rich agricultural crops of northwestern France directly to the age-old custom of regularly using seaweed fertilizer.

Scottish scientists working at the Institute of Seaweed Research claim that seaweed products "enhance the germination of seeds, increase the uptake of plant nutrients, impart a degree of frost resistance and make the plants better able to withstand phytopathological fungi and insect pests."

From seashore to vegetable plot, author Zimmerman finds free, storm washed seaweed a rich harvest of nutrients and humus for her Cape Breton soil.

An explanation for these surprising finds might lie with potassium, one of the "big three" plant nutrients. Scientists are still not certain of the exact role that potassium plays in plant growth, but theories suggest that it helps build plant resistance to disease, aids in retaining water during dry periods and (of particular interest to northern gardeners) protects plants from damage by cold weather.

TRACE ELEMENTS

Although seaweed is only half as rich in nitrogen and phosphorus as cattle manure mixed with bedding, potassium makes up 4.6 per cent of seaweed's bulk (three times higher than manure and bedding). Too, the components of seaweed can play a vital part in other, little-understood areas of plant growth. Although scientists are still at a loss to explain fully the roles of trace elements, it is known that zinc, manganese, boron, iron, sulphur, copper, magnesium, molybdenum and chlorine are all important to healthy plant growth. Seaweed boasts high concentrations of many of these hard-to-find trace nutrients.

Sand and seashore frequently go hand in hand, and although tourists might delight in this, sandy-earthed gardens, nearly devoid of organic material, can wreak havoc on horticultural activities. Fortunately, nature has seen fit to bless seashores with seaweed — an ideal source of much needed humus, and a source that does not increase acidity (sandy soils are often acidic).

The landlocked gardener need not bemoan the use of seaweed as a benefit belonging only to his maritime counterpart. Reports indicate that those troublesome weeds that clog swamps and shallow areas of fresh water bodies also possess many of the qualities of their saline cousins.

Seaweed can be applied to the garden in three ways. We spread it directly as a mulch between rows, to be plowed under at the end of the gardening season, but others report success by working fresh seaweed directly into the soil or by adding it to compost heaps.

There are those who would doubtlessly balk at the thought of eating an ingredient normally destined for the compost heap, but today we sat down to a delicious lunchtime soup that was both simple and quick to prepare. I gathered the seaweed for it just yesterday during a mulch-collecting spree at the beach.

"Seaweed," you may gasp, "actual pieces of it floating in the soup bowl?" Yes, exactly. Why not eat plants grown naturally in the richest mineral environment on earth?

The seaweed I used in the soup was dulse, a red sea plant which grows along coasts near the low-tide mark. It is especially abundant along the Fundy coast where it is harvested, dried and sold commercially.

And it must be admitted that, although my husband approached the soup with apprehension, he then eagerly finished the bowlful.

One rainy morning this past spring, Georgie Sears, a dynamic New Brunswick woman and mother of 12 who has lived for many years on a relatively self-sufficient farm, sat in my Cape Breton kitchen drinking tea and entertaining several of us with her experiences. An episode about gathering enormous sea clams along the coast prompted her to describe the family's annual camping excursion to the Fundy shore during the dulse harvest.

Rockweeds

DULSE HARVEST

When the tide ebbs (and in the Bay of Fundy it goes out a long way) they all scramble among the slippery rocks picking the plentiful red sea plant by hand. Only when the morning low tide occurs early, do some of the group show any reticence to scurry across the uneven terrain after the dulse. "Collected on a warm, sunny day and spread one layer thick on the dark, hot rocks, the plants dry in just a few hours," our friend explained.

While the tale was being told I got out a 12 ounce box of dulse — there is quite a bit of dried seaweed stuffed into a 12 ounce box — and placed it on the kitchen table.

"Well, what do you do with it; I mean, what can you use it for?" our neighbour, Pete, asked.

"Why you just eat it," she replied, smiling and tearing off a large hunk of the papery, red-purple seaweed.

"Don't you use it in cooking?" Pete persisted.

"Guess you could," our friend agreed, chewing happily. "But around our house it goes too fast just like this. In fact, I know a fellow, if you'd set this box down in front of him, why it'd be gone in 10 minutes. . . 10 minutes."

By then we had eagerly passed around the box and were all quietly munching away and reflecting on the surprisingly good-tasting, salty seaweed. And it grows on you. Really, if you've never tried eating seaweed you have every right to be skeptical, but at least give it a try. If you live by the sea it is free food; if not it is relatively inexpensive eating and certainly a delightful change for any palate.

Although none of the dire effects that could befall the careless mushroom hunter await those who pursue seaweed, *Desmarestia*, a variety fairly common to deeper waters off the Canadian Maritime provinces, can make humans sick if eaten.

According to a Dalhousie University seaweed expert, however, there is very little chance that you would ever consume *Desmarestia*. As soon as it dies, the plant takes on an odour akin to that of rotten hay and a bitter taste that, according to the expert, "is just awful." *Desmarestia*, a three-foot tall bushy plant that is brown when growing but turns green shortly after being uprooted, does not resemble any of the edible seaweed varieties.

I visit the beach below our house regularly, and after storms I generally go prepared with a collecting bucket. While our two dogs race in vain after the always-elusive sandpipers, I poke through the huge piles of plants along the water's edge.

While it's probably best to gather plants soon after they wash ashore because they can rapidly

Kelp

Dulse

Rockweed

Rockweed

Irish Moss

decompose, I have collected Irish moss from the beach several weeks later. By then the seaweed is dried out but can be quickly revived in a bowl of water in the kitchen; it still works perfectly well in making custard. I gather quantities of both dulse and Irish moss for cooking.

For centuries the Japanese have enjoyed eating seaweeds, which even today may still comprise up to 25 per cent of their daily diet. *Kombu*, which is primarily used for soup stock, is made from kelp, a family of brown seaweeds with wide, leathery fronds. Kelp is gathered commercially in deep water off Hokkaido Island in Japan.

Nori or laver is a food product usually made from a red seaweed which is farmed in Japan on nets spread in nutrient-rich, constantly renewed tidal waters. (It is also eaten in Britain, Ireland, New Zealand and Korea.) Sun dried and pressed into thin sheets between mats of bamboo, *nori* is toasted over a flame and eaten in strips with soy sauce or crumbled as a delicate garnish over grain dishes.

Wakame has a romantic story behind it. Although today this food comes from a bottom-growing, feather-like seaweed which is harvested by machines, there was a time not very long ago when highly trained and skilled women free divers (without scuba) plucked these plants by hand from the ocean floor and collected them in floating baskets when they surfaced for air. Many Japanese people savour miso soup with *wakame* for breakfast.

If eating sea plants seems foreign to you, consider its creditable health aspects. All the rivers on earth flow into the sea, where they deposit mineral-laden soil. Trace elements such as copper, iron, boron, zinc, iodine and manganese, as well as significant quantities of potassium and magnesium are found in sea plants as in no other natural vegetable; the importance of a natural source of iodine, essential in each of our diets to avoid goiter, should be emphasized.

Also, to quote from the U.S. Fish and Wildlife Services, "Seaweeds are a good source of vitamins — beta-carotene (which is converted to vitamin A by the body), thiamin, riboflavin, niacin, panthotenic acid, vitamin B-12, vitamin C, and vitamin D."

Obviously, sea plants can add new dimensions of colour and delicate flavours to your diet and there can be no doubt as to the nutritional value of these natural vegetables. If you live by the sea or if you plan a visit this summer, why not stroll on the beach or scramble over a rocky shore in quest of sea plants. Whether for food or fertilizer, seaweeds are worthy of your investigation.

If you're landlocked without a hope of seeing the sea in the near future, all the seaweed foods mentioned in this article are available at well-stocked natural food stores. Also, fertilizers from the sea are obtainable commercially. They have gained a measure of acceptance in commercial organic greenhouse operations, but are still quite expensive.

Recipes

Savoury Seaweed and Noodle Soup

1 large handful dried dulse,
   chopped (or kombu)
1 quart cold water or stock
1 bunch onion tops or
   chopped onion
2 Tbsp. oil
1 large handful
   whole wheat noodles (or
   cooked grain instead)
1 tsp. mugi miso (soybean
   and barley paste — optional)

Soak seaweed in the water. Sauté onions in the oil, add seaweed and water and bring to a boil. Add noodles and miso and simmer until noodles are done.

Sea Vegetable Soup with Dumplings

2 large handsful dried dulse,
   chopped (or kombu)
2 quarts cold water or stock
3 Tbsp. oil
2 onions, chopped
3 garlic cloves, minced
2 carrots, diced
1 stalk celery, diced
1 egg
3 Tbsp. milk
2/3 cup flour
¼ tsp. salt
1 tsp. baking powder

Soak seaweed in water. Sauté vegetables in the oil, add seaweed and water and simmer. To make dumplings: Beat egg until light. Add the milk and mix. Add the dry ingredients and mix until paste-like. Drop by teaspoonsful onto boiling soup and turn heat down to simmer. Don't peek and cook covered 15 minutes.

Vegetables (including Seaweed) and Rice

½ cup dried dulse or hiziki,
   chopped
2 Tbsp. oil
½ pound fresh mushrooms, sliced
2 onions, chopped
2 garlic cloves, mashed

1¼ cup or more stock or water
1 cup carrots and/or
   squash, diced
½ pound green beans and/or peas
   and/or broccoli, chopped
1 Tbsp. or more tamari soy sauce
1½ cups dried brown rice
3 cups stock or water

Soak the seaweed in the 1¼ cup stock or water. Heat oil in a wok or heavy skillet and add mushrooms, onions and garlic. Add the vegetables and quickly coat with oil, add seaweed and water and cook until the vegetables are tender. Meanwhile, cook the rice in 3 cups water or stock. Add soy sauce to the vegetables and serve over the rice.

Tossed Salad with Sea Plants

1½ quarts crisp lettuce
   or spinach
½ cup dried seaweed,
   chopped finely
1 cup celery, chopped
2 carrots, diced
1 green pepper, chopped
1 cup grated cheddar cheese
3 hardboiled eggs, chopped

Dressing

1/3 cup safflower oil
3 Tbsp. apple cider vinegar
1 Tbsp. chopped fresh dill
1 egg
¼ tsp. sea salt
dash pepper

Toss salad and dress.

Irish Moss Blancmange

The Cookbook *Out of Old Nova Scotia Kitchens* by Marie Nightengale says of Irish moss:

"One of the popular 'company' desserts of the long ago was blancmange. It was made with milk and Irish moss. The moss, which is still collected on the Fundy shore, was well washed and dried in the sun. Consisting chiefly of vegetable gelatin, it was considered a very nourishing food for invalids.

1/3 cup Irish moss
4 cups milk
¼ tsp. salt
1½ tsps. vanilla

Soak moss for 15 minutes in enough cold water to cover; drain. Pick over, removing any undesirable pieces. Add the moss to the milk and cook in the top of a double boiler for 30 minutes. The milk will only thicken slightly, but if cooked too long the blancmange will be too stiff. Add the salt and vanilla and press through a sieve. Fill individual moulds that have first been dipped in cold water. Chill. Turn onto serving plates and serve with sliced bananas, sugar, and cream."

The New Country Krautmeister

Bringing out the essence of cabbage in aromatic homemade sauerkraut

---

By Robert Mariner

Put the cabbage in the barrel,
Stamp it with your feet.
When the juice begins to rise,
The kraut is fit to eat.

Put it in a pot,
Set it on to bile.
Be sure to keep the cover on,
Or you'll smell it half a mile.

— Old Farmer's Song

One chilly day last winter, after shovelling a foot of snow from the path leading to our loyal outhouse, I came into the steaming kitchen where sauerkraut was cooking, the air dripping with a wonderfully vile, sulfurous odour that drew on my salivary glands until they ached slightly in anticipation. It was, in fact, just such anticipation that led me to shovel the path in the first place — I knew we'd be using it, because sauerkraut is, of course, the best "regulator" to come along either before or since the rediscovery of fibre in food.

Among our ancestors, my wife and I count the family names Schill and Nock and we suspect that the traces of kraut molecules spiralling around in our DNA readily account for our love of fermented cabbage. Sauerkraut is a centuries-old favourite with many European nations and their North American descendants, but it has more than ethnic value. In addition to its pungent taste, sauerkraut has a well-deserved reputation as a source of winter vitamins.

Its retention of vitamin C, in fact, undoubtedly helped spread sauerkraut's good name. Because the fermented product could be kept longer and in a much smaller space than fresh cabbage, sauerkraut was long valued as a scurvy preventative aboard the old sailing ships.

Scurvy victims first became irritable, argumentative and pessimistic, as well they might, given what the future held for them as their bodies' connective tissue deteriorated without ascorbic acid.

The condition of a scurvied ship's crew was not appealing. Old scars would open and new wounds wouldn't heal. Blood vessels would rupture, often starting in the legs, where old, dead blood collected under the skin — hence the name "blackleg." Poor circulation eventually brought gangrene, putrefying gums and death — usually resulting from hemorrhaging of the lungs, digestive system or brain.

In 1863, after winning the battle of Chambersburg during the American Civil War, the victorious Southern troops were suffering both from poor supply lines and scurvy. They demanded 25 barrels of sauerkraut from the conquered city fathers.

Long before that, sauerkraut was recognized by the Dutch for its anti-scorbutic (scurvy fighting) value: spice merchants in the 17th century included it among ships' provisions.

Fear of scurvy is hardly a reason to start making your own sauerkraut today — so little vitamin C is actually required to combat the disease that a daily dose of Jolly Miller Flavour Crystals would easily hold your sinews together. Nevertheless, sauerkraut is a good source of lactic acid and several vitamins. We feel justified in eating it in large amounts as it is one food that provides both bulk and fine flavour without high calorie content (one cup of cooked sauerkraut has but two calories).

Much of today's commercially available sauerkraut is packed in a preservative-laced brine, and we find that we can't trust even the brands that are packed without artificial stabilizers to be free of traces of chlordane, Malathion or other similar treats.

Making sauerkraut is a good outlet for our home-grown cabbages and, furthermore, we

264

Joseph Mahronic

make products that are almost impossible to buy: purplish, crunchy kraut made from red cabbage and tawny, tangy, fermented rutabaga — called turnip kraut by some (including myself). Translated correctly, this would mean "turnip cabbage"; *sauererüben* (sour turnip) is the etymologically correct name for turnip kraut.

It does not take a chemist to make good sauerkraut — the basic process is quite simple. Essentially, you cut up some cabbage, add salt and let it ferment. No water is added — fresh cabbage is 90 per cent moisture, and this is drawn out, along with natural sugars, by the salt.

At the proper fermentation temperatures, bacteria convert the sugars to acids, which in turn prevent the growth of spoilage bacteria. There are, of course, a few tricks to be learned, but no complicated procedures or equipment are needed. Most first-time kraut-makers can be assured of success.

The occasional failures generally result from too little or too much salt, insufficient mixing, very sloppy sanitation, fermenting in a room that is too cold or too warm, or allowing air to reach the cabbage in the working sauerkraut container.

Just as kraut is a culinary paradox — a smelly ambrosia — so the *making* of kraut is a paradox: extremely simple rules help control an extremely complex fermentation process, involving a dizzying array of bacteria. *Leuconostoc mesenteroides*, *Lactobacillus plantarum* and numerous others take their appointed turn to produce the needed ingredients (lactic acid, acetic acid, alcohol, mannitol, carbon dioxide), all in the requisite proportions . . . if conditions are right. It's no wonder that the making of good kraut from one batch to the next has a hint of art as well as science.

IN THE AUTUMN BY THE LIGHT OF THE MOON

Sauerkraut is usually made in the fall when cabbage is harvested and the ideal fermentation temperature range of 65 to 72 degrees (F) is easy to achieve. However, cabbage can be stored and then converted to sauerkraut in the depths of winter, after you've recuperated from the busy harvest, provided the fermenting kraut is kept warm enough.

Those of us blessed with cool summer weather, courtesy of the North Atlantic, rarely suffer

265

from weather that is too hot for kraut-making. I made some in mid-March this year, when I used the last of the cabbages from our root cellar, and then later in June, when we cheated and bought some cabbage. If you try it in warmer months, beware of fruit flies; screen them from the fermenting kraut or they'll introduce wild bacteria and ruin the batch.

Some traditionalists claim that the lowest failure rate results from cabbage shredded and salted during an ascending moon (just after the new moon). Keen-eyed observers testify that the moon pulls the brine over the kraut, protecting it from spoilage. The creed of other kraut-makers specifies salinometers and well-scrubbed containers for best results.

Our household is so hectic that kraut gets made whenever it can and neither my wife nor I generally has time to notice the phase of the moon. On one occasion, one container failed while other containers made at the same time under the same moon bubbled to perfection.

I seem to recall that we made our first batch with great care on the correct lunar schedule, and we're convinced it was the best kraut we've ever made, but that judgement is probably biased by our initial delight in producing an edible product.

The occurrence of partial failures — where most of the sauerkraut is good and only a small amount is lost — is one of the best recommendations possible for not putting all of your kraut in one crock (reminiscent of eggs in baskets and one bad apple). I ferment small quantities in jars, as shown in the accompanying photographs.

The use of jars as fermentation containers also means there is no odour given off and no skimming of surface mould, as there is with crocks or barrels. With no odour, kraut can be made pleasantly anywhere in the house — where temperature is fairly easy to control year-round. So, unless you are enthralled with the idea of using traditional, "old-timey" stoneware crocks or barrels (or are mass-producing kraut for sale or barter), I think you'll find jars ideal.

Making sauerkraut by the jar method is a simple one. After the cabbage has been shredded and the appropriate amount of salt added (see table at the end of this article) the mixture is packed firmly into the jar with a wooden spoon or your hand. If you wait a few minutes for the salt to wilt the shreds, they will co-operate more readily with the packing. Place a piece of cheesecloth or other open-weave material on top of the shreds and compress by wedging wooden sticks (popsicle sticks work fine) into the shoulders of the jar.

The juice will begin to rise after a few hours, but if necessary, the jars may be topped with brine (1½ tsp. pickling salt per cup cold water) — it is essential that the shreds be submerged in liquid.

Keep the jars at 65 to 72 degrees (F) — warmer temperatures will allow spoilage, cooler will prevent fermentation. In the unlikely event that scum forms at the surface, skim it off and change the cloth if it was in contact with the scum. After about seven to 10 days, the brine level will drop suddenly and the kraut is ready to eat or process for long-term storage.

Some of our kraut gets made in wide-mouthed, one-gallon jars (scrounged from restaurants and hot dog stands) and stored in the root cellar throughout the winter. I also use quart-size canning jars for fermentation containers, and the finished kraut requires no repacking. We just pop some jars into the freezer. About an inch of space must be left at the top to allow for expansion and it is wise to test one of your jars to assure that the brand will withstand freezing.

Should you prefer, the canning jars can be processed in a boiling water bath for 30 minutes to allow safe storage at room temperature. Once the fermentation has ceased (no new gas bubbles forming), press the kraut in each jar down firmly to expel bubbles and add brine to bring the level in each jar to within one-half inch of the top. Seal the jars, place them on a rack in a large kettle, assuring that the tops of the jars are submerged. Bring to a boil and keep boiling for 30 minutes.

Perhaps the best advantage to using jars for fermenting sauerkraut is that it encourages experimentation. Looking into the gaping hollow of a huge empty barrel can intimidate a first-time, would-be krautmeister; a quart jar, however, looks tameable. The adventurous may try adding a touch of dill, caraway or celery seed along with the salt, or perhaps even some sliced pears or apples for special occasions.

Of salt, scales and other weighty matters

For successful sauerkraut making, you should borrow or buy a kitchen scale if you don't already have one. While salt is essential to draw forth water and natural sugars from the cabbage, too much salt will inhibit the necessary acid-forming bacteria.

The correct amount of salt is 2 to 2½ per cent, *by weight*, or one pound of salt for each 40 pounds of shredded cabbage in large batches. Because of the variables of shred size, crispness and the extent of compaction, it is impossible for most people to estimate accurately the weight of shredded cabbage. Use a set of scales.

For small batches of sauerkraut, the salt is measured by volume. The following chart gives the correct amount of salt for a number of shred weights:

SALT REQUIRED

Shreds	Salt
1 lb.	2 tsp.
2 lb.	4 tsp.
5 lb.	3 Tbsp.

Krautmaker's apparatus — note the cabbage-filled jar with cloth and wooden wedges in place to keep all strands well-submerged.

After quartering and removing core, cut cabbage in shreds the thickness of a dime. Core is then grated and added to the mixture.

Old-style kraut cutter, in which cabbage is placed in moveable box and repeatedly slid across sharp blade in wooden track below.

Jars should be packed firmly; a few minutes in the salt mixture causes the cabbage to wilt and co-operate with the packing operation.

8 lb.	4 Tbsp. + 2 tsp.
10 lb.	6 Tbsp.
25 lb.	15 Tbsp.

If you have, for example, 17 pounds of shreds and wish to mix all the salt at once, *do not* simply take the 2 tsp. per pound figure and multiply. The slight error in the recommended amount for a small batch would become significant when multiplied. Instead, add the measures required for 10 pounds plus five pounds plus two pounds (10 Tbsp. plus 1 tsp. salt).

We find that approximately two pounds of shreds are required for a one-quart jar (and eight pounds for a gallon).

Remember that most table salt is iodized, generally a good idea, but not for sauerkraut. Use pickling salt to assure proper fermentation and to avoid having the brine become dark or cloudy.

Salt brine will react with some metals to produce an off colour or taste. Particularly avoid old zinc canning lids, because zinc is often contaminated with cadmium (a dangerous heavy metal which may be released by the salt solu-

tion). Glass, wood and unchipped enamelware are all fine for sauerkraut-making.

Old German Recipes

Although the rules are far from hard and fast, sauerkraut in its homeland is often served alongside some manner of pork or sausage, and there are those who believe that nothing makes a better lunch or light dinner than sauerkraut with a spicy knockwurst or Polish sausage.

SAUERKRAUT NATURELL

Sauté a few sliced or diced onions and apples in a pot for a minute or two. Place sauerkraut on top, cover and let cook until just tender.

Both onions and apples reduce the sharpness of the kraut and add a delicate flavour.

WINE SAUERKRAUT

Kraut with class: Add the desired amount of your favourite dry white wine to the uncooked sauerkraut and simmer gently until thoroughly heated.

SAUERKRAUT GARNIERT

2 lb. sauerkraut
8 German frankfurter sausages
½ lb. ham, sliced
½ lb. bacon, sliced
2 carrots, cut in strips
1 large onion, chopped
1 cup white wine
6 or 7 juniper berries
salt and pepper
3 or 4 boiled whole potatoes

Put bacon and onion in the bottom of a medium-sized casserole. Place sauerkraut on top, then carrots and juniper berries. Add the wine, and, if necessary, a little water or broth. Cover and bake at 350 degrees (F) for 1¼ hours.

Boil the potatoes and sausages and fry the ham.

Turn the casserole upside down on a platter and garnish sides and top with potatoes and meats.

SAUERKRAUT AND POTATO DUMPLINGS

German carbohydrates at their best — a bit time-consuming, but the results are well worth the effort.

For the dumplings: Boil 4 medium-sized potatoes and mash them with a little milk. Let cool.

Grate 6 medium-sized potatoes and strain them through a cheesecloth until all the water is removed. (Retain the water. The starch that sinks to the bottom will be used later.)

Mix together the raw and the cooked potatoes, and the potato starch. Add an egg and ½ tsp. salt. Form dumplings with your hands and place 2 or 3 croutons in the centre of each. Set prepared dumplings on a floured board.

Meanwhile boil some water in a large pot and add a little salt. Gently drop the dumplings one by one into the salted water and simmer slowly. The dumplings will sink to the bottom at first and then rise to the top. Cook for about 15 minutes after they have floated.

Remove to a platter and serve covered with plain sauerkraut, Sauerkraut Naturell or Wine Sauerkraut.

Tallow Ho!

The ingredients have changed little in two millenniums, but home soapmaking today is not the steaming-over-a-hot-cauldron task you might imagine

By Merilyn Mohr

Engraving by Francois Millet (1814-1875) Bettman Archive

A healthy indifference to soap, if you have made any recent forays into the drugstore or supermarket, is becoming less and less easy to maintain. At 50 cents or more for a bath-size bar of the most popular brands, soap has become one of those modest necessities that quietly but effectively engorges the household expense budget.

Those truly attracted to primitive living may decide that the only way out is to give up the habit. Soap, after all, is a relative newcomer on the historical hygiene scene. Primitive peoples still use the juices of saponaceous plants, seeds, roots and bark as cleansing agents. Unfortunately, neither the yucca plant nor the soapberry is native to Canada.

In the spirit of presenting alternatives, we must note that, before the discovery of soap, enterprising ancients used a method of cleansing that is both economical and within reach of us all. Dirty clothes were washed with urine which had been allowed to stand until the alkali separated. It was then warmed and stomped into greasy garments with bare feet. The washday blues had meaning then. (As for personal hygiene, a bone or metal shoehorn-like device was used to scrape away the grime.)

The Romans are credited with the invention of soap. Sapo, a hill outside Rome, was traditionally the site of religious sacrifices and, during heavy rains, the animal fat and ashes around the base of the altar would be washed down the hillside to the banks of the Tiber. Roman women found that this mud, rubbed on their clothes, cleaned even better than urine.

Although no one knows when or how the step was taken from the riverside to the marketplace, a complete soapmaker's shop was uncovered in the ruins of Pompeii.

Merilyn Mohr removes bars of homemade oatmeal soap from moulds to be stacked and cured for several weeks.

Add pure lye — not drain cleaner — slowly, to soft water, stirring constantly but using care not to slosh.

Test the temperature of the two ingredient solutions (lye should be body temperature, fat a bit warmer).

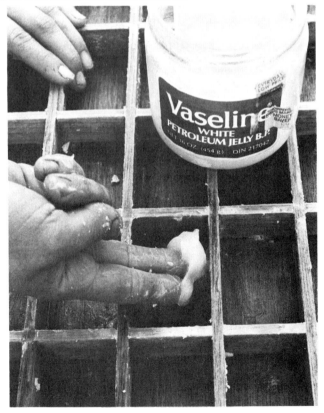

While waiting for the fat and lye solutions to cool, liberally grease moulds with petroleum jelly.

This historical diversion brings us to the only real alternative to using today's pricey, chemical-laced mass-market soaps: make your own. Neither a sacrificial altar nor a downpour is required, but the simple formula has changed very little in these two millenniums.

As the women of the Tiber learned, the only necessary components of soap are fat, lye and water. Homemade soap costs but a trifle to make and one bar will have roughly twice the life span of commercial bars. (The soapmaking reaction is correctly known as *saponification* which creates soap and glycerin. Most big soap companies now use industrially processed fatty acids rather than natural fats, and their soaps do not contain the glycerin of homemade soap and are thus a bit more harsh on the skin.)

One thing is certain: the soap you produce will not make you look 10 years younger, nor is it guaranteed to attract hairy-chested jocks or wet-lipped maidens. Unless you want it to, home-made soap will not leave you smelling of gardenia hedges or Irish whittlers in the springtime. Quite simply, it will do what it is supposed to: keep you and yours clean.

STEAMING MAGIC

Any reluctance about making soap usually involves an image of Grandma steaming over a bubbling vat of rancid fat. This less than titillating vision is a misconception.

In truth, soapmaking is neither difficult nor time-consuming. The ingredients are simple and commonly available, no special equipment is needed, and using the following recipe, success is virtually guaranteed. (My very first batch was made solo with surprising success.)

Especially attractive is the fact that an afternoon of routine-breaking labour and an investment of about $2.00 will easily provide a half-year supply of soap for any family. Best of all, soapmaking is exciting. There is an element of magic and alchemy involved in combining two such unlikely allies as caustic lye and greasy fat to create a firm, white lathering bar of soap.

Once the basic steps are mastered, soapmaking has a way of spurring your creative soul: wild herbs, spices and distilled natural oils can be added for special effects, and you can easily make different soaps for your face, hair, clothes, dishes and pets. And, if it matters to you, there is the satisfaction of knowing that you are breathing some life into an almost moribund tradition.

The tools of the soapmaker's trade are simple. You will need two containers (one each for fat and lye), a wooden stirring stick or spoon, and a mould for the finished product. Rubber gloves are a good idea for beginning soapmakers.

Because lye is extremely caustic and remains active until the soap is set, it is advisable to use only glass, cast iron, ceramic or enamel containers. *Never* use aluminum. It is also wise to dedicate these tools expressly to soapmaking, as even the glass and ceramic glazes will become etched by the lye.

The lye used in all the following recipes is Gillett's Lye, which can be found in most hardware stores. Be sure the lye you use is specifically labelled as "pure" and not a combination product (for unplugging drains). The lye crystals come in nine and one-half-ounce plastic bottles (for about 95 cents).

The water used for making soap should be soft. In hard water, part of the lye will combine with the soluble minerals and leave less to react with the fat. The resulting soap will be greasy and soft. If you cannot get rain water, add washing soda *(Arm and Hammer* is one brand) to soften your hard water.

Fat is the most flexible and interesting of the three ingredients, but in some ways the least pleasant. The kind of fat you use will determine, to a large degree, the quality of your soap. Lard is pork fat and it produces a very soft soap. It should not be used alone, but rather combined with beef fat.

Beef fat (tallow) is the traditional base for soapmaking and it can be used alone. Tallow can usually be obtained from your local butcher, although you should be prepared for raised eyebrows, an indulgent smile or even some reluctance to part with this seemingly useless by-product. (Don't settle for anything but fresh or fresh-frozen tallow — rancid fat being rendered is one of life's less pleasant olfactory sensations.)

Many butchers have contracts with large rendering companies who pick up their tallow, so be prepared to pay five to 10 cents a pound for the fat. By visiting a smaller shop, however, you may find a friendly soul who will give it to you free of charge. (Return with several bars of soap in a few weeks and you may cement the relationship.) About one pound of clean fat would be ample for your test batch, 10 or 11 pounds for the large recipe.

The tallow must be trimmed of every speck of blood and meat before you begin rendering. This next step might be done out-of-doors if you wish to avoid a kitchenful of greasy odours (think of a truck-stop diner during the breakfast rush).

Cut the trimmed tallow into small squares and heat slowly in a heavy kettle until all the fat is extracted. Use care not to scorch the melting tallow — the burned smell and colour will be carried into your finished soap.

When only crispy cracklings remain, strain the fat through several layers of cheesecloth and refrigerate or freeze it at least overnight, or until you are ready to make your soap. (Throw the cracklings to the birds — they love them.)

Since the rendering step is a bit tedious, it is a good idea to process a fairly large quantity at one time to avoid the wasted effort and energy of doing small batches. Invite a friend and make a day of it.

TEST BATCH

With your equipment and ingredients at the ready, try your hand at a small batch. The basic formula I use is simple.

1 cup rendered fat
1 ounce lye
½ cup water

(This recipe changes slightly in bigger batches, as you will see.)

First, melt your previously rendered fat, measure, and set aside to cool.

Add the lye (one ounce equals two tablespoons) to the water, stirring evenly with your wooden spoon. *Do not* underestimate the power of lye. It burns skin. The vapours of this solution are equally caustic if inhaled, so stir near an open window or in a thoroughly ventilated area and stand well back.

The lye will heat the cold water until it steams (a nice show of magic). Set this solution aside to cool. The trick now is to force yourself to wait. Both the fat and lye mixture must cool to lukewarm. Feel the base of each container: the lye solution, when ready, should be about body temperature, the fat slightly warmer. Read a book, brew a cup of tea, but keep checking until the temperatures feel right.

While waiting, you may prepare your moulds. This test batch will yield about four average-size bars of soap, so choose your containers accordingly. A plastic soap dish, the bottom of a plastic bottle, a small firm cardboard box, milk carton, or muffin tin — anything other than aluminum can be used to form the shapes of your soap cakes. Grease the mould liberally with petroleum jelly (Vaseline) so the finished soap will release easily.

Once the moulds are ready and you are convinced that the two solutions are cool enough, proceed. (If you misjudge the temperature, it is no tragedy. If too warm, it will simply mean more stirring before the soap sets. If too cool, it will set very fast, perhaps before you get it into the mould.)

Now add the lye to the fat, pouring slowly and evenly and stirring constantly. The mixture will become cloudy. Continue to stir until your soap begins to set. When the consistency is similar to creamy honey, and your spoon will stay upright by itself for a moment, it is time to pour into the moulds.

At this step the gooey, fudge-like mixture looks good enough to eat. It is not. Remember that the lye remains active until the soap is hard. Keep curious fingers out.

Congratulations, you have now made soap. All that remains is to allow it to harden (about 24 hours). If you have poured the whole batch into one shallow mould, cut it into bars after a few hours or when it reaches the consistency of butter.

Once hard, it can be pressed from the moulds and set in a well-ventilated area to cure for two-to-three weeks (do not allow it to freeze during curing). It will improve with age: a well-cured soap lasts longer and lathers better than a fresh bar.

Following this recipe, it is difficult to have a failure. Almost any result will be useable. There are, however, many gradations of good soap. A truly superb bar should pare smoothly, producing a continuous curl that holds its shape without crumbling. A side view should reveal no separation of layers. A greasy deposit on top means there was too little lye or that your water was hard. If the soap separates into layers, it could be from too little stirring, pouring too early or too late, or from mixing at the wrong temperature.

With this one success behind you, larger batches are now within reach. The basic proportions for making soap in quantity are:

13 cups rendered fat
13 ounces lye
5 cups water

The basic procedure is the same, but the cooling times are, of course, longer. Experimenting with the type of fat will give endless variations on the basic recipe. Vegetable oils — olive oil, safflower, peanut or corn — have their own distinct qualities, but all have superfatting qualities and make for an enriched bar of soap. Up to 20 per cent vegetable oil can be included in the rendered-fat measure. Coconut oil will produce a profusion of suds, but may prove drying to your skin.

Vaseline, cold cream or glycerin (99 cents for 4 ounces) can be added, just before moulding, to enrich the soap.

A mildly abrasive bar can be produced by adding cornmeal or oatmeal after the lye and fat are mixed, and, if you are feeling exotic, stir in a little crushed cucumber to act as an astringent. Whipping air into the mixture just before pouring into moulds will result in floating soap.

Dozens of natural essences sold at health food stores or craft shops can add both fragrance and antiseptic qualities to your soap, among them citronella, eucalyptus and lanolin. Sandalwood contributes an especially pleasant scent to soap, as do wintergreen, cloves, lavender and lemon. These oils vary in price, but the more expensive cost about 90 cents for a half-ounce vial — enough for a year's supply of soap. (Never use commercial perfumes in homemade soap — if they contain alcohol the ingredients may separate.) Add any scents after the lye and fat are well-mixed.

Once you are producing your own toilet soaps, you will probably graduate, as I did, to making laundry soap. My grandmother used her own soap all her life, and her wash was the envy of the block.

A combination of lard and tallow seems to work best, prepared in the same manner as above. The soap is poured into a large enamel pan and cut into bars. When it has completely set and cured, I grate the bars into a powder and use it like any detergent. Since I wash in cold water, I dissolve the soap in warm water before adding it to the laundry. Besides giving the kind of whites that would send a Madison Avenue soap ad man into ecstasy, homemade laundry soap leaves a wash water that is ecologically benign.

Caveat Emptor Tractoris

A Greenhorn's Guide To Buying A Small Tractor

By Barry Estabrook

Stuart Connolly

There were those who flatly predicted that a mechanical contraption would never replace the ever-faithful draft horse as North America's primary source of agricultural power.

Although the clanking and sputtering machines rather quickly proved them wrong and drove the Percherons, Clydesdales and Belgians into near oblivion, we have no evidence that the horse trader allowed himself to be dragged into obsolescence behind his merchandise.

Known for their resourcefulness, among other things, it is quite possible that many dealers in horse flesh saw what the future held, hung up their crops and currycombs and took a seat behind the desk of a farm implement dealer.

"Not that all tractor dealers are bad," a neighbouring farmer explained recently when I approached him on the subject of buying a tractor. "Matter-of-fact, 90 per cent may be perfectly honest. . . but, well, they're only in business for one reason. . ."

Something about the way he was eyeing me as I stood there in his barnyard in still-stiff boots and unfaded jeans suggested that someone like myself would prove a frightful temptation for

273

even the most honest person trying to sell a used tractor.

Nevertheless, there comes a time when many new residents of the countryside find themselves ready to venture into tractor ownership. Assuming that a brand-new machine is out of your price range, buying a used one takes just a bit of cautiousness to come through with your eye-teeth and reputation as a grass-roots business-man intact.

Many are the "hobby farmers" who have been taken in by a tractor with a fresh coat of paint and a dealer's spiel that "these things never die."

The truth is that tractors built up to the end of the 1950's were basic, simple machines without unnecessary frills. Unlike many mechanical devices, tractors then were made to last, and with a little care even the greenest newcomer to the land should be able to secure one of these low-priced dependable machines.

Before heading off to a farm auction or even leaning over the fence to ask a neighbour about that old Ford sitting unused behind the barn, sit down and seriously assess your needs. What jobs will you be doing regularly? Once-in-a-lifetime work, such as excavating or clearing land, can be done by custom operators, who can be found in all agricultural areas. Too, farm equipment can be rented for special jobs.

If plowing, tilling and snow removal are your major needs, think twice before investing in a tractor. A good rototiller and snowblower or two-wheeled plow might be cheaper and easier to maintain in the long run.

On the other hand, small riding garden tractors are not really practical for anyone whose agricultural ambitions are grander than tending a well-groomed lawn and postage stamp-sized garden. These machines are designed to do a limited job well, but even their dealers will tell you that pushing them beyond a narrow limit will result in excessive wear and eventually in breakdowns. Furthermore, at $3,000 for a large garden tractor with a few accessories, they are by no means inexpensive.

If you feel your operation is not large enough to justify the expense of a tractor which will pass most of its days gathering dust in the barn, but still find the need for a tractor arising periodically, do not overlook co-operative buying.

There is a good possibility that you have a neighbour in a similar position, and your combined needs might justify buying a tractor. By all means know this neighbour as a true friend, and take peak periods of use into account. Many friendships have been destroyed as one partner in a joint equipment ownership stood by watching his neighbour plant that crop that he himself would never get in, due to the lateness of the season.

Knowledgeable sources agree that the best type of tractor for the back-to-the-lander is a small (20 - 30 horsepower), pre-1960 model.

"I would never buy one of these new tractors," says Andrew Ptak, a former repairman for a Case dealership in Ontario. "Up until the late '50s they were all made with heavy cast iron and straight steel, but then everyone started cutting corners and making lighter machines with alloys.

"The newer tractors, ranging in price from $5,000 to $8,000 for the more modest sizes, also achieve their horsepower rating by higher engine speeds (RPMs). The older tractors turned over much more slowly and therefore would last much longer. A slow engine with a heavy fly-wheel can develop the same power as a light engine at high RPMs, and last longer," says Ptak.

Among the various manufacturers, none seems to have produced a notorious "lemon" between World War II and the late 1950s. Any of the Fords (the 8N, Jubilee and Major are all popular with small-scale farmers), Cockshutt, Allis-Chalmers, Case, David Brown, Oliver and John Deere are all well respected for the quality and durability of their machines. "The Massey-Harris was also a good tractor," says Ptak, "until the Ferguson interests came onto the scene and began producing cheaper, lightweight machines."

Popular names like Ford, Ferguson, John Deere and International usually have an available supply of parts, but by all means check first for local availability, or your tractor might well serve out its days as a rather bizarre piece of sculpture nestled among the weeds beside the barn, long immobilized because you can't obtain that very small part which just happens to be crucial to operation.

To determine if the tractor is mechanically sound ask someone who knows the machine (and has no personal interest) about its condition. Try to do a compression test on the motor and have the hydraulic pressure checked. Check the plugs for excessive carbon deposits and also the oil for blackness and metal fragments. Play in the front wheels signifies faulty bearings, and dirty air filters or worn fan belts point to a lack of care on the part of the former owner. If there are still doubts, take the machine to a mechanic and have him go over it.

Buying a tractor at a farm sale can have some definite advantages. Often the tractor was really used by a real farmer until the day he died or quit, so it probably is in decent condition (then again, he may have quit partly out of frustration with the tractor).

Depending on the auction, you may have time and opportunity to inspect the tractor while the rest of the farm is being sold (tractors are always auctioned off last). If you feel nervous about it, this is the time to bring along a friend or pay a mechanic to examine the tractor. The drawback is that farm auctions allow no chance for actually trying the machine under field conditions.

Listen to the old-timers, for they may know about the tractor. Take the local gospel on tractors with a healthy measure of skepticism — in this area a Cockshutt, for example, is usually talked down by the old-timers. It is, however, a fine, heavy tractor that can be bought cheaply simply because it is regionally unpopular.

One important thing to take into account is the condition of the rear tires. If they are worn to the point of needing replacing, count on spending upwards of $150 apiece, a price which could gobble up a few years' profit on garden produce without difficulty. Don't automatically reject a tractor with poor tires, however. If the price is low enough, it may be very well worth your while to replace them to get a good tractor. (Cracking, excessive or irregular wear and gouges all spell tire trouble.)

One important choice you will face is whether or not to buy a tractor with a three-point hitch. This feature, a major advance in tractor design in the 1940s, allows equipment (such as a mower) mounted on the rear of the tractor to be raised and lowered by hydraulic power. It is recognized by two arms that can be raised and lowered, with a third upper point of attachment.

The advantages of three-point hitches are that they will take a very wide array of accessories, which are now standardized to fit all three-point hitches. Using the hydraulics, it is much easier to mount and remove equipment and it is also possible to raise equipment (such as a mower blade) to avoid hitting rocks while in motion.

Straightforward tow-type tractors are, however, much less expensive and may suit your needs perfectly. Furthermore, accessories for these older models can be picked up very cheaply at auctions everywhere. With this type of hitch, the implement is simply attached to the tractor's drawbar.

While all three-point hitch tractors have hydraulic systems, there also exist several older tractors that have hydraulics without the three-point hitch. Check where the hydraulics are located. Some are tucked well under the machine and extremely unhandy to reach (the hydraulic system can be used for a front-end loader or to lift tow-type equipment with a remote cylinder).

PLOWS & HARROWS

Decent plows, disk harrows and drag harrows are three basic pieces of equipment for ground preparation, and all can usually be had at auctions for less than $50. These older, modest pieces of equipment will serve the small farmer adequately, and full-time farmers are interested only in larger, more modern implements.

For small acreages, a two-bottom plow is about right. Check plow points and bolts for

looseness or breakage. The coulter, a metal disk which cuts the edge of the furrow wall, should not be chipped or badly dented and should fit snugly in its bearing.

Disks on a disk harrow must be concave, and should be of uniform diameter. Remember that a disk harrow is made to turn the soil, not merely slice it, and if disks are not properly concave, they will do an inefficient job. Also check the boxings on the axle; there should be very little play between the axle and the boxing.

Although most farmers have given up on the old diamond peg harrow, in favour of chain drags, these older harrows are still useful tools and the lack of demand today makes them reasonably priced.

If you are looking at a peg harrow, check each peg to assure that it protrudes on the top side of the harrow. This protrusion means the peg can still be tightened with a hammer if it loosens. If the top of the peg is flush with the top of the harrow, it can no longer be tightened and will eventually fall out, resulting in the perfect spike to puncture a tractor tire on a later pass over that field.

The whole process of buying, fixing and maintaining a tractor is immensely more pleasurable if you know a good mechanic close-by. Tractor mechanics seem to have followed the lead of their cousins in the auto repair business. Many know nothing about the older tractors — indeed, many large new tractor dealerships don't want anything to do with a tractor more than 10 years old. If you are not especially mechanical yourself, check with the small country garages in your area — some have people who grew up with the same tractor you are buying and will provide a wealth of knowledge and help. (Asking around might also turn up a farmer who repairs tractors as a sideline).

More and more boards of education and community colleges are offering basic courses in mechanics, and you might wish to enroll in one of these.

A few basic tools should be obtained and kept around the workshop, along with a supply of parts which are likely to break, fall out or otherwise malfunction.

For the rest, don your filthiest pair of rubber boots, shove a wad of tobacco under your lip, and get out and start looking for that machine.

Bill Milliken

Buyer's Check List

The tractor can be the heart of a homestead's power needs. The right machine can ease your workload, enable you to alter the face of your land for the better, give you spare time for family and friends and generally be a positive force in your home.

The right tractor will do much more than field work — it is also a tool for land clearing, excavating, road building, snow plowing, pond building. It can be a portable power source for a compressor, generator and cordwood saw.

If your income is limited, remember that you will be buying and maintaining a machine that is usually operated by people who, because of their income, can justify the cost. The wrong machine can eat dollars and time, both of which should be spent more wisely.

Can you afford to replace a $40 starter one day and be faced with another expense the following day? Even with the best machine it can happen. On the other hand, you may have to spend little on a tractor that gives many hours of trouble-free service.

CHECK LIST

1. Oil pressure.
2. Compression.
3. Check oil in the sunlight for presence of filings, dirt, blackness.
4. Ease of starting.
5. Ease of shifting.
6. Listen to idle — is it smooth? Remember the 2 and 3 cylinder tractors sound very rough — that's their nature.
7. Check tires, there should be no cord showing or replacement will be necessary. Odd sizes on the rear will blow the tractor's differential.
8. Check play in wheel to the king pins. There will be some in an old machine — decide what you can live with.
9. Work the machine before you buy it. See how it feels under load — does it handle the work, or does it shudder and moan? Check that the governor cuts in when it hits load.

If You Buy From a Dealer

He is getting the top dollar for the tractor, and you can insist on more. Have a compression test and hydraulic pressure check done *in your presence.*

If You Buy At An Auction

Beware! There is very little that you can check.

If You Buy Privately

Take the tractor around the farm. Do some work with it.

Above all, be sure there is a dealer handy who can get parts for that particular model and year tractor (be sure you can determine the model type and, preferably, the serial number — buying parts can be hell if you don't know what you own). For example, in this area a Minneapolis-Moline would be a foolhardy purchase, because there are no dealers. Out west it could be a very good buy.

Finally, don't buy a tractor just because it is a well-respected make or model — Ford makes a good tractor, but an abused Ford is useless, regardless of the manufacturer's name.

Andrew Ptak,
Centreville, Ontario

Thunderstrokes And Firebolts

The phenomenal power of lightning — and how to weather the storm

By Janice McEwen and the Harrowsmith *Staff*

Miller Services

Imagine the chagrin of a Renfrew, Ontario farmer who pulled on the handle of a recently repaired barn door one morning following a thunderstorm only to have the door crumble into a heap of individual boards at his feet.

The man was left sheepishly wondering about his carpentry skills until a local lightning protection contractor examined the door and explained that, unknown to the farmer, a bolt of lightning had hit the barn during the storm. Leapfrogging from nail-to-nail along the Z-shaped bracing boards that supported the door, the lightning made its way to the ground. In the process the heat produced by the bolt reduced the nails to dust.

For the 118 passengers aboard an Air Canada DC-8 jetliner, lightning had much more serious consequences: the ill-fated plane crashed in a swampy field shortly after take-off from Montreal in November, 1963. When investigators finished sifting through the rubble of what is still Canada's worst airplane disaster, lightning was high on the list of probable causes.

Lightning is the most awesome of nature's weather phenomena — a single stroke of lightning produces more electricity than the combined output of all electrical power plants in the United States. The average cloud-to-ground lightning bolt averages only six inches in diameter, but attains a core temperature of about 50,000 degrees Fahrenheit — five times the temperature at the surface of the sun.

Each day some 44,000 thunderstorms break out around the globe, the greatest concentration of them within the belt extending 30 degrees north and south of the equator. As you read this there are 1,800 electrical storms raging throughout the world, and by the time you finish this sentence, lightning will have struck earth 100 times.

Too frequently, lightning strikes spell disaster. Each year several hundred North Americans are killed by lightning, and others die in the fires that follow in the wake of electrical storms. Ten thousand forest fires and more than 30,000 building blazes are caused by lightning. Damages to property and loss of timber are estimated at more than 50 million dollars annually.

Yet the scientific study of lightning is still in pioneering stages, leaving unexplained many aspects of the complicated series of events that take place in the five-thousandth of a second required for the average lightning bolt to strike.

Scientists are, for example, at a loss to explain "ball" lightning, a rare occurrence in which an orb about 20 centimetres in diameter forms at the lightning impact point. This blinding ball of energy is able to move around at a speed of several metres per second and is said to be accompanied by a hissing sound. Ball lightning is able to pass through closed windowpanes and often disappears with an explosion.

Little wonder that this astounding natural force has always aroused man's curiosity and fear.

For our ancient forefathers, there was no doubt about what caused lightning: various gods were flamboyantly expressing their disapproval of somebody's actions.

Zeus, as legend would have it, was particularly

keen to use a handy supply of lightning bolts to express his frequent outbursts of rage. Unfortunate were the troops that attacked friends of this surly deity — Zeus would often step in when his side was losing and tip the tides of battle with a few well-placed bolts among the enemy ranks.

SPARK OF LIFE

But recent findings by Nobel Prize winner Dr. Harold Urey suggest that the ancients may not have underestimated the nearly divine role lightning plays in terrestrial life.

Through laboratory reconstruction of the atmosphere of the young, lifeless earth — an atmosphere composed of ammonia, methane, hydrogen and water — students of Urey found that when electrical sparks, much like lightning, were passed through this medium, amino acids were created — the first building blocks in the evolution of life.

Recent findings also suggest that we can thank lightning (at least partially) for giving the world plants. Although nitrogen makes up 80 per cent of the earth's atmosphere, in its pure state it is useless to plants.

It has been found that lightning causes atmospheric nitrogen to combine with oxygen, forming nitric-oxide gas. This gas dissolves in rain and falls to the earth in useable nitrates. Some scientists estimate that hundreds of millions of tons of these nitrates are produced by lightning each year. It's enough to make a purveyor of bagged 20-20-20 weep.

Benjamin Franklin, that portly Renaissance man of the eighteenth century, made the first real breakthrough in man's understanding of lightning by determining that it was, indeed, a huge electrical spark. But it is ironic (in light of his factual discoveries) that one of the most prevalent schoolboy myths still surrounding lightning features Mr. Franklin as its main character.

Everyone has heard about Franklin's kite flying antics. What few people realize is that his kite was never struck by lightning. Had it been, either the string would have burned and Mr. Franklin would have lost his kite, or the experimenter himself would have been struck, and the world would have lost an able scholar and statesman.

What happened during this famous flight was that there was enough difference in the electrical charge between the earth and the air at the level of the kite to create a small finger-tingling flow of electrical current through Mr. Franklin's string.

Today we know that conditions leading to electrical storms begin when a strong negative charge builds in rain (cumulonimbus) clouds. How this charge develops is still a matter of scientific debate, but an accepted theory is that air turbulence in the clouds creates a build-up of negatively-charged electrons.

Free electrons on the earth directly below the cloud are repelled by the huge numbers of electrons above, and therefore the charge of the earth becomes more positive.

Because opposing charges are attracted to each other, the electrons in the cloud yearn to get to the positive earth.

Air, however, is a poor conductor of electricity. As the cloud matures, the charge continues to build until pressure becomes great enough to permit the electrons to leap through the insulative layer of air.

The first tentative electrons probe toward the earth in a series of steps that gives a lightning bolt its irregular shape. These first electrons clear a path for those in the cloud, and as soon as the first electrons connect with the ground, an avalanche of electricity surges from the sky.

Lightning has struck.

Lightning bolts range from 1,000 to 9,000 feet long, and can attain speeds over 60,000 miles per second.

A lightning bolt seeks the route offering the least electrical resistance in its journey from cloud to ground. Almost any solid object offers an easier path for electricity than air: it could be a tree, a utility pole, a high patch of ground; it could also be your barn, one of your outbuildings — or your house.

Lightning is a hazard deserving special attention from rural dwellers. Grim statistics show that nine out of 10 lightning-caused deaths occur outside city limits. Fire authorities estimate that lightning causes up to 37 per cent of all rural building fires.

G. A. Pelletier, chief of technical services in the Ontario Fire Marshall's Office and one of Canada's foremost authorities on lightning, attributes part of this phenomenal loss of life and property in rural areas to people being misinformed about this frightening natural force.

"Most people are totally unaware of what lightning is, how it behaves and what it can do," he said. "Take the old wives' tale about lightning never hitting the same place twice — a common enough belief. It's totally false. As a matter of fact, if a place has been hit once, it shows that it is a prime site for future strikes."

Pelletier also says that many people believe their homes to be safe from lightning because of the proximity of tall trees or a high television aerial. Neither is necessarily true.

We can thank Ben Franklin's inquisitive (and financially long-sighted) mind for the protection we now have against destruction of property caused by lightning.

"It has pleased God in His goodness to mankind, at length to discover to them the means of securing their habitations and other buildings from mischief by thunder and lightning," wrote Franklin in the 1753 edition of *Poor Richard's Almanack*. He went on to outline a system that not only worked, but which remains, almost unchanged, as the most efficient form of lightning protection.

The heart of a lightning protection system is a series of rods extending at least 12 inches above a structure at lightning vulnerable places: peaks, gable ends, chimneys, etc.

These lightning rods (or "air terminals" in the jargon of lightning experts) are connected to each other by a woven copper cable roughly one-half inch thick. The cable, in turn, is grounded on at least two sides of the building to rods driven 10 feet into the earth, although the depth will vary somewhat in accordance with soil conditions. It is often said that a lightning rod gives protection within a circle whose radius is equal to the distance between the tip of the rod and the ground. Unfortunately, lightning does not always adhere to this rule, but the Canadian Standards Association says that "a properly installed lightning rod system, if not 100 per cent effective, will ensure that in nearly all cases of lightning strikes to buildings, little or no damage will result."

Fire statistics support these claims: in 1975, the most recent year for which figures are available, only 91 of the 2,559 structural fires started by lightning in Ontario occurred in buildings protected by lightning rod systems.

Pelletier explains that a properly working lightning rod system creates an easy route for the electrical charges to follow, diverting them away from the building and allowing them to dissipate harmlessly in the ground.

This, of course, is preferable to the unprotected alternative — where the bolt strikes the roof of the building and passes through the structure itself, leaping through walls, appliances, plumbing fixtures, radiators (and in some cases human beings) en route to the earth.

Peter Burchell, a lightning protection contractor, claims that the most frequent cause of malfunctioning lightning protection systems is also

*The beginner need not approach do-it-yourself lightning protection with trepidation. Experts agree that installing a system on a simple home such as the one illustrated, **above**, lies within the capabilities of the average home handyman. Materials, however, must be purchased from a licensed lightning protection firm. Note that the three air terminals have been placed on lightning-vulnerable gables and the chimney. If the house had no chimney, a terminal would still have been placed between the two end terminals. Guidelines state that there should be no more than 25 feet between air terminals. The terminals are usually copper rods 12 inches high and affixed to the roof by metal connections and supports usually held in place by simple screws. A copper cable runs between air terminals and is also attached to aerials, downspouts and other exterior metal objects. The cable is neatly coursed over roof parapets and ridges and takes as direct a path as possible from terminal to terminal. Ground electrodes are usually one-half-inch diameter copper and steel rods, connected to the cable and driven 10 feet into the earth. (This depth will vary according to conditions: where soil is shallow the main ground electrode is placed in a two-foot trench 12 feet long). Ground electrodes are placed at diagonal corners of the building. A rule of thumb states that two ground electrodes are required for every six air terminals.*

*The house illustrated, **below**, although elaborate, can be protected using the same principles as those for the simpler home. Notice that air terminals have been placed on the chimney and the dormer, potential lightning targets.*

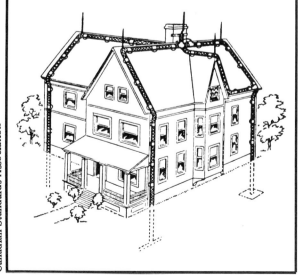

Canadian Standards Association

the easiest to remedy: the copper cable leading from the air terminals to the ground rod has simply become detached at the point where it connects to the ground rod — usually the result of careless lawn mower manoeuvres or a misdirected swipe of a scythe.

The situation can be remedied by obtaining a simple ground rod clamp from a lightning protection company (four or five dollars). Installation of the clamp requires only a wrench and the most basic handyman skills.

Placement of air terminals in an already-installed system is another area worth checking. Uppermost points of the roof (particularly chimneys — favourite gateways for lightning bolts) should have rods. This also holds for ends of roof peaks, gables, etc. Eaves troughs, downspouts, aerials and other metal objects on the roof should also be connected to the system.

Although lightning protection systems have proved their effectiveness, the sad truth is that the vast majority of rural homes are improperly protected or simply not protected at all. Burchell, who has examined systems on thousands of rural homes, estimates that 75 per cent of country homes are ill-protected.

One explanation for the lamentable condition of rural lightning rod systems is that their installation predates 1919, the year when the Canadian Standards Association set down guidelines for installation of lightning protection systems. Before this, hucksters proclaiming themselves lightning protection experts roamed country byways peddling systems that were often useless.

Too, the story of a rural home is often one of haphazard additions every time family fortunes or family numbers expand, and when the nucleus of a home might have been properly outfitted, subsequent additions have often been overlooked.

Assuming that your home or outbuilding does have a lightning protection system (the foot-long air terminals affixed to the roof are tell-tale signs), a brief check-over is in order to ascertain if the system is functional.

Expect to spend from $400 to $1,000 to have a commercial lightning protection firm install a complete system on an unprotected house. If such prices make you balk, you needn't place the security of your home in the hands of insurance companies or that deity who controls lightning bolts.

Dr. Terry McComb, a lightning expert with The National Research Council in Ottawa, claims that it is "definitely possible for someone who is reasonably handy with tools to install a lightning protection system on his home or outbuilding."

Do-it-yourself installation will result in savings of 50 per cent or more.

McComb suggests that someone wishing to install lightning rods on his home begin by obtaining a general book on the subject, such as R. H. Golde's *Lightning Protection* (Arnold), which can be found at public libraries.

The Canadian Standards Association (178 Rexdale Blvd. Rexdale, Ontario M9W 1R3) publishes a useful guide on lightning protection (publication B72-1960) which sells for five dollars. This booklet is well-illustrated and contains specific information on exactly what sort of system should be installed under different conditions.

Should you be leery about the effectiveness of your homemade system, a commercial contractor will come out (for a fee) and check it over.

Barns and other farmyard outbuildings are key targets for lightning, so much so that one insurance company, Farmers' Mutual, offers premium reductions of seven per cent on policies covering farm outbuildings protected from lightning.

Your chances of being killed by lightning this summer are roughly one in a million — certainly no reason to cancel plans for boating, picnics and hiking during the warm season, but reason enough to implement precautions.

An electrical storm that swept the New York City area took a typical toll of human victims. A golfer whose foursome had sought refuge from the rain beneath a tree (a common mistake that accounts for one-third of thunderstorm fatalities) died when lightning struck the tree. His companions were unharmed. The storm's next victim was a fisherman holding a metal casting rod. Lightning leapt from the rod to his jacket zipper. His single companion was injured but recovered. The final victim, a young man, died while standing near a beachhouse.

All of these deaths could have been prevented had the victims followed commonsense safety measures.

A car is perhaps the safest place to be during an electrical storm. There have been few, if any, substantiated cases of lightning striking an automobile, but laboratory experiments show that the charge would pass harmlessly over the metal shell of the car and then leap from the undercarriage to the pavement.

Second only to a car (and virtually 100 per cent safe) is a dry building protected by lightning rods. When the first signs of thunder make themselves manifest, the sensible thing to do is go straight to the shelter of a protected building. Two-thirds of lightning-caused deaths occur outdoors.

When you are caught by a storm in an open area, do not, under any circumstances, take shelter under an isolated tree. If you cannot reach a protected building, seek a low-lying area of open land.

Trees are favourite targets for lightning, and electrical charges that surge from the base of a struck tree can kill for a considerable distance. In one instance, a single bolt of lightning struck a tree in a Utah pasture and killed 500 sheep. There are recorded cases of cattle being killed while standing 100 yards from a struck tree.

Few people are killed by direct lightning strikes. If someone were directly hit, he would be severely burned. In most cases, the lightning victim is not burned but dies because currents cast off from a nearby lightning strike pass

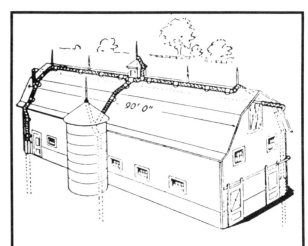

That it's all over whenever an unprotected barn is hit by lightning is an accepted fact known by all country residents. Notice here that the silo and vent tower, as well as all exterior metal objects (vulnerable to lightning strike), have been connected to the system. A wooden sleeve is also recommended to protect the area where the cable joins the ground electrode from breakage by livestock.

Miller Services

through his body, stopping his breathing and heartbeat.

Lightning frequently strikes water and electrical charges travel freely through this medium. Boats are high on the list of undesirable places to be when there is an approaching electrical storm. If you are in a boat, get to shore immediately and move some distance inland; shoreline trees are prime candidates for lightning strikes.

Swimmers, too, are in danger of being injured or killed by electrical charges that surge through water as a result of lightning.

If you find yourself in a protected house at the outbreak of a thunderstorm, take heart; you are safe. Still it is wise to stay away from sinks and bathtubs — your plumbing system is connected to a metal vent pipe protruding through the roof and is a potential lightning target. Avoid touching refrigerators, stoves and other large metal objects. Do not use telephones or other electrical appliances, and stay away from stovepipes, chimneys and fireplaces. Windows and doors should be closed.

If your home or one of your outbuildings is struck by lightning, an immediate check-over is due to insure that no hidden fires have started. (Old-timers often referred to hot and cold lightning — the former causing fires and the latter merely hitting with one explosive bolt.) When lightning fells a human, it is often possible to revive him with prolonged artificial respiration. Many victims have recovered fully, while others were left with sight or hearing impairments.

But even when nestled in the security of a snug, lightning-protected house, there are still some people who find themselves quivering under the bed with the dog at the faintest rumble of thunder. This unfortunate segment of the

Miller Services

population might consider moving to the Arctic or Antarctic — areas which see only one thunderstorm per decade.

If relocation does not fit your plans, we can only offer the slim comfort of words spoken by one lightning protection expert: "If you heard the thunder, the lightning did not strike you. If you saw the lightning, it missed you, and if it did strike you, you would not have known it."

Tales Of The Great White Gardener

There is a certain blissful ignorance among the non-gardening elements of our society, for whom all things good and green appear cellophane-wrapped in the supermarket vegetable crisper, sent there by the Great Market Gardener in the Sky.

I recently encountered one of these non-agrarian types over coffee, and happened to mention my growing desire to effect the cold-blooded murder of a certain ground hog that has been ravaging my lettuce patch.

Hardly able to finish his bran muffin, my friend received this news with a pale disgust only barely covering his desire to jump up and call the Humane Society.

"How could you do that to a helpless little animal?" was his argument.

"With glee," was the answer, "and be careful about calling this woodchuck helpless, thank you."

Actually, this well-intentioned person bears the common misconception that the path taken by a box of young lettuce plants in getting from greenhouse to garden to salad bowl is one filled with warm sunshine and gentle rains.

Obviously this non-gardener had never spent successive spring weekends moving stone, tilling, raking, hoeing and planting only to come out one morning to find a row or two of the garden defoliated by a ground hog.

I should now hasten to explain that I am not

one of the world's recognized experts on woodchucks or woodchuck hunting, and repeat this saga merely to show how a gentle gardener intent only on growing a few zucchini can become a lusty killer of rodents.

Upon finding two rows of young lettuce neatly bitten off at ground level recently, I immediately purchased a packet of poison gas sticks from the local farm co-op. When used properly, the wrapper said, these sticks would kill all manner of "varmints." Just chase them down the hole, toss in a burning poison candle, cover and start planning those Chef Salads.

However, various attempts to use this decidedly non-organic product proved futile, because: (A) this particular woodchuck I am seeking turns out to have a condominium-like maze of entrance and exit holes, and (B) he appears to relish the noxious fumes. Like a stiff drink before dinner, this gas appears only to whet his appetite.

Desperate at the loss of vegetables one Saturday morning (he had by now learned about the spinach and beet greens), I made a fast, angry run to Canadian Tire for one of their precision discount .22 rifles.

This I bought (no questions asked) with a box of "mushroom" shells designed to expand upon impact and cause maximum physical mayhem and quick death for the victim.

Not all *that* humane, I thought, but fair enough for a voracious woodchuck with no feelings for young veggies.

No more had I pulled back into our farm laneway when I spotted a familiar brown form about a third of the way up a row, leaving a barely visible trail of green stumps where pea vines had just stood.

Thus ensued a scene previously only thought possible in a sequel to *The Gang That Couldn't Shoot Straight.*

Much lead flew in the general vicinity of the startled animal, which finally gained cover in a pile of old split rails, aided no doubt, by the fact that the new gun's sights hadn't yet been adjusted.

When he finally emerged later that evening, a patiently planned long shot just missed him. All human caution was then thrown to the wind and the full arsenal put into use. The garden soon resembled a French battlefield in the First World War, with poison gas wafting out of various ground hog holes and a crazed Great White Gardener running about with a rifle, trying to match wits with a ground hog.

By noon the next day two enemy had fallen, with a third missing and presumed gone, but the big game had apparently got away.

Now much more attuned to the ways of this well-cultivated one-acre restaurant, the old woodchuck now makes his visits with the greatest of caution.

While the temptation to admire his sly intelligence is there, I must consider the continuing loss of lettuce and peas. It is all-out war now, and my main desire in life is to see a certain brown blot blown somewhere into the next county.

Just how far this has gone can be seen from the newest tactic, whereby I now can be found of an evening sitting in the hayloft of the barn, overlooking the garden through a crack in the boards.

I admit to slight guilt pangs at taking what is clearly a sniper's position. (When was the last time a good guy in a western or war movie ever fired a shot from the bell tower?) It is a Hun trick, but it should do the job.

—*JML*

Index

Personal Subscription Order Form

Please start a *Harrowsmith* subscription in my name.

☐ One year (8 issues) for $12.00
☐ Two years (16 issues) for $22.00

☐ Payment enclosed.
☐ Please bill me.

Name

Street/Rural Route

Town

Province/State   Postal Code

Payment must be made in Canadian or U.S. funds. All subscriptions outside North America, $16.00 per year Surface Mail; $28.00 per year Air Mail.

Gift Subscription Order Form

Please enter a Gift Subscription in the name of the person listed below.

☐ One year (8 issues) for $12.00
☐ Two years (16 issues) for $22.00

☐ Payment enclosed.
☐ Please bill me.

Gift Name

Street/Rural Route

Town

Province/State   Postal Code

Your Name

Street/Rural Route

Town

Province/State   Postal Code

A tasteful gift card will be sent to you for personalization and forwarding to the gift subscription recipient. Should the recipient of this gift already be a *Harrowsmith* subscriber, his or her subscription will be extended by the number of issues ordered.

Gift Subscription Order Form

Please enter a Gift Subscription in the name of the person listed below.

☐ One year (8 issues) for $12.00
☐ Two years (16 issues) for $22.00

☐ Payment enclosed.
☐ Please bill me.

Gift Name

Street/Rural Route

Town

Province/State   Postal Code

Your Name

Street/Rural Route

Town

Province/State   Postal Code

A tasteful gift card will be sent to you for personalization and forwarding to the gift subscription recipient. Should the recipient of this gift already be a *Harrowsmith* subscriber, his or her subscription will be extended by the number of issues ordered.

BUSINESS REPLY MAIL

No postage stamp necessary
if mailed in Canada

POSTAGE WILL BE PAID BY

MAGAZINE
CAMDEN EAST, ONTARIO
CANADA K0K 1J0

BUSINESS REPLY MAIL

No postage stamp necessary
if mailed in Canada

POSTAGE WILL BE PAID BY

MAGAZINE
CAMDEN EAST, ONTARIO
CANADA K0K 1J0

BUSINESS REPLY MAIL

No postage stamp necessary
if mailed in Canada

POSTAGE WILL BE PAID BY

MAGAZINE
CAMDEN EAST, ONTARIO
CANADA K0K 1J0